St Antony's Series

Series Editors

Halbert Jones
St Antony's College,
University of Oxford,
Oxford, United Kingdom

Matthew Walton
St Antony's College,
University of Oxford,
Oxford, United Kingdom

The St Antony's Series publishes studies of international affairs of contemporary interest to the scholarly community and a general yet informed readership. Contributors share a connection with St Antony's College, a world-renowned centre at the University of Oxford for research and teaching on global and regional issues. The series covers all parts of the world through both single-author monographs and edited volumes, and its titles come from a range of disciplines, including political science, history, and sociology. Over more than thirty years, this partnership between St Antony's College and Palgrave Macmillan has produced about 200 publications.

More information about this series at
http://www.springer.com/series/15036

Paul Betts • Stephen A. Smith
Editors

Science, Religion and Communism in Cold War Europe

palgrave
macmillan

Editors
Paul Betts
St Antony's College
Oxford University
Oxford, UK

Stephen A. Smith
All Souls College
Oxford University
Oxford, UK

St Antony's Series
ISBN 978-1-137-54638-8 (hardcover)   ISBN 978-1-137-54639-5 (eBook)
ISBN 978-1-349-71401-8 (softcover)
DOI 10.1057/978-1-137-54639-5

Library of Congress Control Number: 2016938686

© The Editor(s) (if applicable) and The Author(s) 2016, First softcover printing 2018
The author(s) has/have asserted their right(s) to be identified as the author(s) of this work in accordance with the Copyright, Design and Patents Act 1988.
This work is subject to copyright. All rights are solely and exclusively licensed by the Publisher, whether the whole or part of the material is concerned, specifically the rights of translation, reprinting, reuse of illustrations, recitation, broadcasting, reproduction on microfilms or in any other physical way, and transmission or information storage and retrieval, electronic adaptation, computer software, or by similar or dissimilar methodology now known or hereafter developed.
The use of general descriptive names, registered names, trademarks, service marks, etc. in this publication does not imply, even in the absence of a specific statement, that such names are exempt from the relevant protective laws and regulations and therefore free for general use.
The publisher, the authors and the editors are safe to assume that the advice and information in this book are believed to be true and accurate at the date of publication. Neither the publisher nor the authors or the editors give a warranty, express or implied, with respect to the material contained herein or for any errors or omissions that may have been made.

Printed on acid-free paper

This Palgrave Macmillan imprint is published by Springer Nature
The registered company is Macmillan Publishers Ltd. London

Contents

1 Introduction   1
Stephen A. Smith

Part I   Religion and Social Science in Eastern Europe   33

2 Piety by the Numbers: Social Science and Polish Debates About Secularization in the 1960s and 1970s   35
Jim Bjork

3 The Shepherds' Calling, the Engineers' Project, and the Scientists' Problem: Scientific Knowledge and the Care of Souls in Communist Eastern Europe   55
Patrick Hyder Patterson

4 Romanian Spirituality in Ceauşescu's 'Golden Epoch': Social Scientists Reconsider Atheism, Religion, and Ritual Culture   77
Zsuzsánna Magdó

Part II   Science, Religion and the Paranormal   103

5   Inculcating Materialist Minds: Scientific Propaganda
     and Anti-Religion in the USSR During the Cold War   105
     James T. Andrews

6   Tsiolkovskii and the Invention of 'Russian Cosmism':
     Science, Mysticism, and the Conquest of Nature
     at the Birth of Soviet Space Exploration   127
     Asif Siddiqi

7   Witchdoctors Drive Sports Cars, Science Takes
     the Bus: An Anti-Superstition Alliance Across
     a Divided Germany   157
     Monica Black

Part III   The Socialist Life-Cycle: Between
            Science and Religion   177

8   Writing Rituals: The Sources of Socialist Rites
     of Passage in Hungary, 1958–1970   179
     Heléna Tóth

9   In Search of Rationality and Objectivity: Origins
     and Development of East German Thanatology   205
     Felix Robin Schulz

Part IV   Socialism and the Problem of Religious Heritage   225

10   Religion and *Nauka*: Churches as Architectural
     Heritage in Soviet Leningrad   227
     Catriona Kelly

11 The Antireligious Museum: Soviet Heterotopia
    between Transcending and Remembering
    Religious Heritage   253
    Igor J. Polianski

12 Religion, Science and Cold War Anti-Communism:
    The 1949 Cardinal Mindszenty Show Trial   275
    Paul Betts

List of Contributors

James T. Andrews Modern Russian and Comparative Eurasian History, Iowa State University, Ames, IA, USA

Paul Betts Modern European History at St Antony's College, Oxford University, Oxford, UK

Jim Bjork Department of History, King's College London, UK

Monica Black Modern European History, University of Tennessee, Knoxville, TN, USA

Catriona Kelly New College, Oxford University, Oxford, UK

Zsuzsánna Magdó Department of History, University of Illinois, Champaign, IL, USA

Patrick Hyder Patterson Modern European History, University of California, San Diego, CA, USA

Igor Polianski Universität Ulm, Ulm, Germany

Felix Robin Schulz School of History, Classics and Archaeology, Newcastle University, Newcastle-upon-Tyne, UK

Asif A. Siddiqi Department of History, Fordham University, New York, NY, USA

Stephen A. Smith All Souls College, Oxford University, Oxford, UK

Heléna Tóth Otto-Friedrich-Universität, Bamberg, Germany

List of Figures

Fig. 10.1	Foucault's Pendulum in the Museum of Atheism, St. Isaac's Cathedral, early 1930s.	237
Fig. 10.2	Metropolitan Grigory of Leningrad, late 1940s	242
Fig. 10.3	Boris Piotrovsky (1908–1990) photographed in 1965	246
Fig. 11.1	Torture chamber, Diorama	262
Fig. 11.2	Leo Tolstoy as Judas before the Final Judgment. Wall painting, church of Tazovo	263
Fig. 11.3	R. Rakauskas, 'The shadow of the past'. Third prize in the atheistic photo competition 1962	264
Fig. 11.4	P. Mikhailov, 'Thinking', Oil painting 1964	265
Fig. 11.5	'Working in Soviet Society', Mstyora panel, fragment	268
Fig. 12.1	Cardinal Mindszenty, Budapest Trial, Feb. 8, 1949	285

List of Tables

Table 8.1 Rites of passage in Hungary and Czechoslovakia in 1967   196
Table 8.2 Rites of passage in Hungary, 1963–1987   198

CHAPTER 1

Introduction

Stephen A. Smith

This volume of essays explores the inter-relationship of Communism, science and religion in the Soviet Union and the states of Central, Eastern and Southeastern Europe that were incorporated into the Soviet bloc after the Second World War. These regimes, which I shall refer to interchangeably as 'Socialist' or 'Communist', were committed to the application of science, technology and instrumental reason to all areas of social life, and a key assumption of their vision was that religion would wither away as societies moved towards full-blown Socialism. From our current vantage point, more than a quarter of a century after the fall of the Soviet bloc, organized religion, rather than falling into desuetude, seems to have proved surprisingly resilient through the Communist era, and in some regions of the former Soviet bloc religion has undergone a significant resurgence. This, obviously, raises the question of what, if any, impact state efforts to eliminate religion and disseminate science actually had. The scholarly literature has tended

The volume arises from a conference held at St. Antony's College, Oxford, 16–17 May 2014, entitled 'Science, Religion and Communism in Cold War Europe'. We would like to thank St. Antony's College, that college's Modern European History Research Centre, and All Souls College Oxford for their financial support.

S.A. Smith (✉)
All Souls College, University of Oxford, Oxford, UK
e-mail: Stephen.smith2@history.ox.ac.uk

© The Editor(s) (if applicable) and The Author(s) 2016
P. Betts, S.A. Smith (eds.), *Science, Religion and Communism in Cold War Europe*, St Antony's, DOI 10.1057/978-1-137-54639-5_1

to treat Communism and religion and Communism and science as distinct fields. The historiography on Communism and religion has mainly focused on state repression, on the official promotion of 'scientific atheism', on collaboration of the churches with the state, on the role of religion in fostering political resistance and on the emergence of civil society in Eastern Europe.[1] Relatively little work has been done on the lived experience of believers, although that is now changing.[2] So far as the Cold War more broadly is concerned, it is striking how little attention has been paid to religion as a factor shaping that conflict, despite an enormous historiography on relations between the superpowers in the post-war period.[3] Scholarly work on Communism and science has focused mainly on the Soviet Union and has tended to concentrate on the role of the state in promoting science during the Cold War, especially military technology, on the negative impact of official ideology on scientific endeavour and on the contribution of science and technology to industrialization in the economies of Eastern Europe.[4] What the present volume seeks to do is to connect these two

---

[1] D. Pospielovsky (1984) *The Russian Church under the Soviet Regime, 1917–1982*, two vols. (Crestwood, NY: St. Vladimir's Seminary Press); John Anderson (1994) *Religion, State and Politics in the Soviet Union and Successor States* (Cambridge: Cambridge University Press); D. E. Powell (1975) *Antireligious Propaganda in the Soviet Union: A Study of Mass Persuasion* (Cambridge, MA: MIT Press); S. P. Ramet (1998) *Nihil Obstat: Religion, Politics and Social Change in East-Central Europe and Russia* (Durham, NC: Duke University Press); G. Weigel (2003) *The Final Revolution: The Resistance Church and the Collapse of Communism* (New York: Oxford University Press); B. von der Heydt (1993) *Candles Behind the Wall: Heroes of the Peaceful Revolution that Shattered Communism* (London: Mowbray).

[2] C. Wanner (2007) *Communities of the Converted: Ukrainians and Global Evangelism* (Ithaca, NY: Cornell University Press); C. Wanner (ed.) (2013) *State Secularism and Lived Religion in Soviet Russia and Ukraine* (New York: Oxford University Press); M. D. Steinberg and H. J. Coleman (eds.) (2007) *Sacred Stories: Religion and Spirituality in Modern Russia* (Bloomington: Indiana University Press); P. Betts (2010) *Within Walls: Private Life in the German Democratic Republic* (Oxford: Oxford University Press), ch. 2; B. Berglund and B. Porter-Szucs (eds.) (2010) *Christianity and Modernity in Eastern Europe* (Budapest: CEU Press).

[3] D. Kirby (ed.) (2002) *Religion and the Cold War* (London: Palgrave); P. E. Muehlenbeck (ed.) (2012) *Religion and the Cold War: A Global Perspective* (Nashville: Vanderbilt University Press); L. N. Leuştean (ed.) (2010) *Eastern Christianity and the Cold War, 1945–91* (London: Routledge).

[4] L. R. Graham (2004) *Science in Russia and the Soviet Union. A Short History* (Cambridge: Cambridge University Press); J. L. Roberg (1998) *Soviet Science Under Control: The Struggle for Influence* (London: Macmillan); D. Hoffmann and K. Macrakis (1998) *Science under Socialism: East Germany in Comparative Perspective* (Cambridge, MA: Harvard University Press); I. V. Bromlei (1986) *Main Features of Science Organization in Socialist Countries* (Moscow: Institut istorii estestvoznaniia i tekhniki).

historiographies and to challenge the implicit assumption that science and religion were counterposed, i.e. that the advance of science was necessarily at the expense of religion. The volume seeks to explore the contradictory and often surprising interplay of 'science' (including the social sciences), state-backed atheism and religious belief, seeking, for example, to show how between the 1940s and the 1980s religion and atheism were subject to a continuous process of reconstitution as objects of social science, or how the natural sciences were drawn to occult themes. It seeks to emphasise the modes of interpenetration and coexistence of science, atheism and religious belief and the difficulties of drawing clear boundaries between them, pointing, for example, to the role of scientific expertise in formulating policy in both atheistic education and religious heritage. A secondary theme of the volume is to connect the usually separate historiographies of the Soviet Union and of Eastern Europe (the volume looks at Poland, East Germany, Hungary, Romania, Yugoslavia and Czechoslovakia) and to encourage comparison of the different ways in which Communism, religion and science were triangulated across the different states of the Soviet bloc. The essays demonstrate that although the governments of the Eastern Bloc adopted the tenets of Stalinist anti-religious policy in the late 1940s, they rather quickly adapted them, according to the specific configuration of the religious field in their country, to its sociological make-up, its intellectual and cultural traditions and the extent to which it had undergone secularization in the interwar period.

The main themes that emerge from the 11 essays of the volume relate to the following broad areas. First, by the 1960s, Communist states faced increasing difficulties explaining the persistence of religious belief, notwithstanding the socio-economic modernization and educational advances that were assumed to erode religiosity. Scholars began to distance themselves from the vulgar Marxism of the 1950s and turn towards the social sciences and psychology to find ways of understanding the resilience of religious belief and practice. Those charged with propagating scientific atheism were increasingly forced to concede that its limited impact on the populace was due to its failure to tackle existential questions concerning human purposes, meanings and values to which religion had traditionally offered answers. One response was to accelerate the process of creating secular rituals to replace religious rituals that marked the key rites of passage and to seek to engage with the inescapable reality of death. Second, this was the era when, especially in the Soviet Union, great achievements in science, notably in the space race, afforded the government a certain legitimacy. Achievements in science and technology aroused public enthusiasm, especially in the context of Cold War competition, and the popularization of science made great strides. Somewhat paradoxically, the Cold War also spurred professional scientists to take an interest in phenomena that seemed to resist explanation in terms of dialectical materialism, an interest that extended to the occult. Third, scientism could easily take on quasi-religious forms, as institutions such as the anti-religious museum—never as numerous in Eastern Europe as in the Soviet Union—suggest. Fourth, the Christian Churches, for their part, showed increasing willingness to endorse the claims of science and, correspondingly, to reduce the sphere of knowledge in which religion had traditionally claimed authority. In addition, the involvement of the Russian Orthodox Church in the World Council of Churches or the Second Vatican Council in the Roman Catholic Church called into question the leninist shibboleth that all religious institutions were bastions of reaction. Finally, the material heritage of the religious past, once held in contempt, became increasingly valued from the 1960s for its symbolic potential to connect the Socialist present to the historic culture of the nation, as many Socialist states increasingly asserted a national identity.

The volume is not concerned with the efforts of the Soviet Union and its client states to combat institutionalized religion or with church–state relations as such. Yet the often tense relationship between church and state cannot be ignored, since it provides the institutional framework within which the contest between science, religion and official ideology played out. It should also be mentioned that the focus is very much on the Christian churches, since these dominated the religious landscape of Eastern Europe, but we should not forget that in the Soviet Union the attack on religion extended to Islam, Judaism, and Buddhism, and that in the Socialist bloc as a whole a decimated Judaism and—in parts of Yugoslavia, Albania and Bulgaria—a hardy Islam were also elements of the religious landscape.

THE SOVIET STRUGGLE AGAINST RELIGION

The Bolsheviks stood in an Enlightenment tradition, reinforced by nineteenth-century philosophical materialism, which championed reason as the source of social progress and looked to the application of scientific knowledge to nature and society as the key to furthering human advancement. In their hostility to religion, however, the Bolsheviks went much further than the Second International. In Germany the Social Democratic Party (SPD) had endorsed a generalized secular humanism, confidently believing that science would bring intellectual emancipation from religion, and had encouraged worker activists to read classic materialist texts such as Ludwig Büchner's *Kraft und Stoff: Empirisch-naturphilosophische Studien* (1855) (Force and Matter: Empirico-philosophical Studies) or Ernst Haeckel's *Die Welträtsel* (1895–1899) (*The Riddle of the Universe*, 1901). Yet the SPD upheld freedom of religion and avoided vituperative anti-religious rhetoric.[5] By contrast, the Bolsheviks revelled in hostility to religion, not least because the Russian Orthodox Church was such a pillar of the old regime. It was also the case the Russian intelligentsia, more than its counterparts in western Europe, had espoused atheism as intrinsic to progressive politics. As Victoria Frede observes, questioning the existence of God took on a 'peculiarly intensive, existential quality in Russia', in contrast to France, Great Britain, and Germany, where 'metaphysical doubt gradually resolved itself into polite agnosticism'.[6] After October 1917, with Orthodox hierarchy and clergy opposed to them, the Bolsheviks determined to bring the Church to heel, not only instituting the separation of Church and state that had become standard in Europe since the French and American revolutions, but also expropriating it economically and radically curtailing its freedom to operate as an independent organization. Much has been written about the fierce campaigns against the Church and the efforts of bishops, priests and laity to resist them. Suffice it to say that notwithstanding its institutional battering at the hands of the state, it was clear by the time of the Second World War that the Orthodox Church would not go under anytime soon.

At the time the Bolsheviks took power, the concept of secularization in its modern sense was not yet in circulation. Like many Russian intellectuals in the nineteenth century, however, including some churchmen, they took for granted that socio-economic modernization—whether construed

---

[5] S. Prüfer (2002) *Sozialismus statt Religion. Dei deutsche Sozialdemokratie vor der religiösen Frage 1863–1890* (Göttingen: Vandenhoeck & Ruprecht); T. H. Weir (2014) *Secularism and Religion in Nineteenth-Century Germany* (Cambridge University Press).

[6] V. Frede (2011) *Doubt, Atheism, and the Nineteenth-Century Russian Intelligentsia* (Madison: University of Wisconsin Press), p. 3.

as urbanization, migration, industrialization, the decline of face-to-face communities, or the rationalization of social life—would bring about a decline in the social influence of religion. The Bolsheviks, however, were never content to let modernization take its course. They believed that the state must take conscious, determined action to undermine religion and promote science, using the full panoply of its repressive, ideological, pedagogical and legal powers. They perceived religion in all its forms as inimical to Socialism, since it instilled values at odds with the activism and collectivism required of Socialist citizens, encouraging fatalism and humility, acquiescence in the status quo, sectarian dogmatism, a concern with individual salvation and a general orientation to the world beyond this.[7] Socialist society, they believed, would be atheist or it would be nothing, a perspective exported to the countries of Eastern Europe in the late 1940s.

The positive counterpart to Bolshevik anti-religious propaganda was the vigorous commitment to propagate a scientific understanding of the natural and social worlds. The assumption was that each advance of scientific knowledge, together with increased control of nature that it enabled, would expose the falsity of religious claims. Different agencies of state, therefore, including the agitprop sections of the Communist parties, schools, the Red Army, scientific institutions, the press, publishing houses and the radio, were enlisted into combining anti-religious propaganda with the popularization of scientific knowledge. 'Scientific-atheist propaganda', as Igor Polianski shows in his article on anti-religious museums, meant not only refuting the claims of religion about the origin of man and the universe in the light of modern science but also using historical criticism to debunk stories in the bible and demonstrate the age-old subservience of religious institutions to the ruling classes. A taste of this heady brew can be perceived in a training course compiled by the League of Militant Godless in 1937, which called for demonstration of the impossibility of miracles, the refutation of the 'unscientific representation of nature' in the Bible, Koran and Talmud, and the exposure of the ignoble history of the Roman Catholic Church in persecuting scientists such as Bruno and Galileo.[8]

Though the Bolsheviks were loud in their denunciation of religion, they never came close to fashioning a coherent and consistent strategy for anti-religious work. Even in respect of the Orthodox Church, frontal attacks between 1922 and 1925 and again during the 'cultural revolution'

---

[7] Powell, *Antireligious Propaganda*, pp. 7–8.
[8] State Archive of the Russian Federation. GARF, f. A-2306, op. 39, d. 65.

(1928–1931), gave way to periods of standoff, when the regime sought to win the cooperation of senior clerics. Behind the twists and turns of policy lay fundamental disagreements within Bolshevik ranks about the relative importance of the struggle against religion as a policy objective; on whether the struggle against religion was a matter of long-term education or whether it could be advanced through 'administrative methods'; on whether positive promotion of scientific knowledge was more effective than anti-religious propaganda; and on whether religion could be left to wither away as the social conditions of the masses improved. Initially, the League of Godless, founded in 1925 as a non-state organization, advocated the use of propaganda and education in respect of religion in opposition to the so-called 'priest eaters' of the Komsomol, who favoured a confrontational approach, revelling in antics such as setting pigs loose in church or staging 'anti-festivals' at Easter and Christmas. The Union itself was split between a majority of *Kulturträger* and a minority who favoured a 'class struggle' approach to religion. During the 'cultural revolution', the Union assumed a more combative stance, accusing both the Komsomol—which by this stage was pursuing a more moderate approach to anti-religious work—and the Commissariat of Enlightenment of excessive compromise with the reactionary proponents of religion. Despite occupying a 'prominent and even noisy position within the Bolshevik propaganda apparatus', the League of Godless was in almost constant crisis, its obsession with recruitment, training cadres and selling magazines, leaving little time to engage seriously with the concerns of believers.[9]

The census of 1937, the first and last Soviet census to ask about religious belief, revealed that 42.9 % of adults in the USSR claimed to be non-believers, made up of 55 % of men and 33 % of women, with non-believers concentrated in the 20–29 and 30–39 age groups.[10] This almost certainly underestimated the extent of religious belief, for we know that some were reluctant or afraid to admit that they were believers.[11] Nevertheless, the pattern revealed of a decline in religious belief in the younger generation, especially among men, seems plausible. The census suggested that the percentage of believers had almost halved between the generation that

---

[9] D. Peris (1998) *Storming the Heavens: The Soviet League of the Militant Godless* (Ithaca: Cornell University Press), p. 226.

[10] *Naselenie Rossii v XX veke: istoricheskie ocherki*, tom. 1 1900–1939 gg. (Moscow: Rosspen, 2000), p. 184.

[11] C. Merridale (1996) 'The 1937 Census and the Limits of Stalinist Rule', *Historical Journal*, 39:1, 233.

lived through the revolution—roughly, those in the 30–34 and 35–39 age groups—and the 'first soviet generation' (those in the 16–17 and 18–19 age groups).[12] It was this 'first socialist generation' whose worldview the Soviet state was seeking to mould through schooling, propaganda, youth organizations and military service. Doubtless family and community continued to be influential in the socialization of these children, yet this was a period when the hallowed patterns of rural life were being shattered by collectivization and pell-mell urbanization, and there is evidence that parents had come to feel less confident that traditional beliefs and values could equip their children for the Socialist future.[13] That said, we should not exaggerate the impact of anti-religious propaganda. We know, for example, that anti-religious education in schools was poorly organized and underfunded.[14] Nevertheless, young people were directly exposed to soviet institutions where the culture was resolutely secular and where the verities of village life were constantly challenged. Whether this amounted to what party leaders confidently called a 'process of departure from religion' (*protsess otkhoda ot religii*) may be doubted, but a process of secularization—uneven, to be sure—does seem to have been underway.

The Second World War represented a massive watershed in the development of Soviet society. Following the battle of Kursk in July 1943, when the Red Army gained a strategic advantage that would last for the rest of the war, Stalin's government engineered a rapprochement with the Russian Orthodox Church, notably by allowing the restoration of the Moscow Patriarchate, in an effort to bolster Russian patriotism. A further motive in areas of Ukraine that had been under German occupation seems to have been to work with the Church to counter the threat from the Greek Catholic Church.[15] This proved to be the start of a period of tense cooperation between church and state that would last, with some hiccups, until 1958. In these years, the number of

---

[12] *Naselenie*, p. 143.

[13] S. A. Smith (2007) 'The First Soviet Generation: Children and Religious Belief in Soviet Russia, 1917–41', in S. Lovell (ed.), *Generations in Twentieth-Century Europe* (Basingstoke: Palgrave), pp. 79–100.

[14] Larry Holmes concludes that: 'From 1917 to 1941, Soviet Russia's schools contributed almost nothing to a direct assault on religion. The educational system, like society itself, proved more tradition-bound and inert than expected'. L. E. Holmes (1993) 'Fear no evil: schools and religion in Soviet Russia, 1917–1941', in S. P. Ramet (ed.), *Religious Policy in the Soviet Union* (Cambridge: Cambridge University Press), p. 125.

[15] S. M. Miner (2003) *Stalin's Holy War: Religion, Nationalism, and Alliance Politics, 1941–1945* (Chapel Hill: University of North Carolina Press).

parishes grew and the Church was allowed to raise income and rent and to build property. There were even reports that local soviets made use of priests to promote economic and political campaigns, organize grain deliveries and encourage subscription to state loans and participation in soviet elections. From 1947, however, hardliners in the party leadership regularly expressed concern about the extent to which the Church (and certain Protestant confessions) were regaining the ground they had lost from the 1920s.[16] Church attendance, especially on feast days, was increasing, more people were baptizing their children (at one least one-fifth of infants in the Russian Federal Republic in the 1950s), more people were getting married in Church and pilgrimages to sacred shrines were growing in popularity.[17]

With the death of Stalin and the access to power of Nikita Khrushchev in 1954, many of Stalin's policies were reversed. Khrushchev's desire to revitalise the commitment to creating a Communist society entailed a reinvigoration of efforts to create the world's first atheist society, to show the Soviet public, as Khrushchev put it, the 'last priest'. On 7 July 1954, against a background of agricultural crisis and religious revival in the countryside, the Central Committee of the Communist Party of the Soviet Union issued a decree entitled 'On mistakes in carrying out scientific-atheist propaganda among the population', which seemed to revive the *Sturm und Drang* anti-religious policies of the pre-war era: 'Churchmen and sectarians are seeking different methods to poison the consciousness of the people with the drug of religion'. Almost immediately, however, it proved divisive in a society still coming to terms with the trauma of Second World War and the repudiation of Stalinism, and on 10 November 1954 the campaign was brought to a halt. The decree announcing the halt claimed that 'instead of systematic and deep propagation of natural-scientific knowledge and broad explanatory work, based on the reality that believers are consciously leaving behind the influence of the Church and other religious prejudices, some officials have begun to apply administrative methods to religious ministers and believers and to insult their feelings'.[18] For the next few years, a tussle went on behind the scenes between hardliners, led by M. A. Suslov, second

---

[16] E. Zubkova (1998) *Russia after the War: Hopes, Illusions and Disappointments, 1945–1957*, trans. and ed. H. Ragsdale (Armonk, NY: M. E. Sharpe), pp. 69–70.

[17] M. V. Shkarovskii (1999) *Russkaia Pravoslavnaia Tserkov' pri Staline i Khrushcheve* (Moscow: Krutitkoe Patriarshee Podvor'e), p. 357.

[18] M. V. Shkarovskii (2010) *Russkaia Pravoslavnaia Tserkov'v XX veke* (Moscow: Veche), pp. 350–53; J. D. Grossman (1973) 'Khrushchev's Anti-Religious Policy and the Campaign of 1954', *Soviet Studies*, 24:3, 374–86.

secretary of the CPSU, and an opponent of Khrushchev, and the Council for the Affairs of the Russian Orthodox Church, the body charged with overseeing the hierarchy, handling complaints and informing party leaders of developments on the ground.[19] The hardliners won out in 1958 when a new campaign against the Church was launched that would last until 1965. During the campaign, over 6000 churches and over 1000 non-Orthodox places of worship were closed down, monastic communities disbanded, restrictions placed on parents' right to teach their children religion and a ban enforced on the presence of children at church services (beginning in 1961 with Baptists, and then extending to the Orthodox in 1963).[20]

Meanwhile since the 1930s relatively little finance or energy had been put into anti-religious propaganda, indeed the League of Militant Godless had ceased activity with the onset of the war. In 1954 the Znanie Society—the organization tasked with popular enlightenment and the subject of James Andrews's article—called for the creation of a special journal to propagate atheism.[21] In the same year, 'scientific atheism' was formally consecrated as a dimension of Marxism–Leninism, construed as 'integral yet relatively independent'.[22] However, it was only with the renewal of Khrushchev's anti-religious campaign that a journal devoted to the propagation of scientific atheism, *Nauka i religiia*, appeared. In 1959, institutions of higher education were required to teach a course on 'The Foundations of Scientific Atheism', although it only became mandatory in 1964, and in secondary schools evolution and the origins of life began to be taught intensively. Yuri Gagarin's failure to discover God on his orbit of the earth in 1961 was milked to the full in anti-religious propaganda. With the end of Khrushchev's anti-religious campaign in 1964, however, the Council on Religious Affairs immediately softened its propaganda against the Orthodox Church, reserving its opprobrium

---

[19] T. A. Chumachenko (2002) *Church and State in Soviet Russia: Russian Orthodoxy from World War II to the Khrushchev Years*, ed. and trans. E. E. Roslof (Armonk: M.E. Sharpe), ch. 3.
[20] Anderson, *Religion, State and Politics*, chs. 2 and 3.
[21] V. Smolkin-Rothrock (2014) 'The Ticket to the Soviet Soul: Science, Religion, and the Spiritual Crisis of Late Soviet Atheism', *Russian Review*, 73, 177.
[22] J. Thrower (1983) *Marxist-Leninist 'Scientific Atheism' and the Study of Religion in the USSR* (Berlin: Mouton), p. 149.

for so-called 'sectarians', particularly unregistered religious groups such as Baptists, Jehovah's Witnesses and Seventh Day Adventists.

SOCIALISM AS SCIENCE

In nineteenth-century Russia, notwithstanding the country's extraordinary literary, musical and artistic efflorescence, it would be fair to say that science held a place of pre-eminence in the cultural field. For the intelligentsia the natural sciences, in particular, possessed the key to the transformation of socially and economically backward Russia. A. I. Herzen was typical: 'Without natural science there can be no salvation for modern man; without that healthy nourishment, that vigorous elevation of thought based on facts, and that proximity to the realities of life ... our soul would continue to be a monastery cell, veiled in mysticism spreading darkness over our thoughts'.[23] The Bolsheviks were the inheritors of this belief in the capacity of science to emancipate humanity from 'superstition'. The Twelfth Congress of the All-Russian Communist Party (Bolshevik) in 1923 stated that 'systematic work must be done to create in the new generation a serious urge to master science and technology'.[24] In 1926, the programme for workers' anti-religious circles explained: 'Natural phenomena have a law-governed character and are independent of the desires of man. As human society studies the laws of nature, it subordinates natural phenomena to its will'.[25]

At the same time, millions of Russians, mainly in the countryside, lived in a culture that had not yet instantiated the epistemological distinctions characteristic of the post-Galilean world—between man and nature, between the natural and supernatural orders, between a natural realm governed by physical laws of cause and effect and a metaphysical realm of the spiritual. There was thus an urgent need for a crash programme to disseminate knowledge of science and technology. Through lectures, exhibitions, pamphlets and primary schools, different agencies set about explaining phenomena such as

---

[23] A. Vucinich (1989) *Darwin in Russian Thought* (Berkeley: University of California Press), p. 8.
[24] A. E. Gorsuch (2000) *Youth in Revolutionary Russia* (Bloomington: Indiana University Press), p. 19.
[25] *Antireligioznik* (1926), no. 10, 53.

thunder and lightning, germs and basic hygiene, electricity and the internal combustion engine. For the more curious, popular books and pamphlets were published on a gamut of topics from astronomy, evolution, biology and geography to agronomy. In 1921 a poor peasant who had fought in the Red Army wrote a letter to the peasant newspaper: 'Send me a list of books published on the following subjects because I am interested in everything: chemistry, science, technology, the planets, the sun, the earth, the planet Mars, world maps, books on aviation, the number of planes we possess, the number of enemies the Socialist Republic has, books on comets, stars, water, the earth and the sky'.[26] After 1928, as James Andrews has shown, the Bolsheviks came to favour a more utilitarian approach to science education that emphasized its application to industrial and agricultural production.[27]

The period of the First Five-Year Plan and forced collectivization (1928–1932) was one in which science was subject to heavy ideological interference. 'Red' experts were favoured over 'bourgeois' experts and scientific research was oriented away from pure research to production. Debates among Deborinists, mechanists and others took place about the class nature of science and its relationship to base or superstructure in Marxist theory. A new round of ideologization of science occurred between 1948 and 1952, as Lysenkoist biology, based on the idea that organic nature is infinitely malleable and subject to human manipulation, reached its apogee. Trofim Lysenko's rise, in fact, had less to do with ideology than with his extravagant promises of increased agricultural yield, yet he used a language of class to justify his work. Increasingly, however, Stalin came to accept that science dealt with laws that were beyond the ability of humans to control.[28] In 1948, for example, he crossed out an entire section of Lysenko's draft speech to the Lenin Academy of Agricultural Sciences, entitled 'The Fundamentals of Bourgeois Biology are False'. Opposite Lysenko's comment that 'any science is class-oriented by its very nature', Stalin wrote 'Haha!!! And what about mathematics? What about Darwinism?'[29] This did not, of course, mean that science was freed from

---

[26] R. Robin (1991) 'Popular Literature of the 1920s: Russian Peasants as Readers', in Sheila Fitzpatrick et al. (eds.), *Russia in the Era of NEP* (Bloomington: Indiana University Press), p. 261.

[27] J. T. Andrews (2003) *Science for the Masses: The Bolshevik State, Public Science, and the Popular Imagination in Soviet Russia, 1917–34* (College Station: Texas A&M University Press), p. 3.

[28] E. Pollock (2006) *Stalin and the Science Wars* (Princeton, NJ: Princeton University Press).

[29] K. O. Rossianov (1993) 'Stalin as Lysenko's Editor: Reshaping Political Discourse in Soviet Science', *Configurations*, 1:3, 451.

ideological pressure—in the early 1950s Soviet science was still suffused with rhetoric about combating idealism, cosmopolitanism and obsequiousness to the West—but it marked a step towards greater objectivism..[30]

Following Stalin's death, it became acceptable to discuss foreign science and technology in relatively positive ways. *Nauka i Zhizn'* saw the number of articles attacking western science fall from 14 in 1953 to 2 in 1957.[31] Science was no longer presented as a product of class-specific ideology and more as an international pursuit based on objective enquiry.[32] The context of the Cold War, however, generated fierce rivalry between the USA and the Soviet Union on the terrain of Big Science and Big Weaponry, rooted in the relentless arms race and race into space. On both sides of the Iron Curtain the 1950s saw a huge increase in state and military spending not only in weapons research and the physical sciences but also in the biological sciences and geoscience (though in the Soviet Union there was much less interest in the social sciences than in the USA, where modernization theory was yoked to the ends of US foreign policy). From the late 1950s, Laika (the first dog in space), photographs of the far side of the moon and, above all, Yuri Gagarin's triumphant flight into space in 1961 sparked enormous public interest in the 'new frontier' of the cosmos and fostered genuine pride in the achievements of the Soviet system. Triumphs in space served to highlight the dynamic and future-oriented character of the Soviet project, justifying massive spending on technology that had few obvious public benefits. Internationally, the Soviet Union exploited the propaganda value of its scientific achievements to the full, sending cosmonauts on foreign tours and organizing international exhibitions.[33] As James Andrews shows in his chapter, the popularization of science improved considerably. In 1958, for example, 394 science documentary films were produced, including a regular cinema journal *Science and Technology* and a quarterly children's cinema journal, 'I want to know everything'.[34]

---

[30] A. Kozhevnikov (1998) 'Rituals of Stalinist Culture at Work: Science and the Games of Intraparty Democracy circa 1948', *Russian Review*, 57:1, 26.

[31] M. Froggatt (2005) 'Science in Propaganda and Popular Culture in the USSR under Khrushchev (1953–64)' (Oxford University, DPhil), 74.

[32] Froggatt 'Science in Propaganda', p. 135.

[33] J. T. Andrews and A. A. Siddiqi (2011) *Into the Cosmos: Space Exploration and Soviet Culture* (Pittsburgh: University of Pittsburgh Press).

[34] Froggatt, 'Science in Propaganda', p. 35.

Socialism and the Paranormal

The practice of science always takes place in a social context in which political, social and cultural pressures shape lines of investigation and even the philosophical assumptions that underpin scientific enquiry. Asif Siddiqi reminds us that the achievements of the Soviets in space were shaped in part by a philosophical tradition that many might see as antithetical to the rational objectivism of scientific enquiry. He offers an illuminating discussion of the roots of the Soviet space programme as they lay in the essentially mystical tradition known as Cosmism. By the 1960 Soviet scientists were at the forefront of world research into paranormal and occult phenomena, all indirectly stimulated by competition with the USA. Dr. Leonid L. Vasil'ev, chair of physiology at Leningrad University, headed a special laboratory for parapsychology and predicted that 'The discovery of the energy underlying extrasensory perception will be equivalent to the discovery of atomic energy'.[35] Paul Betts's chapter on the trial of Cardinal Mindszenty in Hungary in February 1949 shows that in the USA, too, the Cold War stimulated the interest of the Central Intelligence Agency in 'dark psychology' and the paranormal, although by the 1960s the USA state felt it was falling far behind the Soviet Union in the field of paranormal research. In the Soviet Union by 1967 some 20 centres specializing in such research were in operation, with a budget of well over 13 million dollars. In May 1968, Moscow hosted the First International Conference on Parapsychology.[36] Other Eastern Bloc countries were also involved in such research. In Czechoslovakia, which had a tradition of psychical research that went back to the beginning of the twentieth century, scientists developed psychotronic generators to store and use the bio-energy they believed was behind such phenomena as psychokinesis.[37] In her chapter, however, Monica Black reminds us that within the Socialist bloc there were those who were deeply sceptical of what they considered to be a turn towards the irrational or mystical. She tells an engaging tale about two comrades-in-arms active in what she calls the 'superstition wars' on opposite sides of the West/East German border. She suggests that in

---

[35] S. Ostrander and L. Schroeder (1970), *Psychic Discoveries behind the Iron Curtain* (Englewood Cliffs, NJ: Prentice-Hall), p. 7.

[36] Ostrander and Schroeder, *Psychic Discoveries*, p. 300.

[37] Ostrander and Schroeder, *Psychic Discoveries*, p. 366.

this case it was the experience of Nazism more than the Cold War that inspired heir antipathy to pseudoscience.

For the still largely rural populations of Eastern Europe the blurred boundary between science and 'superstition' was most evident in the sphere of healthcare, where 'folk medicine', or more properly ethnomedicine, came under challenge from modern biomedicine. In Russia peasants had traditionally made a loose distinction between illnesses with a 'natural' cause, such as chills or bruises, and those with a 'supernatural' cause, such as infections, nervous diseases and internal ailments (with a mediating category of illness brought on by moral transgression). The *znakhar'* (from the Russian verb 'to know') and, more commonly by the twentieth century, his female counterpart, the *znakharka*, were healers who used therapies based on spells, charms and herbal remedies to drive out illness, restore mental well-being, adjust broken limbs, stop haemorrhages, cure snake bites or undo curses and bewitching. These therapies were associated with the operation of white magic.[38] In the pre-war era, the Bolsheviks held such folk medicine in contempt. 'What absurd methods will the wizards, wise women, znakhari and folk healers not resort to? They use spells and spit on the sick, blow on the sick part of the body or clean it with water that has been used to wash a corpse. They stick pieces of paper onto the sick person inscribed with secret signs, they give the sick person live fleas to drink in water or even urine'.[39] By the 1960s, modern health provision, based on biomedicine, had spread across the Soviet Union, and there was little official tolerance of ethnomedicine, although it is noteworthy that in some major cities homeopathy clinics existed, in spite of a 2-year study in 1939 that concluded that homeopathy was entirely ineffective.[40] Nevertheless among the populace, and not only in the countryside, a sick person was still likely to consult a folk healer, depending on the nature of their illness or the effectiveness of their initial treatment, or simply as a form of double insurance. Now some researchers sought to differentiate between the 'rationality' of herbal medicine and the 'superstition' of curative magic. This trend, which

---

[38] L. Ivanits (1989) *Russian Folk Belief* (Armonk, NY: M. E. Sharpe), p. 111.

[39] A. Rostovskii (1925) *Pop, znakhar' i vrach: religiia i meditsina* (M: Izd-o Narkomzdrava), p. 4.

[40] L. I. Min'ko (1971) *Znakharstvo: istoki, sushchnost', prichiny bytovaniia* (Minsk: Nauka i tekhnika), p. 103.

had its parallel in the rise of alternative therapies in the West, continued into the 1980s, by which time it had become fashionable for a now far healthier population to turn back to folk medicine, as the extraordinary popularity of the *Almanac of Russian Folk Medicine* (Russkii narodnyi lechebnik) by P. M. Kurennov, a naturopath who had fled to Harbin after the October Revolution, showed.

In the course of the 1950s, universal health case was instituted across the Eastern Bloc, and here, too, the expectation was that the medicalization of healing would drive out folk medicine. Yet Bulgaria, offered a dramatic counter to that expectation, showing the capacity of the paranormal to vie with biomedicine. In the late 1960s and early 1970s, Mother Vanga, a blind clairvoyant born in 1911, transformed herself with the assistance of the Bulgarian government into a kind of living saint, successfully connecting her visions, premonitions and spiritual experiences to the language of biomedicine. Drawing on a cross-class clientele, which included members of the political elite, Mother Vanga offered cures for physical illness and solace from bereavement and loss.[41] Her most famous disciple was Lyudmila Zhivkova, only daughter of Todor Zhivkov, the long-time party leader and head of state. At the height of her political career—she died at age 39 years—Zhivkova was a member of the Politburo and head of the State Commission for Science, Culture and Art, a committed Theosophist and anti-Soviet by temperament. The Institute of Suggestology, active from 1966 to 1986, began a large-scale study of Vanga's gifts, under its director Dr. Georgi Lozanov, who held a Ph.D. in Psychology from Moscow State University. He defined suggestology as applied parapsychology and saw it as a way to build an individual's extrasensory perception.[42] In the words of Galina Valtchinova, the appropriation of Mother Vanga represented a 'remaking of religion under Socialism (and) amounted to taking it outside ecclesiastic institutions, giving it a rational face, and putting it into a "scientific" mold, thereby dissolving "religiosity" into the blurred "magical-religious-medical field"'.[43]

---

[41] G. Valtchinova (2009) 'Between Ordinary Pain and Extraordinary Knowledge: The Seer Vanga in the Everyday Life of Bulgarians during Socialism (1960s–1970s)', *Aspasia*, 3, 106–30.

[42] Ostrander and Schroeder, *Psychic Discoveries*, pp. 286–8.

[43] G. Valtchinova (2010) 'State Management of the Seer Vanga: Power, Medicine and the "Remaking" of Religion in Socialist Bulgaria', in B. R. Berglund and B. Porter-Szucs (eds.), *Christianity and Modernity in Eastern Europe* (Budapest: CEU Press), p. 261.

Church, State and Socialism

As mentioned above, until recently scholars have neglected the importance of religion in the Cold War, especially prior to the 1960s. The USA presented the struggle against Communism as a moral crusade of godliness against godlessness, and more Americans identified Communism with atheism than with nationalization of the means of production. Presidents Truman and Eisenhower seized on the absence of religious freedom as a symptom of the wider subjugation of the masses in the Eastern Bloc. Pope Pius XII, an inveterate enemy of Communism, identified the Christian cause with that of the West and discussed containment with Truman.[44] Paul Betts's chapter reveals how in Hungary in February 1949 the trial of Cardinal Mindszenty became a cause célèbre of religious persecution, human rights and Soviet tyranny. For its part, the Soviet Union and its allies cast the Catholic Church, in particular, as a citadel of obscurantism and a bulwark of fascism and reaction. The Orthodox Churches, especially the Russian, were pressed into service on the Cold War stage by their respective governments, particularly through their international activities in support of the Soviet-backed peace movement. The Christian Peace Conference, for example, founded in 1958 by the Czech pastor Josef Hromádka, is said to have received Soviet funding. And the involvement of the Russian and Romanian Orthodox Churches in the World Council of Churches from 1961 may have helped ensure that that body did not unite to condemn religious persecution behind the Iron Curtain.[45] During the 1960s, however, the specifically religious dimension of the Cold War eased somewhat as the World Council of Churches and then, following the Second Vatican Council (1962–1965), the Roman Catholic Church took an increasingly progressive stance on peace and justice issues. These developments probably helped strengthen doubts in Communist circles about the satisfactoriness of their anti-religious policies.

Despite the image of the Eastern Bloc as being under the thumb of Moscow, policies towards religion were rather varied and reflected national differences. In the late 1940s, Stalin imposed the Bolshevik model of separating

---

[44] D. Kirby (2006) 'The Cold War, the hegemony of the United States and the golden age of Christian democracy', in H. McLeod (ed.), *Cambridge History of Christianity Volume 9: World Christianities c.1914–c.2000* (Cambridge: Cambridge University Press), pp. 285–303; A. Preston, 'Introduction: The Religious Cold War', in Muehlenbeck (ed.), *Religion and the Cold War*, pp. xi–xxii.

[45] Lucian Leuştean, 'Introduction', *Orthodoxy and Cold War*, p. 8.

church and state (where this had not already happened), nationalizing ecclesiastical property, banning religious instruction in schools, curbing the public activity of religious organizations and restricting their access to mass media. Almost universally, church leaders reacted hostilely to the Communist takeovers, although some priests and laity were more positive, and this precipitated the arrest and imprisonment of the most vocal ecclesiastical critics. Following the Mindszenty case, the Vatican advised Catholic bishops to pursue a nonconfrontational course but not to collaborate actively with the communist authorities. Communist governments proceeded to implement a tactic of divide-and-rule, favouring some confessions while treating others harshly. Over time, however, a modus vivendi between churches and the state was reached, some clerics accepting that some engagement with the state was inevitable if resources were to be accessed, others endorsing the regimes for opportunistic, nationalist or social-justice reasons. By the 1960s overt conflict between state and religious organizations had diminished in most countries, a reflection of the fact that the Eastern Bloc governments were increasingly inclined to pursue courses independent of Moscow and that the churches themselves were less anti-Communist than they had been in the 1950s. One sign of the general easing of church–state tension was Christian-Marxist dialogue, which had its heyday between 1966 and 1971. Finally, we should note—although the effect is hard to measure—that rapid modernization in many Socialist countries from the late 1950s served to erode the high levels of religiosity that had been evident at the end of the Second World War.

These generalizations aside, what was crucial was the diversity that characterized the religious fields of the different countries incorporated into the Socialist bloc, a diversity that influenced the religious policies practised by the different governments. This diversity was manifest in, first, the historic divide between Western Christian confessions and Eastern Orthodox Churches, the latter being organized along ethnic lines of autocephaly. Secondly, there were countries on both sides of the Iron Curtain that were largely monoconfessional, notably the overwhelmingly Roman Catholic countries of Poland, Croatia and Lithuania and, on the Orthodox side, Bulgaria and Romania. In these countries religion served to underpin national identity. Thirdly, and more common, were countries that were religiously mixed, where one confession might have a plurality of adherents yet exist alongside other confessions with significant social support. In Hungary, for example, Roman Catholics were the largest denomination but there was a significant

Reformed confession and a small Lutheran minority. Albania was largely Islam (Sunni, but with a Bektashi minority) but had an Orthodox minority, and in the north a small Catholic minority was strong.[46]

Following the death of Stalin in 1953, religious policies began to diversify. In Poland the authorities initially sought to contain the influence of the Catholic Church in public life, but after 1956 restrictions on religious education, the press, seminaries and monasteries were lifted, and in Lublin the one Catholic university in the Eastern Bloc functioned. Jim Bjork's nuanced chapter, however, highlights the importance of regional variations in the strength of Polish Catholicism and points to the extent to which secularization was taking place up to the mid-1970s. In 1978 mounting intellectual and working-class opposition to the regime was given impetus by the election of Archbishop Karol Wojtyła of Krakow to the papal throne. The return of John Paul II to his native land in the following year firmed up anti-Communist sentiment and catalysed the events that would lead 10 years later to the fall of the regime. In East Germany, the only country in the Socialist bloc where Protestantism was dominant, state policy towards the Evangelical Church went from being repressive in the first couple of decades to becoming one of the most liberal after 1969, when the Evangelical Church broke ties with its counterpart in West Germany and declared 'critical solidarity' with the regime. In return, the government made concessions with regard to church property, tempered atheist propaganda and, by the mid-1970s, allowed the Lutheran churches to appoint bishops without state interference and made provision for an alternative to military conscription.[47] In Yugoslavia, too, the authorities pursued a relatively liberal course, while remaining vigilant about the association of different confessions with different ethnic groups (the Serbian Orthodox Church with Serbia, the Catholic Church with Croatia and Slovenia, and Islam with Bosnia). By the 1960s religious orders and welfare organizations operated freely, large religious gatherings were permitted, and the state did not interfere in seminaries and church life.

In Romania up to the 1970s religious policy also appeared relatively liberal but this hid the fact that government support for the Romanian Orthodox

---

[46] This overview is based mainly on Ramet, *Nihil Obstat*; P. Walters, 'The Revolutions in Eastern Europe and the Beginning of the Post-Communist Era', in H. McLeod (ed.), *The Cambridge History of Christianity: vol.9: World Christianities, c.1914–c.2000*, pp. 348–65; Leuştean (ed.), *Eastern Christianity*.

[47] Betts, *Within Walls*, pp. 77–8.

Church entailed its almost complete cooptation. It suited the Romanian Communist Party, which was bereft of nationalist credentials, to court the Church while clamping down on other confessions. Orthodox clergy were paid a salary by the state (something that did not happen in the Soviet Union or Bulgaria), churches were restored, and religious publications were fairly free to publish. Following the 'divorce' from Moscow in 1964, Nicolae Ceauşescu set about creating an autarkic, militaristic form of 'national Stalinism' but did not baulk at appropriating religious symbols to bolster his legitimacy.[48] He oversaw the restoration of many churches, tolerated Orthodox baptisms, marriages and funerals, and even publicized his father's burial with Orthodox rites in 1972. From this time on, however, there was a slide to a more repressive policy, with churches closed and atheist propaganda stepped up, developments discussed in Zsuzsanna Magdo's fine chapter.

In a religiously diverse country such as Czechoslovakia, where peasant Slovakia was largely Catholic and the Czech lands Catholic, Hussite, Evangelical and Uniate, the government was inclined to a divide-and-rule policy. With the brief exception of the Prague Spring, the Czechoslovak authorities clamped down on all religion but especially on the Catholic Church, whose clergy were licensed by the state and paid official salaries and whose bishops were subject to government approval. By the early 1980s, 8 out of the 13 Catholic dioceses had no bishop and over one-third of Catholic parishes had no priest.[49]

If politics was paramount in determining the fate of the churches, levels of socio-economic modernization were not irrelevant. In Czechoslovakia and East Germany, the two most industrially developed countries of the Socialist bloc, a very significant degree of secularization took place during the Communist period, with falling church attendance and observance of Christian life-cycle rituals.

Atheism Reconfigured

By the mid-1960s, the resilience of religion in the Soviet Union was causing concern in party circles. On the one hand, two generations had grown up within the Soviet system, largely without any connection to organized religion, and they tended to present themselves as nonbelievers,

---

[48] V. Tismaneanu (2003) *Stalinism for All Seasons: A Political History of Romanian Communism* (Berkeley, CA: University of California).

[49] S. Bruce (1999) *Choice and Religion: A Critique of Rational Choice Theory* (Oxford: Oxford University Press), p. 102.

if not committed atheists. On the other, there was a widespread if cautiously expressed conviction among political and scientific elites that the drive to eliminate religious belief was failing. The exploiting classes had long disappeared, industrialization, urbanization and the collectivization of agriculture had transformed Soviet society, creating the conditions that secularization theory predicted would erode religious belief. Yet religion displayed a capacity to adapt to conditions of modernity. Those tasked with propagating 'scientific atheism' were especially aware that it barely resonated with believers. In November 1963 the ideological commission of the Central Committee established an Institute of Scientific Atheism, tasked with coordinating research on the topic. Its journal, *Voprosy nauchnogo ateizma*, published some serious research 'hidden among the ideological genuflections and the considerable quantity of hackwork'.[50] The number of articles in the press on scientific atheism rose sharply and the new ideological commissions that were attached to party organizations were exhorted to improve atheist education.[51] The editorial board of *Nauka i religiia*, a journal that had been going for barely 5 years, was overhauled in summer 1964. The journal, it was said, must cease to see science and religion as opposed and 'address issues that extend far beyond the relationship of science and religion, (such as) the meaning of life, happiness and solace'.[52] Rather than underscoring the negative aspects of anti-religious propaganda, the journal should speak to themes such as moral norms and appeal to readers' emotions and everyday concerns. Increasingly, some of those associated with the journal began to argue that fundamental to atheism's failure was its refusal to confront existential questions.[53]

All the regimes of the Socialist bloc espoused scientific atheism, yet the energy invested in this varied from state to state. In 1954 Walter Ulbricht declared that East Germany 'will carry on its campaign against superstition, mysticism, idealism and all other unscientific worldviews', and the launch of Sputnik in 1957 encouraged his government to step up atheist education in schools.[54] In Romania a Society for the Popularisation of Science and Culture was established in 1949 to 'propagate among the labouring masses political and scientific knowledge to fight obscurantism, superstition and

[50] Anderson, *Religion, State and Politics*, p. 39.
[51] Anderson, *Religion, State and Politics*, p. 42.
[52] Cited in Smolkin-Rothrock, 'Ticket', p. 176. S. Luehrmann (2011) *Secularism Soviet Style: Teaching Atheism and Religion in a Volga Republic* (Bloomington: Indiana University Press).
[53] J. Thrower (1992) *Marxism-Leninism as the Civil Religion of Soviet Society* (Lampeter: Edwin Mellen), p. 47, pp. 49–50.
[54] Betts, *Within Walls*, pp. 59, 66.

mysticism and all other influences of bourgeois ideologies', but, as Magdo mentions, few resources were devoted to this end.[55] Albania was the only state in the Socialist bloc to declare itself atheist, the Hoxha regime unleashing a drive in 1967 to close every mosque, church and monastery in the country. Article 37 of the 1976 Constitution proclaimed: 'The state recognizes no religion and supports atheistic propaganda in order to implant a scientific materialistic world outlook in the people'. The following year the penal code imposed terms of 3–10 years imprisonment for 'religious propaganda and the production, distribution, or storage of religious literature'. Interestingly, the motivation seems to have been largely nationalist, with the Albanian Party of Labour, claiming that the country's three confessions were destroying the 'national uniqueness' of Albania.

In Czechoslovakia, atheist education was pursued with rigour although, unusually, the government allowed clergy into primary schools after school hours to give religious education. In December 1957, the Central Section of Scientific Atheism of the Czechoslovak Society for the Dissemination of Political and Scientific Knowledge organized a conference to discuss the poor state of anti-religious work in the country's schools. In 1958 and 1959 mandatory courses in atheism were introduced in all institutions of higher education.[56] In 1960 the new constitution laid down that education be conducted 'in the spirit of the scientific worldview of Marxism-Leninism'.[57] In the mass media film, radio and television produced work with anti-religious themes, especially hostile towards a Catholic Church seen as deeply reactionary. What is striking is that the number of Czechoslovak citizens claiming to be Christian fell dramatically from 94.6 % in 1950 to about 30 % in 1985, though how far this derived from anti-religious policy, the secular nature of public institutions, or the secularizing effects of socio-economic modernization is hard to determine.[58] It is probably not without significance that Czechoslovakia had a long tradition of secularism and freethinking, organized around mainly middle-class associations, which facilitated the rapid decline of religious belief under Communism. Whereas religion has revived since 1989 in many parts of the former Socialist bloc, including the Soviet Union, the proportion of

---

[55] L. N. Leuştean (2007) 'Constructing Communism in the Romanian People's Republic. Orthodoxy and State, 1948–49', *Europe-Asia Studies*, 59:2, 303–29.

[56] Ramet, *Nihil Obstat*, p. 125.

[57] Ramet, *Nihil Obstat*, pp. 129, 130.

[58] Bruce, *Choice*, p. 102.

self-defined atheists in the Czech lands (i.e. excluding Catholic Slovakia) has remained extremely high at 64.3 %.[59] It is also significant that levels of professed atheism remain high in Estonia and former East Germany, both countries, like the Czech lands, with a predominantly Protestant tradition. Neither of these countries, however, was subject to the same level of systematic anti-religious work as Czechoslovakia, although East Germany had a similarly long tradition of radical freethinking. Clearly this, combined with a Protestant, individualist culture and relatively high levels of socio-economic modernization, provided fertile soil for rapid secularization.

In the Brezhnev era Soviet sociologists of religion were enlisted to help explain the persistence of religion. Like sociologists of religion everywhere, they grappled with the standard problem of finding appropriate measures of religiosity. Did declining church attendance, for example, indicate a decline in religious belief?[60] One of the interesting avenues of research they pursued was investigation of the religiosity of women, since women were clearly the major vector for the reproduction of religious belief across generations. Why, these sociologists asked, did women comprise at least two-thirds of Orthodox congregations? They explored factors such as the loss by millions of women of husbands and potential spouses as a result of the war, low levels of education, isolation in the city, marginalization from the world of employment and—the stand-by since the 1920s—their 'political backwardness'. In addition, a few cited the 'heightened emotional perception' of women compared with men and their 'desire for beauty'. Certain factors, however, could still not be talked about, such as the perception on the part of many rural women that Soviet institutions represented a threat to the moral order of the family and community.[61]

Elsewhere in the Socialist bloc, official concern about the refractoriness of religious belief also led to the enlistment of the social sciences. Jim Bjork discusses the relatively well developed field of quantitative surveys of religious practice in Poland in the 1960s. Patrick Hyder Patterson offers

---

[59] I. Borowik, B. Ančič, R. Tyrała (2013) 'Central and Eastern Europe', Stephen Bullivant and Michael Ruse (eds.), *The Oxford Handbook of Atheism* (Oxford: Oxford University Press), ch. 39.

[60] Powell, *Antireligious Propaganda*, p. 132.

[61] J Anderson (1993) 'Out of the Kitchen, Out of the Temple: Atheism and Women in the Soviet Union', in S. P. Ramet (ed.), *Religious Policy in the Soviet Union* (Cambridge: Cambridge University Press), pp. 206–30; I. Paert (2004) 'Demystifying the Heavens: Women, Religion and Khrushchev's Anti-Religious Campaigns, 1954–64', in M. Ilič, S. E. Reid and L. Attwood (eds.), *Women in the Khrushchev Era* (Basingstoke: Palgrave), p. 206.

an illuminating comparison of Hungary and Yugoslavia—regimes both known for their ideological flexibility—where social scientists distanced themselves from Marxist materialism in a bid to develop more refined understanding of religious commitment. Rather than perceive religion as a problem to be dealt with by scientific education and social engineering, they engaged with it as a response to the dilemmas of human existence, although in more explicit fashion than their Soviet counterparts, pointing to such factors as a widespread sense of helplessness in the face of personal and social problems. In the rather curious case of Romania, as Zsuzsanna Magdo shows, Ceauşescu's liberalization in the second half of the 1960s and very early 1970s, saw religiosity recognized as a 'complex phenomenon' that needed to be investigated empirically through ethnography, psychology and sociology, elements of which might even be appropriated to create a distinctively Romanian spirituality. From the 1960s, then, in Eastern Europe in particular the models and methods used by sociologists of religion differed little from those of their counterparts in the West, a reminder that where intellectual traffic was concerned, the Iron Curtain was not an impermeable barrier.

Forging Socialist Rituals

It was with the creation of a complex of life-cycle rituals that the Soviet Union and the Socialist bloc countries ventured furthest in turning Marxism–Leninism into ersatz religion. Yet in contrast to the indifferent success of anti-religious propaganda, the efforts to supplant Christian rituals with secular ones appear to have been rather successful. The Bolsheviks had always recognized that there was more to religion than creedal propositions and they acknowledged the place of ritual in cementing communities and marking key rites of passages. In the 1920s they made a rudimentary attempt to craft substitutes for Christian (and to a much lesser extent, Islamic) life-cycle rituals, but these were vitiated by crude anti-religious blather. A book that outlined a model ceremony for 'octobering' a newborn child, a kind of Socialist baptism, explained that after welcoming the arrival of the new citizen into the Socialist commonwealth, the master of ceremonies should:

> paint a bright picture of the crudity and barbarism of baptism, tell how they blow and spit on the child to drive away the devil, how they imperil the health of the baby by immersing it in cold water etc. They should consult

a church calendar for the Russian translation of saints' names and use them to pour scorn on the baptismal ritual—e.g. the name Klavdiia means 'lame'; Spiridon, 'a basket'; Varvara, 'rude'.[62]

As the militant godless movement dwindled during the 1930s, these secular rituals, never remotely popular, fell into disuse. Only in the late 1950s was there a new attempt to create such rituals and, interestingly, the Soviet Union played a follower rather than a leader role, certain countries in the Socialist bloc taking the initiative. Significantly, within the Soviet Union it appears to have been the Baltic states that pioneered these new rituals. In spring 1957, for instance, the Central Committee of the Estonian Communist Party proposed an alternative to the Lutheran confirmation ceremony. In 1959, Leningrad became the first city in the Soviet Union to open a special palace of weddings and in 1965 it opened a baby palace where a special rite for newborn babies was performed.[63] Only in 1963 did the Ideological Commission of the Central Committee call for the 'more active introduction into the life of the Soviet people of non-religious holidays and rites'.[64]

The new rituals seemed to filled a social need. With civil weddings much of the traditional symbolism—wedding gowns, flowers and so forth—was retained, but the couple was reminded of their obligations to Socialist society and to stable family life. In Leningrad the official in charge of a wedding pronounced 'The Soviet family is a union of people warmly loving one another who have decided to go through life together. It is the most important cell of our state'.[65] Civil weddings became so popular by the 1970s that in some major cities it was difficult to book a slot in the local palace of weddings. In Ukraine by 1978 there were 100 palaces and houses, 400 rites bureaux, 8500 halls, 1500 choirs and 1150 orchestras dedicated to enacting the new rituals.[66] Between 1975 and 1982, secular weddings in Ukraine rose from 64 % to 82.9 % of all weddings and secular funerals rose from 6 % to 49.5 % (though in the late 1980s there was a revival of religious rituals).[67] The ritual to mark the birth of a new baby

[62] I. V. Stepanov (1929) *Kak vesti antireligioznuiu propagandu v derevne* (Leningrad: Priboi), p. 72.
[63] Anderson, *Religion, State and Politics*, p. 46.
[64] Anderson, *Religion, State and Politics*, p. 48.
[65] Lane, *Rites of Rulers*, p. 76.
[66] Anderson, *Religion, State and Politics*, p. 120.
[67] Anderson, *Religion, State and Politics*, p. 120.

began to grow after 1963. This might entail a communal ceremony in which new parents and their relatives attended a Day of Family Happiness, where they were presented with certificates, medals or symbolic gifts such as a Young Pioneer scarf; or it might comprise a simple visit to a family by a local official bearing flowers and a birth certificate. Officially designated as a 'ritual of name-giving', it was often called in popular parlance a *krestiny*, a christening (the equivalent term circulating in Hungary from the mid-1950s).[68] In contrast to the rituals of the 1920s, these late-Soviet rituals were neither explicitly anti-religious nor marked by a heavy emphasis on duty to the Socialist state. They were focused on the family, and the purpose was to provide an emotionally fulfilling and aesthetically pleasing ceremony to mark a rite of passage. At the same time, of course, the hope of the authorities was that by tapping into the emotional freight of these ceremonies, the individual and family would be bound more tightly into the Socialist collective.

In novel fashion Heléna Tóth explains in detail how new life-cycle rituals were created in Hungary from the mid-1950s on, focusing on the role of Zoltán Rácz of the Institute of Public Education. As she argues, the process of creating rituals that would inject values of Socialist humanism into daily life required innovation and imagination. Rácz looked particularly to the Czechoslovak experience (as did his East German counterparts), reminding us of the extent to which ideas and practices were borrowed and adapted by Socialist states. As in the Soviet Union, secular weddings were especially popular, with more than 70 % of weddings civic by the 1980s. The name-giving ceremony was not quite so popular but by the mid-1970s it extended to about 30 % of newborn children. In East Germany secular weddings and name-giving ceremonies also rose steadily in popularity, as did the Jugendweihe, or youth dedication ceremony. The latter faced dogged resistance from the churches, given its close resemblance to Christian confirmation, and especially after the 1953 uprising—and again in the late 1950s—the regime projected it very much as an attack on Christian values, aimed at 'imparting to young people useful knowledge in basic questions of the scientific world-view and Socialist morality, at raising them in the spirit of Socialist patriotism and proletarian internationalism'. Nevertheless, by 1958, Jugendweihe participation was up to 40 %, and by 1959 up to 80 %

---

[68] C. Lane (1981) *The Rites of Rulers: Ritual in Industrial Society—The Soviet Case* (Cambridge: Cambridge University Press), pp. 68–9.

of East German youth; while Christian confirmation dropped from 75 % to less than 33 % by the second half of the 1950s.[69]

In contrast to this was the much slower uptake of secular funerals. In his innovative chapter Felix Schulz discusses the ideological problems that Marxist–Leninists faced in coming to terms with death and in creating an emotionally rewarding Socialist funerary culture. As materialists, they rejected the idea of an afterlife and insisted that death marked the end of existence. However, in the 1960s and 1970s philosophers struggled to fill an ethical vacuum, pointing to the need to celebrate the service of the deceased to the collective and his or her success in living a life according to the ideals of Socialist humanism. Czechoslovakia was again the pioneer in creating a robust Socialist funerary culture. In 1958 the Communist Party took the conduct of funerals into state hands, redesigning cemeteries, promoting cremation, creating provision for the interment of ashes, designing secular funerals and training orators.[70] In East Germany, by contrast, some 60 % of cemeteries remained in church ownership (the figure for Hungary was even higher, at almost 70 %).[71] There secular funerals borrowed oratory, music and—later silent commemoration—from Christian sources and orators were trained to deliver speeches in a 'thoughtful and emotionally convincing' manner, using 'clear, good German without excessive pathos', and instructed to 'take into consideration the individuality and personality of the deceased, in both his life and work'.[72] Despite this, the advance of secular funerals was fairly slow. By the 1970s, moreover, there were complaints about the empty, formulaic quality of many such funerals and the shabby state of some crematoria. So although by 1986 only 7 % of the population remained members of the Evangelical Church, 30–40 % opted for a church funeral.[73]

Diversity in the ways that Socialist bloc countries handled death is neatly captured in respect of cremation. In December 1918 the Bolsheviks had legalized this and opened the first crematorium in Petrograd in 1920, though it quickly malfunctioned, and it was not until 1927 that the first crematorium started to operate in Moscow. Bolshevik discourse around cremation followed that of the international cremation movement, emphasizing the secular, hygienic and economically rational advantages of cremation over inhumation, but adding to

[69] Betts, *Within Walls*, p. 58, p. 72.
[70] F. R. Schulz (2013) *Death in East Germany, 1945–1990* (Oxford: Berghahn), pp. 204–5.
[71] Betts, *Within Walls*, p. 76
[72] Betts, *Within Walls*, p. 76.
[73] Betts, *Within Walls*, p. 82.

this strong anti-religious invective.[74] Yet if the propaganda about cremation was upbeat, practical efforts to promote it were feeble. As late as 1996, there were only eight crematoria in what had been the Soviet Union—three in Moscow, two in Ukraine (Kyiv and Khar'kiv) and one each in Minsk, St. Petersburg and Ekaterinburg—and a majority of these had only been opened in the 1980s. A widely cited figure for the Soviet Union is of a national cremation rate of around 45 %, but it seems to be based solely on cities with crematoria.[75] In fact, as in all Orthodox countries, there was strong popular opposition to cremation. This was rooted in Orthodox doctrine that at the Last Judgement the bodies of the dead will be resurrected and reunited with the soul, as well as in more demotic beliefs that the soul does not leave the body immediately after death and that the earth is the Mother to which all of us must return.[76] Where Orthodoxy was the dominant confession, rates of cremation were extremely low, though slightly less so in Serbia than in Bulgaria and Romania. In Romania it was a nearly decade into the Communist era before the number of cremations reached the pre-war level, and even in 1989 there were only 1689 cremations out of a population approaching 23 million.[77]

By contrast, Czechoslovakia built upon a robust pre-war cremationist movement, and in line with its centralized drive to promote a Socialist funerary culture, invested resources into promoting cremation and building crematoria. In 1962, the national cremation society in Czechoslovakia claimed a membership of about half a million out of population of 12 million, an astonishing figure if true. Between 1956 and 1997 the cremation rate rose from 16 % to 76.2 % (the latter figure relates only to the Czech Republic), most of that increase taking place from the 1970s.[78] East Germany had an even older cremation movement than Czechslovakia: in 1964 it had 53 crematoria, only two of which had been built since 1950. The cremation rate rose from 24.4 % in 1956 to 63.2 % in 1985, making it one of the highest in Europe, and a stark contrast to West Germany, where the rate was 28 % at

---

[74] A. Rostovtsev (1931) *Za ideiu krematsii* (Moscow: OGIZ), p. 38.

[75] C. Binns (2005) 'Russian and Soviet Transitions' in D. J. Davies (ed.), *Encyclopedia of Cremation* (Aldershot: Ashgate), 370–1.

[76] S. A. Smith (2009) 'Spasenie dushi v Sovetskoi Rossii' [The Salvation of the Soul in Soviet Russia], *Neprikosnovennyi Zapas*, 2 (64) http://magazines.russ.ru/nz/2009/2/ss16.html.

[77] M. Rotar (2013), *History of Modern Cremation in Romania* (Newcastle: Cambridge Scholars), pp.275–77.

[78] Schulz, *Death*, pp. 134, 141.

the end of the 1980s.[79] Felix Schulz points out that the rise in the cremation rate in Britain was actually greater than in East Germany, which leads him to be cautious in seeing Communist ideology as the major driver of change. Instead he stresses local variation in the cremation rate across East Germany and stresses the economic and hygienic factors that made cremation attractive.[80] Certainly, in the last third of the twentieth century cremation developed rapidly in the capitalist as well as the Communist worlds, and factors such as cost, land shortage, the decline of communal and kinship ties, religious tradition (countries such as East Germany and the Czech lands with Protestant traditions being more accepting of cremation than those with Roman Catholic or Orthodox traditions), all influenced the level and speed of adoption of cremation. All of this serves to remind us that we should be cautious in assuming that Communist political culture was determinative of social developments across the Socialist bloc.

Religious Heritage and Nationalism

Jorn Rusen has suggested that the 'disenchantment of the world', which Max Weber saw as constitutive of modernity, brings an 'inescapable diachronic comparison' between past and present that fosters a drive to re-enchant the past. 'Rationalization is only one side of the coin of modernization. There is always a reaction against it, a re-enchantment in the relationship to the past which at least compensates for the loss of sense and meaning brought about by rational methodologies'.[81] As Communist countries modernized, the perception of increasing distance from the nation's often blood-soaked past fostered complex feelings towards the nation's history that included—even in party circles—a desire to embrace elements of the past that had been expunged from official memory. In May 1964, 6 months before Khrushchev's ouster, students in the Soviet Union founded the Rodina (Motherland) club 'to promote the study of historic monuments and to study ancient art and history'.[82] In July 1965 the Council of Ministers of the RSFSR set up the All-Russian Society for the Preservation of Historical and Cultural

[79] Schulz, *Death*, pp. 126–7, 136, 149.
[80] Schulz, *Death*, pp. 133, 153.
[81] J. Rusen (1996) 'Some Theoretical Approaches to Intercultural Comparative Historiography', *History and Theory*, 35:4, 21.
[82] Y. M. Brudny (1998) *Reinventing Russia: Russian Nationalism and the Soviet State, 1953–91* (Cambridge, MA: Harvard University Press), p. 68.

Monuments (VOOPIK), a mass heritage organization under the loose control of the party. This preservationist impulse is commonly tied by scholars to the emergence of conservative Russian nationalism, and though the connection is strong—e.g. those around *Molodaia Gvardiia* argued that neglected churches were a sign that Russia was losing her 'civilization of the soul'[83]—many activists in VOOPIK were apolitical, concerned solely to safeguard and reconnect to Russia's cultural heritage. In a previous essay Catriona Kelly has shown how historic churches in Leningrad, once seen as symbols of counter-revolution, were configured in the Brezhnev era as 'national heritage'.[84] In this volume she explores the salience of *nauka*—'science' in the sense of the full range of intellectual and academic activity—in determining which churches and architectural styles were deemed worthy of preservation. In a perceptive analysis of how policy emerged out of unequal interaction between believers, architect-conservators and atheist-minded officials—a model of interaction between differentially empowered groups that may well have salience for other areas of policy making—she shows how the re-inscription of churches as heritage facilitated the preservationist movement. Nowhere was this new sensibility more apparent than in the Russian North, where the astonishing wooden churches of Arkhangel'sk, Karelia and Vologda were being left to rot as people deserted the countryside. Preservationists campaigned to create a museum-reserve out of the multidomed churches on Kizhi island in the middle of Lake Onega in Karelia, which in 1966 was designated a museum of national importance and in 1990 a UNESCO World Heritage site. The guidebook of the new museum described the churches of Kizhi as a 'bright reflection of the artistic genius of Russian people ... All were created by the hands of the simple people, although they were ordered to build them by princes, monasteries and churches'. Comments in the visitors' book, however, revealed a different sensibility: 'Without faith, we cannot live. It is the foundation that makes life and makes the building of life steady and keeps it safe from catastrophe'.[85] Kelly reflects on the frequent claim that the Soviet project of secularization was a 'failure',

[83] Brudny, *Reinventing*, p. 76.
[84] C. Kelly (2013) 'From "Counter-Revolutionary Monuments" to "National Heritage": The Preservation of Leningrad Churches, 1964–1982', *Cahiers du monde russe*, 1, 1–30.
[85] S. A. Smith (2015) 'Contentious Heritage: The Preservation of Churches and Temples in Communist and Post-Communist Russia and China', in Paul Betts and Corey Ross (eds.), *Heritage in the Modern World* (Past and Present Supplement, 10), p. 189.

pointing to the striking success of Leningrad's architect-conservators in pushing through their own vision of how historic churches should be maintained for their aesthetic and architectural features, if not for their religious history.

From the 1960s in many parts of the Socialist bloc, regimes still formally loyal to Moscow began to pursue a more independent, 'nationalist' course. Communist states, like all twentieth-century polities, based themselves on the territorial and social space of the nation–state, and one of the most effective ways in which these often deeply unpopular regimes could build legitimacy was to focus the identities of their citizens upon the nation (often defined in ethnic rather than civic terms) and to present the state as the institutional repository of the nation. We have seen that in Orthodox countries, particularly, Socialist governments were not slow even in the Stalinist years to tap into the cultural capital of the autocephalous churches. As a family of largely self-governing churches, technically 'in communion' with one another, but not overseen by any centralized authority akin to the papacy, the autocephalous churches possessed symbolic capital that Communist states could seek to appropriate. Lyudmila Zhivkova, the 'uncrowned princess' of Bulgaria, created a permanent exhibition of Bulgarian medieval religious icons in the crypt of Alexander Nevski Cathedral in Sofia and saints Cyril and Methodius, bringers of Christianity to the Bulgarian people, were vigorously promoted to the point where they became icons of Bulgarian nationhood.[86] In Romania Zsuzsanna Magdo tells us that 79 churches were restored at state expense between 1960 and 1977, retaining their religious functions and also serving as sites of national memory (a point that Igor Polianski develops in an unusual way). And even regimes that were hypersensitive to the danger of playing the nationalist card, such as East Germany, could by the 1960s celebrate Martin Luther as a national hero rather than as the 'traitor against the peasants' that he had been portrayed in the 1950s. Thus religion, once seen as mortal threat to the building of a workers' state, could, once the institutional church had been subordinated, be tapped as a source of symbolic capital for nation-building. Lenin would not have been amused.

---

[86] I. N. Atanasova (2004) 'Ljudmila Zhivkova and the Paradox of Ideology and Identity in Communist Bulgaria', *East European Politics and Societies*, 18:2, 291.

PART I

Religion and Social Science in Eastern Europe

CHAPTER 2

Piety by the Numbers: Social Science and Polish Debates About Secularization in the 1960s and 1970s

Jim Bjork

In February 1976, an officer in Poland's Interior Ministry prepared a memorandum discussing the Roman Catholic hierarchy's internal assessment of the level and trajectory of religiosity in the country. The memo, based on reports from informants within the church, described a 'growing crisis of the church', reflected in slipping attendance at mass, abandonment of old devotional habits by migrants from the countryside to the city, and skepticism toward traditional moral strictures among youth. Although Poland's bishops were debating various strategies for overcoming these challenges, the author of the memo concluded that this crisis would likely only 'deepen and accelerate' in coming years.[1]

Few predictions have so badly missed the mark. As we know now, it was the Communist regime that was on the verge of crisis in the mid-1970s, a crisis that would deepen and accelerate over the following decade and eventually lead to its collapse. Poland's Catholic church, by contrast, was about to enter its golden age, as the election of a Polish pope spurred

[1] Informacja dot. ocen hierarchii kościelnych nt. stopnia religijności niektórych środowisk i grup społecznych w Polsce, 9 February 1976, Ministerstwo Spraw Wewnętrznych, Departament IV, Institut Pamięci Narodowej (hereafter IPN) BU 0713/250.

J. Bjork (✉)
Department of History, King's College London, UK

© The Editor(s) (if applicable) and The Author(s) 2016
P. Betts, S.A. Smith (eds.), *Science, Religion and Communism in Cold War Europe*, St Antony's, DOI 10.1057/978-1-137-54639-5_2

an unprecedented surge in religious vocations and devotional activity. But the Interior Ministry memo, while no doubt reflecting a degree of wishful thinking, was not an exercise in failed intelligence. Poland's Catholic hierarchy, the parish clergy, and church-affiliated sociologists of religion did, indeed, take seriously the basic premise of the secularization theory: that industrialization, urbanization, and other processes of modernization inexorably eroded the social significance of religion.[2] And many observers sympathetic to the church, as well as observers critical of it, saw persuasive evidence that these processes were already at work in postwar Poland. Doubts about the secularization story were, to be sure, also very much in evidence. But various objections to the secularization thesis only coalesced gradually and relatively late into a coherent story of a Polish religious Sonderweg, an exceptional immunity to trends prevailing almost everywhere else in Europe. Ironically, the privileging of this particular counternarrative was enabled by the Communist regime's own investment in national consolidation, which demanded that local or regional cultures, including particularistic forms of piety, were viewed as reflective of an underlying, integral Polish-national culture.

The 1960s ushered in a heyday of quantitative investigations into religious practice in Europe, producing a wealth of studies both of rates of participation in religious practices and, through surveys, of religious beliefs and attitudes. Such approaches fell out of fashion by the end of the century, as scholars digesting the linguistic turn looked askance at 'positivist' number crunching.[3] But it would be unfair to dismiss the era's enthusiasm for statistic gathering as reflecting a naïve conviction that a phenomenon as complicated as religion could be reduced to a few numerical indices. It reflected, rather, the more modest proposition that having systematic information about the prevalence of certain devotional behaviors might help scholars move beyond anecdotes and stereotypes toward a broader

---

[2] The first postulate of David Martin's (1978) *A General Theory of Secularization* (Oxford: Oxford University Press), p. 3 was that 'religious institutions are adversely affected to the extent that an area is dominated by heavy industry'. Quoted in R. Finke (1992) 'An Unsecular America' in D. Bruce (ed.) *Religion and Modernization* (Oxford: Oxford University Press), p. 154.

[3] The work of Callum Brown, a leading historian of religion in Britain, is illustrative. After pursuing increasingly sophisticated statistical analyses of data on religious practice in modern Britain, Brown turned in the early twenty-first century to scrutiny of life narratives as a way of understanding secularization. C. Brown (2001) *The Death of Christian Britain* (London: Routledge).

understanding of everyday religious practice. And the fact was that such basic information was the exception rather than the rule across much of Europe until late in the twentieth century. Only belated and painstaking compilations of diocesan-level data started to provide a composite picture of both current-day and, to some extent, historical devotional practices. The dearth of basic data was especially acute in Poland. It was not until 1980 that systematic data on church attendance were first gathered across every diocese in Poland.[4]

It was little wonder, then, that the first moves toward collection of such data, in the 1960s and early 1970s, aroused both considerable interest and considerable controversy. At times the controversies conformed to predictable ideological divides. One mini-debate about devotional statistics emerged in the first months of 1966, during an especially tense period in church–state relations.[5] Edward Ciupak, a sociologist at the University of Warsaw, had published data suggesting that barely 20 % of Varsovian Catholics who were obliged to attend mass had actually done so on typical Sundays in 1962 and 1963. Scholars affiliated with the Catholic church took exception to this low figure. Writing in the Catholic weekly *Znak*—and emphasizing that the earlier findings had appeared in 'an atheist journal' (*Argumenty*)—Father Józef Majka of the Catholic University in Lublin argued that the survey cited by Ciupak had failed to take into account all of the capital's churches and chapels. 'One could assume', he concluded, that if all services at all houses of worship were properly accounted for, fully half of all obliged Catholics in Warsaw attended mass on any given Sunday.[6]

---

[4] Diocesan statistics on church attendance and communion since 1980 can be found on the website of the Statistical Institute of the Catholic Church (ISKK): http://www.iskk.ecclesia.org.pl/praktyki-niedzielne.htm, date accessed 30 July 2015.

[5] The year 1966 marked the millennium of Poland's nationhood/statehood, and the state and Catholic hierarchy had been investing in rival series of commemoration during the run-up to the anniversary. This background tension spiked into more heated confrontation at the beginning of the year, when the Polish bishops addressed a letter to their German counterparts asking for and offering forgiveness for injustices suffered on each side during and after the Second World War. The letter provoked a wave of outrage that was clearly orchestrated by the regime but also reflected some genuine criticism from the broader Polish public. B. Kerski, T. Kycia, and R. Zurek (2006) '*Przebaczamy i prosimy o przebaczenie*': *Oredzie biskupow polskich i odpowiedz niemieckiego episkopatu z 1965 roku, Geneza, kontekst, spuscizna* (Olsztyn: Wspolnota Kulturowa 'Borussia').

[6] Józef Majka (March 1966) 'Jaki jest katoliczym polski?' *Znak* 18:3, pp. 272–93, quote from p. 278; Edward Ciupak (1964), 'Religijność warszawiaków', *Argumenty*, 8:37.

But this kind of empirical standoff, with pro- and anticlerical camps subscribing to competing data sets, was more the exception than the rule. Although state- and church-affiliated scholars tended to have different default methodologies—with the former focusing on survey data and the latter able to conduct direct head counts—each 'camp' accepted and used numbers generated by both methods. And the two sociological camps adhered to strikingly similar frameworks in interpreting those numbers. One point of convergence was a working assumption that Poland was a land where 'traditional' piety was strong. This involved three substantially overlapping understandings: (1) the Polish countryside demonstrated consistently high levels of religious practice; (2) levels of religiosity had been quite high in Poland's recent past; and (3) religious practice in Poland was very high compared to other European countries. In articles in the Polish Catholic periodical *Znak* (cited above) and the sociological journal *Social Compass* and in a chapter on Poland in the edited volume *Western Religion*, Father Majka emphasized the 'very high' rates observance in Poland compared to Western Europe, arguing that few rural localities registered rates of church attendance below 65 %.[7] An Interior Ministry report from 1974 did not quibble with such portrayals. It estimated 75–80 % observance in rural areas and conceded that religiosity in Poland was higher than either in other Soviet bloc states or in the capitalist West, due in part to the 'different cultural heritage' of the country.[8]

A second point of convergence was acceptance of the core of the secularization thesis: that industrialization, urbanization, and associated social and cultural transformation tended to erode traditional forms of rural devotion. State officials and state-university-based scholars operating in a Marxist framework described this as a categorical and inexorable process. The Interior Ministry report cited above insisted that religious observance was 'objectively lower' than in the 1930s, due to the 'participation of the population in the consumption of mass culture' and the 'disintegration of local communities' under the impact of modernization.[9] Ciupak wrote matter-of-factly about the 'undisputed' influence of industrialization on

---

[7] In addition to the *Znak* article cited in fn. 4, see Majka (1968) 'The Character of Polish Catholicism', *Social Compass*, vol. 3–4, and Majka (1972) 'Poland' in Hans Mol (ed.) *Western Religion: A Country by Country Sociological Inquiry*, (The Hague: Mouton). 'Very high' characterization from 'Character', p. 190.

[8] 'Problemy Pracy PZPR w dziedzinie kształtowania socjalistycznej świadomości narodu', May 1974, IPN BU 0713/250.

[9] 'Problemy Pracy'.

changes in religious behavior and the 'current process of dechristianization in our conditions'.[10] Church-affiliated scholars predictably disputed sweeping characterizations of 'dechristianization' in an urbanizing and industrializing Poland, but they generally conceded the link between modernization and secularization. In the late 1970s, Father Władysław Piwowarski, a colleague of Majka's at the Catholic University in Lublin and perhaps the most prolific of Poland's church-affiliated sociologists of religion, summarized studies of religious practice in industrializing communities near Puławy (southeastern Poland), Płock (central Poland), and Nowa Huta (on the outskirts of Kraków) and noted fairly uniformly low rates of Sunday observance (in the range of 27–33 %).[11] While rejecting any straightforward narrative of modernization producing religious collapse, Piwowarski discussed these findings in terms of industrialization and urbanization precipitating decline in religious practice. Even scholars such as Majka, who went furthest in insisting on the relative robustness of religious practice in Polish cities, presented this as an exception rather than the rule; he readily conceded, even emphasized, that Catholic religious practice in urban and industrial areas elsewhere in Europe was quite low and therefore that modernization did normally lead to secularization.

This degree of consensus would not be terribly interesting if the evidence supporting it was, indeed, incontrovertible. But the propositions shared by Marxist and Catholic observers actually sat uneasily with the empirical studies being conducted at the time. The image of a devout Polish countryside certainly matched up quite well with findings from some areas, such as in Małopolska (Galicia), where Sunday observance was confirmed to be high. But it was difficult to reconcile with research conducted across much of the rest of the country. In one local study of a village near Warsaw, Ciupak found church attendance in the range of 23–29 %.[12] In another village in the same region, Sunday observance was 20 %.[13] And these localities were hardly rogue outliers. As diocesan-wide, parish-by-parish surveys accumulated in the 1960s and 1970s, it became clear that levels of religious observance across much of the Polish countryside were quite low. In the predominantly rural dioceses of Lublin, for

---

[10] E. Ciupak (1973) *Katolicyzm ludowy w Polsce* (Warsaw: Wiedza Powszechna), pp. 121–2.
[11] W. Piwowarski (1977) *Religijność miejska w rejonie uprzemysłowionym* (Warsaw: Więź), p. 264.
[12] E. Ciupak (1961) *Parafianie?* (Warsaw: Książka i Wiedza), p. 150.
[13] W. Piwowarski (1966) 'Religijność miejska w dwudziestoleciu powojennym w Polsce' *Znak* 18:3, pp. 294–317, statistic from pp. 303–304.

example, the average rate of observance in 1971 was only 36 %.[14] In an earlier (1965) study of the diocese of Warmia, Father Piwowarski found an average attendance rate of 44 %, with observance under 40 % in 9 of the 23 counties.[15] These levels of practice were not only well below the ostensible 'minimums' ascribed to the Polish countryside but also well below the rates of observance found in many dioceses across contemporary Western and Central Europe.[16]

Given the striking, rapidly accumulating evidence that 'traditional' religious practice in rural Poland was not always robust to begin with, why did Polish commentators subscribe to a narrative in which rural piety only came to be challenged by the novelties of industrialization and modernization? Why, in particular, did Marxist, regime-friendly scholars fail to seize on what would seem to have been tantalizing evidence of weak traditions of popular religiosity across large swathes of the country? I would argue that two ideological factors played a role in this puzzling omission. First, the Marxist framework used by state-affiliated researchers privileged stories of diachronic, evolutionary change. The same inexorable forces of industrialization and urbanization that were ostensibly fueling the construction of a new socialist Poland could be seen as guaranteeing the erosion of Catholicism's social base. Conceding the religious nature of the country's feudal and bourgeois pasts, in other words, provided a useful foil for the triumph of a future secular Poland. Likewise, pointing to high levels of observance in more economically advanced areas of Western Europe would have contradicted arguments about secularization being inexorably driven by economic modernization, instead suggesting that reduction of the social role of religion in the Soviet bloc was driven by political repression.

---

[14] W. Zdaniewicz (1983) 'Stan dominicantes a liczba duchowieństwa w diecezjach polskich' in W. Piwowarski (ed.) *Religijność ludowa, ciągłość i zmiana* (Wrocław: Wydawnictwo Wrocławskiej Księgarni Archidiecezjalnej), p. 108.

[15] W. Piwowarski (1969) *Praktyki religijne w diecezji warmińskiej: studium socjograficzne* (Warsaw: Akademia Teologii Katolickiej), pp. 118, 123.

[16] Piwowarski, *Praktyki*, pp. 19–21. It is worth noting that the official West German average (45 %) was calculated as a percentage of *all* Catholics, while figures from Polish dioceses were calculated as a percentage of *obliged* Catholics (total Catholics minus children and the infirm, usually estimated as 25 % of the population). Average West German church attendance, in other words, was well above that recorded in the diocese of Warmia. Interestingly, Piwowarski's comparative survey includes every country in Western Europe with a significant Catholic population but does not include *any* statistics from other countries in the Soviet bloc, even neighboring ones with mostly Catholic populations.

A second, partially overlapping factor was nationalism, by which I mean both a degree of conscious Polish patriotism and the more generic methodological nationalism built into the social sciences. The patriotic impulse may have come into play in considering the implications of highlighting the weakness of religious practice in certain parts of the country, which would have required delving into the extent of regional difference in postwar Poland. This was a potentially treacherous and divisive endeavor in a country with such a brief history of shared statehood that was still trying to absorb vast swathes of newly resettled territory taken from Germany at the end of the war. The large regional differences in devotional practice evident in most countries of Western Europe, after all, were often reflective of intractable ideological tensions, animosities that had fueled virtual (or, in the case of Spain, actual) civil wars in the recent past. Suggesting the existence of similar deep divisions within Polish society was therefore not a matter of detached scholarly observation but an approach that many Poles would have deemed irresponsible and unpatriotic. Indeed, as scholars such as Michael Fleming have argued, the Communist regime and the Catholic hierarchy in Poland found considerable common ground in relying on religious confession as a ready marker of ethnicity in the postwar project of creating a homogeneous nation–state.[17] The Catholic church's claims to 'post-Protestant' church buildings in the Western Territories were framed as part of the campaign to establish 'Polishness' in the region, while Communist officials were keen to treat any Roman Catholics with any connection to Polish culture as 'recoverable' Poles.[18] It would have been awkward—albeit not impossible—simultaneously to use nominal Catholic affiliation as the defining marker of Polishness while defining the most observant Catholics as un-Polish.

Perhaps more effective than any conscious agenda of patriotic integration, however, was the cumulative impact of social-scientific assumptions

---

[17] M. Fleming (2009) *Communism, Nationalism and Ethnicity in Poland, 1944–50*, London: Routledge), pp. 102–8.

[18] The rhetoric of shared commitment to Polonization comes through clearly in church officials' communications with state officials in the first years after the end of the war. For example, Bolesław Kominek, appointed by the Primate, August Hlond, as apostolic administrator in the Opole region, emphasized to the commandant of the local militia that the church 'adhered consistently to the correct policy of Polonization of the region', and he argued that facilitating the broader distribution of the church-affiliated press would serve that aim. Kominek to Komendant Wojewódzkiej Milicji Obywatelskiej 17 November 1945, ARz 00704, Archiwum Archidiecezjalne w Katowicach (hereafter AAK).

and practices. The default understanding of 'societies' as synonymous with 'nations' made the nation the governing point of reference in the vast majority of social-scientific studies, even if it was not the actual unit of analysis. A given village or a parish was understood, first and foremost, as a *Polish* village or *Polish* parish, an example that would provide insight into national-level phenomena. In a book on popular Catholicism published in the early 1970s, for example, Ciupak devoted a chapter to discussion of religious life in small-city parishes.[19] He drew on statistics from ten parishes in ten different provinces and also included qualitative commentary by the local pastors on the religiosity of each community. But although those commentaries often made explicit references to the importance of divergent local traditions, especially in parishes in the Western Territories where the population was composed of a mix of natives and in-migrants, Ciupak showed little interest in the particularities of local historical experiences. Indeed, the individual parishes were anonymized, identified only as 'Parish A' through 'Parish J', transforming them from specific places into abstract samples drawn from a homogenized Polish national space.[20] Anonymization of smaller communities has been, to be sure, a common practice among sociologists, essentially an extrapolation of efforts to preserve the anonymity of individuals. But regardless of intention, the effect of such practice was unmistakable: Both individual actors and local communities were defined as irreducibly 'Polish', while other aspects of identity were blurred or erased entirely.

Interestingly, although church-affiliated sociologists also treated regional differences gingerly, they were more forthcoming in acknowledging them. Whereas Ciupak referenced regional differences only vaguely and dismissively, Majka's survey of Polish Catholicism referred to 'definite geographical differentiations' and 'distinct types of religiosity' in different parts of the country. He even specified that these differences often reflected the legacy of long-time rule by different empires during the partition era, with former Prussian, former Austrian, and former Russian territories exhibiting distinctive patterns of religious observance. Majka did, to be sure, express some discomfort with an internal

---

[19] Ciupak, *Katoliczym*, pp.107–38.

[20] Individual parishes were, to be sure, still linked to specific województwa. But since provincial boundaries had been deliberately drawn to cut across historic frontiers (of the German, Habsburg, and Russian empires, of interwar Poland, of the wartime General Government, as well as of Roman Catholic dioceses), this obscured as much as it illuminated about historical specificity.

diversity that seemed to correspond so closely with external influences, arguing that regional differences that were traceable to 'various forms of anti-religious pressure exerted by occupying powers' were now 'slowly disappearing'.[21] But what was most distinctive and impressive about the approach of clerical sociologists such as Majka—both reflecting and shaping the broader approach of the Polish episcopate to debates about the trajectory of Polish religiosity—was an ability to see highly idiosyncratic local cases not as subversion of a coherent national narrative but rather as templates for the generation of such a narrative.

This suppleness was apparent in the twist that clerical commentators gave to the story of Poland as a traditionally pious rural society subject to recent modernizing and secularizing pressures. As noted above, church-affiliated scholars had registered a fairly tepid dissent to this secularization narrative, conceding its general applicability in the West while insisting that its implications were not fully evident in Poland. Even this latter claim, evident in Father Majka's plea that 'one could assume' a rate of church attendance close to 50 % in Warsaw if attendees at smaller churches and shrines were tracked down, was strained. In the same article, Majka had admitted that religious practice in most of the country's large urban and industrial centers—Łódź and the Zagłębie conurbation, as well as Warsaw—left much to be desired. His explanation that this was due to 'specific circumstances' in those areas seemed a feeble attempt to edit most of Poland's largest cities out of generalizations about urban Poland.[22] The single case that Majka cited as a strong counterexample, an urban area where religiosity remained unambiguously high, was the Upper Silesian industrial conurbation, situated just to the west of the Zagłębie region. Two years later, in 1968, one of Majka's students in Lublin, Father Rajmund Bigdoń, submitted a doctoral dissertation devoted to the evolution of religiosity in the industrial district of late nineteenth- and early twentieth-century Upper Silesia, specifically in the city of Bytom. This did not seem, by most standards, high-impact research: It was not even published until 2004.[23] And yet it immediately became the linchpin of claims of Polish immunity to industrialization-driven dechristianization. When Majka published his

[21] Majka (1968), p. 197.
[22] Majka (March 1966), pp. 283–4. The claim of 50 % observance in Warsaw seemed especially dubious given findings, discussed earlier, of much lower rates of practice in the more rural areas surrounding the city and elsewhere in central Poland.
[23] R. Bigdoń (2004) *Religijność mieszkańców Bytomia w dobie industrializacji* (Opole: Wydawnictwo Uniwersytetu Opolskiego).

chapter in *Western Religion* in 1970, he insisted that the example of Upper Silesia 'gave the lie to the thesis that industrialization and urbanization will invariably bring about a decline in religiosity'.[24]

Other, more recent examples of resilient religiosity in industrial milieus, such as Nowa Huta (a steelmaking center established in the late 1940s in the outskirts of Kraków) or Gdańsk (largely repopulated after the war following the departure of the German-speaking population), were suggested over the next decade.[25] But these cases simply were not as convincing. Since they involved recent migrants, mostly from devout rural areas and involved in industrial labor for the first time, early signs of robust practice could plausibly be explained as residual habits destined to erode over subsequent decades. And when data from these new urban–industrial centers were gradually accumulated by the 1970s, indicators of piety were respectable rather than spectacular. As noted earlier, Piwowarski found rates of church attendance below 40 % in parishes in Nowa Huta. A head count of church attendance in the diocese of Gdańsk (a small diocese limited to the city and its immediate environs) in 1962 found a rate of observance of 52 %, slipping slightly to 48 % in a follow-up count in 1968.

By contrast, the diocese of Opole, which encompassed the western part of the Upper Silesian industrial region, including the city of Bytom, registered church attendance of 73 % in a head count conducted in 1968. The diocese of Katowice, including the larger, eastern part of Upper Silesia's industrial heartland along with some more rural hinterlands, registered a church attendance rate of 82 % in 1962—by far the highest diocesan rate recorded anywhere in Poland at the time—although this had slipped markedly to 63 % by 1974.[26] Closer scrutiny at the local level revealed that these high overall rates were boosted by exceptionally high rates of observance among long-time residents whose families had witnessed the first waves of industrialization in the nineteenth century. Devotional practice among immigrants from elsewhere in Poland, often entering industrial labor for

---

[24] Majka (1972), p. 414. While referring to 'other Polish working-class milieus' (besides Łódź, Zagłębie, and Warsaw) that ostensibly resisted de-Christianization, the only citation that Majka offered for this claim was Bigdon's dissertation on Bytom.

[25] See the summary discussion of sociological studies in Maciej Pomian-Srzednicki (1982) *Religious Change in Contemporary Poland: Secularization and Politics* (London: Routledge & Kegan Paul), pp. 144–7.

[26] W. Zdaniewicz (1983), pp. 105–28.

the first time, was seen as generally vigorous but patchier and more fragile.[27] Nowe Tychy, a large-scale planned community built out from a small pre-existing town south of Katowice, provides another example of the difficulties of presenting those currently migrating to cities and transitioning to industrial work as exemplars of resilient Polish piety. In an article published on this 'city of the future' in *Znak* in the summer of 1961, Andrzej Woźnicki described the city as divided into various 'ethnic groups', each with a consciousness of its distinctive origins and a tendency toward in-group marriage. The small native Silesian population and repatriates from Poland's lost eastern territories maintained especially powerful senses of group identity. While the author expressed some enthusiasm about the 'growth of a new social organism' in Nowe Tychy as well as some optimism about the potential for new pastoral 'experiments' to spur 'new possibilities for social order', he presented the transitional situation as daunting. The moral level of Nowe Tychy's inhabitants was, he judged, 'very low', with high rates of alcoholism and criminality and an underlying 'hedonism of everyday life'. Frequency of communion in the newer parish in Nowe Tychy was the lowest in the diocese. Church attendance looked a bit healthier, but only if one made exceptional statistical adjustments to take into account the difficulties that the city's disproportionate number of young, working-age parents and young children faced in attending Sunday services.[28] In short, instilling habits of religious observance in the Poland's 'cities of the future' was possible, but it would be an uphill struggle.

Identifying precedents for such a trajectory was therefore crucial for the Catholic church. As Stefan Wilkanowicz observed in an article on contemporary Catholicism in Poland published in the early 1960s, the

[27] Local priests routinely differentiated among parishioners based on their origins, with 'autochthones' (natives of the Western Territories), 'migrants' (from other parts of postwar Poland), and 'repatriates' (from the territories lost to the Soviet Union) understood to constitute the three main demographic groups. A. Sitek (1986) *Organizacja i Kierunki Dzialalnosci Kurii Administracji Apostolskiej Slaska Opolskiego w latach 1945–1956* (Wroclaw), pp. 42–55.

[28] A. Woźnicki (July–August 1961) 'Nowe Tychy—Miasto Przyszłości', *Znak* 13:12, pp. 1071–88, quotes from pp. 1079–85. In a striking extrapolation of principles for omitting from devotional statistics parishioners who were not 'obliged' to attend to attend mass, Woźnicki excluded 1000 mothers of infants, 3000 pre-school-age children, and 2000 others who needed to work on Sundays. With the number of 'obliged' parishioners thus reduced from 15,500 to 9500, he calculated a church attendance rate of 75 % (p. 1087). Using the more straightforward West German methodology (dominicantes = total attendees divided by total Catholics), the figure would have been 45 %.

'popular-worker' type of Catholic was similar to the 'traditional-rural' type but was 'more socially well developed, somewhat deeper from a religious perspective, and more resistant to the destructive processes of social change'. The clear implication was that this was a far more useful model in a country becoming ever more urban and industrialized. But as Wilkanowicz stated matter-of-factly, this was also a type that was 'above all characteristic of Upper Silesia' rather than reflecting a broader national pattern.[29] How, then, was the Upper Silesian exceptional case to be turned into an emblematic Polish case? One key was portraying the region as an extreme example of a familiar theme in Polish history: resistance to foreign domination. This turned an otherwise awkward fact—that prior to the mid-twentieth century, Upper Silesia had not been part of a Polish state since the Middle Ages[30]—into a marker of national authenticity. In a region run by German-speaking political and economic elites, the Catholic church could be presented as a purely plebeian institution, drawn from and in solidarity with a working-class Polish-speaking population. While not typical of the past experience of the rest of Poland, where the church had a closer association with the Polish gentry and was thus sometimes viewed more skeptically by peasant and working-class residents, it could serve as a normative template for a modern Poland now experiencing more widespread urban and industrial development, as well as a new (Soviet) form of foreign domination.[31]

As an historical explanation of religious resilience in Upper Silesia, this account was not entirely convincing. Far from representing a purely Polish frontier Catholicism, locked in perennial confrontation with an ostensibly Protestant Germandom, Catholicism in Upper Silesia in the late nineteenth and early twentieth centuries was more accurately characterized as a hybrid, borderland phenomenon. The region's parish clergy underwent varying degrees of linguistic and cultural germanization, and

---

[29] J. Kłys, S. Wilkanowicz, H. Szczypińska, J. Eska, M. Skwarnicki, J. Susuł, and H. Bartnowska, (December 1961) 'O współczesnych postawach religijnych w Polsce', *Znak* 13:12, pp. 1611–77, quote from p. 1623.

[30] The territory of the diocese of Katowice became part of interwar Poland in 1922, while the territory of the diocese of Opole was taken over by Poland in 1945. Silesia had come under the rule of the Holy Roman Empire, and specifically the Kingdom of Bohemia, in the early fourteenth century.

[31] Pomian-Srzednicki provides a nice summary of this interpretation, with the Zagłębie region serving as an example of the more typical pattern in the lands of the old Polish–Lithuanian Commonwealth: Pomian-Srzednicki, *Religious Change* p. 144.

only a minority expressed sympathy for the Polish-national cause.[32] The portrayal of the local population as passionately Polish-national becomes even more dubious if one tracks them further into the twentieth century. Indeed, even as Catholic authors such as Majka and Bigdoń were invoking the turn-of-the-century residents of Bytom as the marquee example of the Polish church's exceptional navigation of the challenges of modernity, thousands of those residents descendants were declaring themselves ethnically German and emigrating to the Federal Republic.[33] And yet the historical accuracy of this account of Silesian piety was largely beside the point. The narrative linking of high rates of observance in Upper Silesia with Polish patriotism proved to be a potent 'creative misreading' of the region's history, a dubious account of the Upper Silesian past that nonetheless provided plausible inspiration for imagining the Polish future.

Such inspiration and such imagination were sorely needed in the period from the late 1960s to the mid-1970s. In the preceding decade, leading up to the Millennium celebrations of 1966, the 'Great Novena' of special devotional activities spearheaded by Poland's primate, Cardinal Wyszynski, had generated a surge in visible religious activity in the country and even seemed to produce an uptick in the self-reporting of religious practice recorded by government-sponsored surveys.[34] But although scholars such Jan Kubik and Maryjane Osa have seen the Great Novena as a prelude to the successes of the Solidarity era, its immediate aftermath was disappointing for the church. Research conducted by Piwowarski between 1967 and 1970 found that many Catholics could not even identify what the Great Novena was.[35] And as already noted, fresh findings from local studies and diocesan head counts were pointing to at least slow, and perhaps even quite rapid, declines in religious practice in Poland. A nationwide survey

---

[32] J. E. Bjork (2008) *Neither German nor Pole: Catholicism and National Indifference in a Central European Borderland* (Ann Arbor, MI: University of Michigan Press).

[33] Between 1953 and 1964, more than 22,000 residents of Bytom emigrated to West Germany; by 1966, only half of the residents of the area were native to the area. Eugeniusz Klosek (1993) '*Swoi* i '*Obcy*' *na Gornym Slasku od 1945 roku* (Wrocław: Wydawnictwo Uniwersytetu Wrocławskiego), pp. 37–40.

[34] Pomian-Srzednicki, *Religious Change*, p. 143.

[35] J. Kubik (1994) *The Power of Symbols Against the Symbols of Power: The Rise of Solidarity and the Fall of State Socialism in Poland* (University Park: Pennsylvania State University Press), pp. 127–128. M. Osa (2003) *Solidarity and Contention: Networks of Polish Opposition* (Minneapolis, MN: University of Minnesota Press), pp. 75–8. Kubik duly noted Piwowarski's findings, conceding that it 'shattered' more euphoric views of the impact of the Great Novena, pp. 127–8.

of religious practice conducted by Poland's Main Statistical Office in 1969 and again in 1977 indicated a modest but significant drop during this period in Sunday observance and a sharp drop in attendance at weekday services.[36]

This shift from optimism in the early 1960s to anxiety in the early 1970s also reflected growing unease about the repercussions of the Second Vatican Council and trends in the universal—but especially the West European—church. Discussions in the Catholic press in the run-up to the Council had focused on opportunities for renewal in areas ranging from the liturgy to the role of the laity to relations with other religious communities.[37] Within a few years after the Council's conclusion, however, the leaders of the church in Poland were already offering darker evaluations of recent trends within Catholicism and in modern culture more generally. Cardinal Wyszyński reportedly told colleagues at a Christmas party in 1969 that the priests that he saw in Italy 'imitated the mobs on the Roman streets' in not wearing cassocks. While young people in Poland currently exhibited higher 'ideals', he warned that 'among us such a generation could also arise'.[38] By the mid-1970s, discussions within the episcopate featured general unease about whether Poland might be caught up in the post-Vatican II crisis that 'has already engulfed many societies in the world that were previously seen as eminently Catholic'.[39]

The exceptionalist counternarrative, sketched out in Bigdoń's case study and extrapolated by Majka and other church-affiliated authors, offered not only an assurance that patriotic Polish populism could help preserve laypeople's loyalty to the church but also, even more importantly, an affirmation of the efficacy of diligent, effective pastoral care. Parish priests in Upper Silesia had, in Bigdoń's account, dealt with the challenges of industrialization by aggressively subdividing parishes and sponsoring a dense

---

[36] L. Adamczuk (1990) 'Dynamika rytualnej aktywności religijnej Polaków w latach 1969–1976–1984' in W. Piwowarski and W. Zdaniewicz (eds.) *Religijność polska w świetle badań socjologicznych* (Warsaw: Pallotinum), pp. 23–34. Adamczuk's discussion of the reliability of such survey results in a volume published by a church-affiliated press is worth noting, as is his observation that the results might be seen as *less* susceptible to bias than surveys conducted by the 'church apparatus'.

[37] See, for example, the optimistic vision of 'The Church after the Second Vatican Council' sketched out by Father A. Bardecki (June 1962) 'Kościół po II Soborze Watykańskim', *Znak* 14:6, pp. 790–815.

[38] Quoted in B. Porter (2011) *Faith and Fatherland: Catholicism, Modernity, and Poland* (New York: Oxford University Press), pp. 39–40.

[39] 'Informacja', as cited in fn. 1.

network of cultural and social organizations.[40] The take-away message was one of clerical empowerment, of the ability of conscientious pastors to foster resilience in, and even revival of, religious practice. The direct impact of this message should certainly not be exaggerated; if the elaborate devotional program of the Great Novena had failed to produce a self-sustaining virtuous cycle of religious renewal in Poland, reflections on the ostensible successes of pastoral care in Upper Silesia more than a half-century earlier were certainly not about to do so. But I would suggest that identification and investigation of an Upper Silesian model of industrial piety did provide a significant boost for church-affiliated observers at a moment when other indicators of religious practice were looking less than inspiring. The ostensible lessons of this model also buttressed the arguments of Bishop Karol Wojtyła of Kraków and other reformist bishops pushing for a more vigorous pastoral response to the secularizing tendencies of the age.[41]

When Poland's famous religious revival became undeniably visible in the early 1980s, it was signaled by some sudden and impressive shifts in devotional indices. After slipping from 53.5 % to 49 % between 1969 and 1976, reported Sunday observance soared to 62.7 % in 1984.[42] It was more difficult to read trends from the head counts at mass conducted directly by the clergy since, as noted earlier, these were conducted only patchily in some dioceses before 1980, but these data also suggested that a general downward trajectory in the 1970s had been reversed by the early 1980s.[43] Clerical vocations (enrollment in seminaries) had been

[40] Bigdoń, *Religijność*, 286–97.

[41] 'Informacja', as cited in fn. 1. Upper Silesia is not directly referenced in this Interior Ministry memo on debates within the episcopate, but the reported emphasis of Wojtyła and his allies on 'thoroughgoing work of the entire church' and moving away from 'folklore' and 'fanaticism' was in keeping with the model suggested in Bigdoń's work. And of the seven bishops described as supporting Wojtyła's approach, one was a native of Upper Silesia and another had spent much of his pastoral career in the region, further suggesting the familiarity of the example.

[42] Adamczuk, 'Dynamika', 27.

[43] Of the five dioceses where a systematic count of mass attendance had been conducted in the late 1960s or 1970s, four recorded a drop in the subsequent count in 1980. Compare Zdaniewicz (1983), p. 108 with L. Adamczuk and W. Zdaniewicz (eds.) (1991) *Kościół katolicki w Polsce 1918–1990: Rocznik statystyczny* (Warsaw: Główny Urząd Statystyczny, Zakład Socjologii Religii), p. 172. There was a modest surge in recorded church attendance between 1980 and 1982 (from 51 % to 57 %), but levels of observance then returned to a narrow range of 49–53 % for the remainder of the Communist era. Summary data for the 1980s can be found on the website of the Statistical Institute of the Catholic Church (ISKK): http://www.iskk.ecclesia.org.pl/praktyki-niedzielne.htm, date accessed 30 July 2015.

climbing in the mid-1970s after a sharp drop at the beginning of the decade, but these fluctuations remained well within the range evident since the 1950s. By contrast, the further steep rise in vocations at the very end of the decade, followed by the sustaining of this high level through the 1980s, was an unprecedented phenomenon that almost doubled the seminarian–parishioner ratio in the country.[44]

The dramatic change in atmosphere in Poland from the late 1970s to the beginning of the 1980s can also be seen in more qualitative form in some of the most famous cultural productions of the period. Compare, for example, Andrzej Wajda's film *Man of Marble*, released in 1976, with his *Man of Iron*, released in 1981. The first film deals with a young film student investigating the history of a Stakhanovite worker from the Stalinist era, who was lauded as a hero of labor but then denounced as a saboteur. It thus addresses a range of controversial issues, yet it does not make the slightest reference to religion. Even the hero, a first-generation worker coming from the countryside near Kraków, is not portrayed as adhering to any religious practice or belief. Five years later, in *Man of Iron*, which dealt with the contemporary story of the Solidarity-led strikes at the shipyards in Gdańsk, religious references abound. They arise mostly from extended footage of praying crowds and of Catholic memorial services for victims of repression, as well as cameo appearances by devout contemporary figures leaders of Solidarity as Lech Wałęsa. But Wajda's fictional characters, some carried over from the first film, also reflect the change in mood. Agnieszka, the bold and unattached filmmaker from *Man of Marble*, remains strong-willed and independent-minded in her appearance in *Man of Iron* but has now become a wife and mother. In describing her marriage to the son of the subject of her planned documentary, she says that although she had never been religious, she knew that her wedding 'had to be' in a church, that this was 'necessary' and 'essential'.[45]

How can we account for this striking shift between the mid-1970s and the early 1980s? The most obvious dramatic development of that period—the selection of Bishop Karol Wojtyła of Kraków as the head of the universal church in 1978—must be central to any plausible explanation. Indeed, I would suggest that the impact of John Paul II's pontificate has been somewhat *under*estimated in recent accounts of the Polish

---

[44] Adamczuk and Zdaniewicz (eds.) (1991), pp. 33 and 144–5.

[45] The scene of the church wedding comes around the 2 h, 3 min mark in the film. Lech Wałęsa and his wife serve as witnesses at the wedding.

Catholicism during this period.[46] First, John Paul II's return visits to his homeland—the first already in 1979, the second in 1983—provided occasions for mass public gatherings on an unprecedented scale, with crowds regularly topping one million. While not openly oppositional, the pope's first visit provided a precedent and template for the quantum leap in organizational scale achieved by the Solidarity trade union in 1980. The impact of John Paul II's personal charisma is more difficult to quantify. But as Timothy Garton Ash's vivid contemporary descriptions of the pope's visits convey, the emotional interaction between the pope and the crowds that came to see him was galvanizing in a way that the more repetitive rituals of regular pilgrimage to Częstochowa or other sites could never be.[47] References to John Paul II as a uniquely inspirational figure penetrate all the way down to parish-level commentary. In 1987, 4 years after a papal visit to the region, a visitation report for the industrial suburb of Siemianowice in Upper Silesia cited 'the personality of the Pole-Pope' as 'undoubtedly the most positive factor in the development of the faith', crediting John Paul II's example for a growing tendency among parishioners to stand up publicly for their faith and discuss religious issues with friends and neighbors.[48]

A second and related way in which the pontificate of John Paul II transformed Polish Catholicism was its effective erasure of what had previously been persistent tension between Roman-papal and Polish-national sentiments. From Gregory XVI's encyclical Cum Primum (1832), which denounced the contemporary Polish uprising against Russian rule, to the letter sent by Pius XII to the German bishops (1948), which lamented the 'unprecedented' treatment of German refugees driven out of Poland's new western territories, the Holy See had been very easy to portray as a foreign power generally unsympathetic to Polish interests.[49] With the

---

[46] Osa, for example, has favored a more long-term account of successful pastoral mobilization starting in the 1950s, with an emphasis on the dynamics of social movements rather than personalities. But to the extent that individuals are highlighted, Wyszyński rather than Wojtyła would seem to play a more transformative role: M. Osa (Spring 1997) 'Creating Solidarity: The Religious Foundations of the Polish Social Movement', *East European Politics and Societies* 11:2, pp. 339–65.

[47] T.G. Ash (1990) 'The Pope in Poland', in *The Uses of Adversity* (New York: Vintage Books), pp. 47–60.

[48] Kwestionariusz, 9–10 October 1987, AL 2100 (Siemianowice, St. Antoni), AAK.

[49] Porter provides an excellent summary of the tensions between Polish nationalism and the formal teachings of the Catholic church during the partition era: Porter (2011), pp. 158–68. On the germanophilia of Pius XII and the repercussions for Poland, see M. Phayer

ascension of a Polish bishop to the papal throne, this distinction between the 'Roman' church and the church in Poland all but vanished. And this was not just a question of foreign relations. One of the defining features of ultramontane Catholicism was a potential affinity between local/regional particularism and Catholic universalism, an affinity that had a potential to 'leap-frog' over the intermediate level of national identification. Under John Paul II's pontificate, there developed instead a much tighter local–national–universal symbiosis. This transformation was most noticeable in an area like Silesia, where the visit of the pope in 1983 effected what could be described as a gentle nationalization of religious traditions that had previously been more regionalist and ultramontane than Polish-patriotic.[50]

The flipside of the breadth and consistency of John Paul II's charismatic appeal within Poland, even among those whose Catholicism had not previously been closely associated with national identity, was the *limits* of his charismatic appeal outside of Poland. He was, to be sure, a popular figure in much of the world, and his pontificate could plausibly be linked to some evidence of global Catholic revival. Clerical vocations, for example, which had dipped alarmingly in the 1970s, recovered a bit in the 1980s. This included particularly robust growth in Africa and Asia but also a noticeable surge (albeit from a very low level) in neighboring parts of the Soviet bloc, such as Czechoslovakia.[51] But rates of recruitment to the clergy remained largely stagnant in Western Europe and never returned to the levels of the early 1970s.[52] The gap between indices of piety in Poland and those in Western Europe, which had been fairly modest up until the 1970s and, indeed, had sometimes suggested higher levels of religiosity in the West, became a chasm by the beginning of the 1990s. A more convincingly 'Catholic Poland' found its foil in a more convincingly 'secular Europe'.

* * * *

(2007) *Pius XII, the Holocaust, and the Cold War* (Bloomington, IN: Indiana University Press).

[50] For the particular example of Góra Świętej Anny (Annaberg), one of the stops on John Paul II's 1983 visit to Poland, see J. Bjork and R. Gerwarth (2007) 'The Annaberg as a German–Polish Lieu de Memoire', *German History* 25:3, pp. 372–400.

[51] From a global low point of 8.33 seminarians per 100,000 Catholics in 1976, recruitment to the priesthood recovered to 10.5 by 1988. The figure for Czechoslovakia had dropped to 2.39 by 1976 but had risen to 5.25 by 1988. Adamczuk and Zdaniewicz (1991), pp. 32–3.

[52] The ratio of seminarians to the total Catholic population in France, for example, rose from 2.83 in 1976 to 3.24 by 1988, but this was still well shy of the ratio of 4.98 in 1972. In Italy, similarly, there was an uptick from 10.75 to 11.15 between 1976 and 1988, but the ratio of 15.2 from 1972 was never seen again. Adamczuk and Zdaniewicz (1991), pp. 32–3.

This essay has, I hope, demonstrated several benefits of revisiting the history of Catholicism in Poland through the lens of devotional statistics. First, it reminds us of the value of such investigations in recovering empirical information about past human behavior. Rather than simply providing tedious confirmations of conventional wisdom, the collection of data on church attendance, Easter communion, clerical vocations, and other indices of popular piety often generated surprising findings. In Poland, as in other countries of Europe where such studies were being undertaken, one common theme was the discovery of a striking degree of regional and even local diversity, challenging any easy generalizations about national religious norms. For scholars interested in patterns of religious practice in earlier periods, whether as an end in itself or to develop an accurate 'before' picture to allow evaluation of the impact of later trends, the snapshots provided by these earliest statistical studies remain invaluable.

The devotional number crunching of the era was not, of course, a purely 'objective' endeavor. The generation of statistics depended on a host of methodological choices, and those choices sometimes corresponded to the ideological and/or institutional interests of the scholars in question. For example, the practice of calculating church attendance (or Easter communion) statistics as a percentage of 'obliged' Catholics rather than all Catholics—a practice not followed in other countries—resulted in deceptively high headline figures for rates of observance in Poland. More common was garden-variety confirmation bias, in which bits of data were cherry-picked to demonstrate either the resilience of Catholic piety or its erosion, depending on the worldview of the scholar in question. But as we have seen, serious methodological disputes or systematically divergent interpretations of data were actually rather rare. Both clerical investigators and secular, state-university-based scholars shared a broad 'scientific' consensus on the desirability of gathering accurate data and referred to one another's work critically but respectfully.

An even more striking degree of consensus emerges when we look at how different observers embedded their numbers into narratives. Both clerical and secular scholars tended to work within the broad framework of secularization theory: they understood 'traditional' rural communities as inherently religious and 'modern' urban/industrial life as inherently corrosive of religious belief and practice. They broadly subscribed, in other words, to what the sociologist Jose Casanova has described as the 'knowledge regime of secularism', a shared understanding that 'secularity' is a 'sign of the times'. This might seem an odd claim given that Casanova goes on to describe the triumph of this 'knowledge regime' as a 'self-fulfilling prophecy', leading to the dramatic and apparently inexorable erosion of religious belief and practice in Western Europe.[53] This obviously did not happen in Poland. But, as I have argued in this essay, Poland's late twentieth-century devotional *Sonderweg* did not involve a complete rejection of—or immunity from—the overall secularization narrative. Instead, it rested on a creative re-working of it. While the general secularizing tendency of urbanization and industrialization was never really questioned, church-friendly scholars insisted that this tendency was not irresistible. They pointed in particular to one region (Upper Silesia) where mass industrialization had already happened but where levels of religious observance had remained robust. This provided the template for a unique devotional path that the rest of the nation could follow as it underwent a more belated economic modernization.

Such an account of how Polish Catholic exceptionalism emerged in the late twentieth century bears a resemblance to Casanova's suggestion of how a Polish-led pan-European religious revival might conceivably occur in the future. But I would argue that the way that Catholic commentators have understood and narrated the Polish experience—as a fundamentally *national* phenomenon—has had its own self-fulfilling dynamic. Just as it enabled the development of Polish exceptionalism, processing idiosyncratic local stories into representative national stories, it tended to disable alternative framings, such as imagining local evidence as reflective of forms of 'borderland piety' or thinking about the role of old imperial state systems in shaping religious traditions. This triumph of a nationalist 'knowledge regime' is perhaps the most striking, and certainly one of the most frustrating, aspects of past and, indeed, current investigations of religion in the Soviet bloc. For Polish observers, the nation–state was almost always the default unit of study, and comparative analysis tended to be superficial and fleeting. Interest in and knowledge about trends in religiosity in other parts of the Soviet bloc was even more limited than interest in and knowledge about developments in Catholicism in Western Europe. As the diverse contributions to this volume demonstrate, this structuring of past scholarship does not impose inescapable constraints on future scholarship. But narratives of national exceptionalism have shaped, and will continue to shape, how patterns of religious change can be imagined.

---

[53] Jose Casanova, 'Catholic Poland in Post-Christian Europe', *Transit Europäische Revue*, Nr 25/2003, http://www.iwm.at/read-listen-watch/transit-online/catholic-poland-in-post-christian-europe/, date accessed 30 July 2015.

CHAPTER 3

The Shepherds' Calling, the Engineers' Project, and the Scientists' Problem: Scientific Knowledge and the Care of Souls in Communist Eastern Europe

Patrick Hyder Patterson

Especially in the early years, the public face of communism in Eastern Europe was famous, or infamous, for its hostility toward religion. Compounding the assumed complicity of religion in regressive class structures was its rejection of science, and specifically its contradiction of the materialistic worldview propounded by Marx, Engels, Lenin, and their successors. With its purported power to remove the causes of religiosity from interpersonal relationships and from the individual mind, the expert practice of science held promise not just for the development of popular education in scientific atheism but also for the broader socio-political process of overcoming religion. During the socialist period, many academics in both East and West subscribed to an ideal of the social sciences *as science*—that is, as rigorous, empirical, hypothesis-testing methods yielding reliable results—an ideal that today has been abandoned in many quarters among the 'softer' social science disciplines, if not indeed turned into the target of inward-looking critiques of the hegemony of scientific mentalities and discourse. Operating in this spirit, early social-scientific analyses

P.H. Patterson (✉)
University of California, San Diego, CA, USA

of religion by investigators in the communist world often regarded the subject disciplines as practical tools in the effort to build socialism.

Such approaches, however, were not unchanging, uniform, or universal. This essay investigates the extent to which social scientists in parts of communist Eastern Europe proved able to develop more refined understandings of religious commitment that, with time, might move away from the cruder formulations of earlier Marxist atheist ideology toward more sophisticated, nuanced, and temperate methods. My focus here is on Hungary and Yugoslavia, two polities known for relative flexibility in ideology and practice in the years of mature socialism from the 1960s onward. As I show here, there were noteworthy developments in both countries. In Hungary, analysts of religion moderated their combative tone as they confronted an unpredicted persistence of religiosity and tried to adjust their theory to uncomfortable facts. In Yugoslavia the social science of religion changed even more, as many academics responded to delayed secularization with new visions that treated faith less as a problem to be engineered away through scientific training and more as a manifestation of firmly rooted, persistent, and in some sense even 'natural' responses to the travails and dilemmas of human existence. Adaptable as they were, however, members of the Marxist mainstream in both countries continued as dedicated materialists, insisting that religious faith lay within the reach of science, and often relying on psychological concepts of religion that allowed their interpretations to remain, in critical ways, consonant with socialism.

Any effort to evaluate the science of religion in communist Europe—that is, an examination of the purported naturalistic *sources* of religion—leads us to a focus on sociology as the primary site of scientific engagement and, in turn, to the correlate science of psychology. The boundaries between the two fields were not always clear—a fact that, I conclude, ultimately helped reinforce the self-understanding of sociologists as practitioners of an indisputably 'scientific' discipline. A useful, if necessarily rough, guide to prevailing practices comes from Croatian sociologist Nikola Skledar, who endorsed a division between the psychology of religion, focused on 'the sphere of the individual psyche' as manifest in personal experiences and feelings, and the sociological domain, occupied instead with 'the communication of religious experience to others, the exteriorization of religious feeling or its transfer to others'.[1] The

---

[1] Nikola Skledar (1989) 'Sociologija religije kao posebna sociologija', *Revija za sociologiju* 20, nos. 1–2: 87–100, at 98.

intellectual stakes involved in this distinction—and arguably, the political–ideological consequences as well—were higher than might appear at first glance, as psychological approaches were often inclined to identify innate, universal dynamics in the workings of human cognition and emotion. In more rigid forms, such views might run afoul of traditional socialist doctrine on the primacy of social conditioning, as expressed in Marx's insistence that 'the "religious disposition" [*das religiöse Gemüth*] is itself a social product, and that the abstract individual ... belongs, in reality, to a particular form of society'.[2]

In practice, the balance of scholarly attention in communist Eastern Europe often ended up weighted toward sociological inquiry, with psychological phenomena in strict terms receiving relatively little experimental or even conceptual attention, no doubt due in part to what Slovenian sociologist Marko Kerševan noted in 1974 as 'the poorly refined quality of the Marxist psychology of religion'.[3] But this weakness was not confined to Marxism: The pattern in Eastern Europe parallels a contemporary disciplinary shift away from religion among psychologists elsewhere in the world, following an energetic start in the work of William James, Sigmund Freud, Carl Jung, and other pioneers in the field. Even in the absence of strong empirical research into the psychology of religion, however, mental and emotional processes continued to constitute a critical area where science intersected with religious commitment, identity, belief, and practice.

LIFE IS TOO COMPLICATED: THE SCIENCE OF RELIGION IN SOCIALIST YUGOSLAVIA

> Doesn't the modern priest play the role of a psychiatrist? ... So many people come, weary of life, broken by the daily routine, trapped by the confines of a power alien to them, in order to be able to confide in someone, to unburden themselves, to experience catharsis ... And the priest, in a half-darkened space with barely audible music, making sympathetic pronouncements, takes apart the person's already-fragmented personality in order to integrate it later into a whole that will help the person to change his identification, to be convinced that he is not himself, that he is rising into the heavenly blue,

---

[2] Karl Marx [1845], *Thesen über Feuerbach* http://de.wikisource.org/wiki/Thesen_über_Feuerbach.
[3] Marko Kerševan, 'Odnos komunistov do religije', *Komunist*, April 1974, in Kerševan (1984), *Religija v samoupravni družbi* (Ljubljana: Državna založba Slovenije), pp. 75–90, 82.

that he is being pulled away from the conditionality of this world, that grief and loneliness are leaving him, ...

—Yugoslav sociologist Esad Ćimić, 1970[4]

The science of sociology was in many ways ill-equipped to address problems of religiosity under socialism. For decades it suffered from a reputation as a 'bourgeois science', with a long struggle for legitimacy. In the Soviet Union the discipline remained extraordinarily weak, suspect, and peripheralized—so distanced from international methods and standards, in fact, that Liah Greenfeld dismissed it in 1988 as merely 'a branch of social technology, a managerial science oriented towards the promotion of the goals and the increase of the ideological and administrative efficiency of the Soviet government', or in other words, not a genuine manifestation of the social science practiced in the West but instead 'a different branch of intellectual activity which for some reason uses its name'.[5]

In the unorthodox ideological climate of Yugoslavia, however, the basic potentials were different, and a much more open style came to typify examinations of religion.[6] Even as early as the mid-to-late 1950s, there were some signs of a notable departure. Sociologist Ante Fiamengo, for example, had begun promoting a new agenda for concrete empirical research, one remarkable for its radical reduction of overt Marxist determinism. Fiamengo, who exerted a strong influence on the emerging field through his work at the University of Sarajevo and later at the University of Zagreb, suggested that this would be the proper remedy for a sociological practice that, beyond the sheer paucity of its output, was notable for its 'restrained, incomplete, and abstract character' and 'risked ending up in normativism'.[7] Although social scientists working in the Marxist tradition clearly continued to recognize fundamental contradictions between their worldview and that of the religious objects of their study, by the mid-to-late 1960s there was evidence of a major shift in priorities and methods. And so while the prominent Zagreb

[4] Esad Ćimić (July–August 1970) 'Religija kao psihička i kao moralna činjenica', *Pregled: časopis za društvejna pitanja* 60, nos. 7–8: 33–44, at 34.
[5] Liah Greenfeld (1988) 'Soviet Sociology and Sociology in the Soviet Union', *Annual Review of Sociology* 14: 99–123.
[6] See Dragoljub Đorđević (2008) *Uzornici i prijani: skice za portret YU sociologa religije* (Belgrade: Čigoja štampa,); Marco Orsolic [Marko Oršolić] (1973) 'La sociologie de la religion d'inspiration marxiste en Yougoslavie', *Social Compass* 20, no. 1: 73–82; Siniša Zrinščak (1999) *Sociologija religije: hrvatsko iskustvo* (Zagreb: Pravni fakultet).
[7] Ante Fiamengo (1956) 'Sociologie et religion en Yougoslavie', *Archives de sociologie des religions* 2, no. 2: 116–120, 117–118.

philosopher, social critic, and member of the Marxist-humanist *Praxis* group Branko Bošnjak was insisting that trying to reconcile socialism with the continued existence of religion 'would mean that this kind of socialism is not founded on a rational or scientific comprehension of the world and history', Fiamengo was cautioning (indeed, in the same 1969 volume on religion and society) that 'life is too complicated for us to be able to classify its constituents into two opposing camps'.[8] Empirical complexity on the ground dictated a change in approach.

When the change came it was striking. By the early 1970s, Esad Ćimić could flatly state that Yugoslavia's reformist socialism had no place for the old 'bureaucratic mentality' that viewed religion 'exclusively as a social evil which it is necessary to suppress but—because of its power—to tolerate temporarily'.[9] Ćimić, who held positions in various institutions around the country and especially in Croatia, was another central figure in shaping the sociology of religion, and the stance he advocated here became more or less a standard line among Yugoslav specialists.

Although the Yugoslavs' change in tone and practice was remarkable, worries about the restraints of orthodoxy never quite subsided, right up through the last years of the communist period, and there were occasional sanctions against sociologists who proved too radical.[10] Notwithstanding the big differences that emerged in the practice of the science of religion by self-identified Marxist academics in Yugoslavia, there is some divergence of opinion as to just how free of political considerations their approaches could become. Typical of one view is the praise offered by another shaper of the field, Slovenian sociologist Zdenko Roter, for the achievements of specialists working in his own Ljubljana and in Split, Sarajevo, Zagreb, and Belgrade, who from the 1960s onward had developed, Roter said, 'an empirically based sociology of religion … as a complete scientific project on an autonomous level' with output comparable in quality to that of academics in the West.[11] That judgment contrasts somewhat with the

---

[8] Branko Bošnjak (1969) 'Kritika religije', in *Religije i društvo* (Zagreb: Stvarnost), pp. 7–35, p. 11; Fiamengo, 'Crkva i politika u samoupravnom društvu', in ibid., pp. 143–162, p. 160.

[9] E. Ćimić (1971) 'Aktualnost Leninovega odnosa do religije', *Teorija in praksa* 8, no. 3: 438–447, at 446.

[10] See Sergej Flere (1994) 'The Development of Sociology as a Contested Science in Post-World War II Yugoslavia', in Mike Forrest Keen and Janusz Mucha, ed. *Eastern Europe in Transition: The Impact on Sociology* (Westport, Conn.: Greenwood Press), pp. 113–124.

[11] Zdenko Roter (2013) *Padle maske: od partizanskih sanj do novih dni* (Ljubljana: Sever & Sever), p. 239. Roter notes similar evolutions in Poland and, somewhat surprisingly, Czechoslovakia.

ambivalent evaluation offered by Ćimić just after the end of the communist monopoly on power, when he concluded that the narrowness of Marxist views could and did distort sociological work throughout the socialist period—and admitted that his own studies were not immune.[12] One Catholic theologian has advanced a radically different reading of the history, criticizing recent efforts by Roter and his colleague Marko Kerševan to promote a secularist vision of post-communist church–state separation by arguing that their advocacy amounts to little more than an *aggiornamento* of the past program of 'the so-called "sociologists of religion"', who in this view 'exploited the power of totalitarian authority, the public media, and other means of compulsion, for indeed they were put in their positions in order for them to do away with religion'.[13] The battle continues.

On balance, the record suggests that many Yugoslav social scientists moved rather quickly to conform to international disciplinary conventions and avoid treatments that merely confirmed the laws of historical materialism. Even as they did so, however, they held onto other established ways of thinking about, and thinking through, the science of religion. Investigations of the past, present, and future of religion became more subtle and less burdened with dogma, diverging widely from those in most other places in communist Europe. Nevertheless, there was no doubt that the dominant perspectives stayed firmly rooted in epistemology and methodology understood to be reliably scientific in their ability to yield valid results. The underlying conceptions that remained, and in particular the psychology of religion, proved quite hospitable to the older Marxist readings.

The most common linkage to psychological science came in connection with the prevailing assumption that religion arose, at least in part, through its capacity to fulfill human needs. This stress on the satisfaction of psychological imperatives was, for example, a framing principle in the influential methodological textbook on the scientific study of religion written in 1970 by Vuko Pavićević, who had served as the director of the Center for Sociological Research at his home institution, the University of Belgrade. Pavićević emphasized the emotional sources of religious belief.

---

[12] Ćimić (1991) 'Ateizam i suverenost', in Štefica Bahtijarević, ed. *Prilozi izučavanju nereligioznosti i ateizma: zbornik* (Zagreb: Institut za društvena istraživanja Sveučilište u Zagrebu), pp. 109–142, p. 135.

[13] Janez Juhant (2014) 'Nasilje in sočutje v ideologijah in religijah in slovenska tranzicijska resničnost', *Bogoslovni vestnik* 74, no. 2: 175–189, at 184. Juhant branded Roter as part of 'the repressive apparatus of the party'. Ibid., 186.

These included, above all, the 'depressive feelings' connected with the fundamental human condition of powerlessness, such as fear, dread, and helplessness. In this view, the generative role of emotions and related psychic drives and constructs was critical. But emotions were not the end of the story. 'The ultimate psychological source of religious ideas and religious actions (prayers, making offerings)', Pavićević asserted, 'lies in the pursuit of security, happiness, a full life, immortality', and consequently, a truly comprehensive scientific analysis also had to take account of the 'psychological paths and forces' that actualize religious belief in human life, including, 'first and foremost, symbolic thinking, fantasy, and the propensity toward the personification of objective forces'.[14]

This perspective, common among Yugoslav sociologists, embraced an underlying psychology of religion that, conveniently enough, paralleled familiar Marxist ideas about the origins of religion as a way of managing a hostile, often overwhelming world. As Pavićević put it, belief in supernatural powers represented a method by which the human being 'conquers his own feelings of powerlessness and despair', although he cautioned against a wholesale dismissal of this as merely a form of self-deception: 'religious ideas are *means* for the *psychological* mastery of existential problems, and therefore it is inadequate to designate them as "false" or "erroneous"; it is enough to ask whether they are *efficacious* or not, *useful* or not useful'.[15] Similar ideas about the origins and functions of religion were an important motif in the work of Ćimić, who frequently spoke in explicitly psychological terms, as when he stressed religion's role in providing at least 'an illusory overcoming of dependency', a solution for the basic absence of human freedom, a way of dealing with death and the awareness of one's personal finiteness, and a compensation for the problems that accompanied individualism and social isolation.[16]

In keeping with spirit of the times and the widespread mid-century concern for the existential dilemmas of human existence, the search for meaning and fulfillment was a central element of the psychology that

---

[14] Vuko Pavićević (1970) *Sociologija religije sa elementima filozofije religije* (Belgrade: Univerzitet Zavod za izdavanje udžbenika Socijalističke Republike Srbije), pp. 16–17.

[15] Pavićević, *Sociologija religije sa elementima filozofije religije*, 10–11 (emphasis in original).

[16] E. Ćimić (1970) *Socijalističko društvo i religija: ispitivanje odnosa između samoupravljanja i procesa prevladavanja tradicionalne religije*, 2d ed. (Sarajevo: Svjetlost), pp. 40–42. See also E. Ćimić (1971) *Drama ateizacije: religija, ateizam i odgoj* (Sarajevo: Zavod za izdavanje udžbenika); Ćimić, 'Religija kao psihička i kao moralna činjenica'.

Yugoslav sociologists explicitly or implicitly deployed. As Roter observed in 1964, just as the new reformist orientation among Yugoslav sociologists was emerging, 'whether it is a question of primitive or contemporary religions, they are all at once the attempt of the human being to find an answer to his existential problem', and as such, Roter maintained, religion involves 'the creation of an illusion' that might serve as 'compensation' for 'the social condition of dependency, uncertainty, frequent lawlessness, humiliation, incompleteness' and other essential problems of human experience.[17] An 'open Marxist approach' sensitive to these dilemmas, Nikola Skledar argued, 'sees the sources of religion in the onto-anthropologically determined quality of the person as a finite, deficient being, with the need for meaning and completeness'.[18] As such comments suggest, for scholars conditioned by the contemporary cultures of both Marxism and social science to prize law-like consistency, the recognition of a psychological science behind religion promised insights that would be all the more useful for their universal validity—that is, for their naturalistic, reliable, 'determined' applicability to the human condition.

Whereas Marxist interpretations often tended, naturally enough, to stress the importance of psychology in its interpersonal, social dimensions, it is clear that many recognized the importance of psychology in its personal, interior forms as well. A connection with the science of the individual mind, Ćimić asserted, was important for understanding the origins and sources of faith, or as he put it, 'the emergence of religion from definite points of the psychological structure of the human being'.[19] Psychological dynamics were, in Ćimić's view, vital to a richer, more complex, yet still authentically Marxist sociological explanation of the unanticipated persistence of religiosity even after four decades of socialist conditions. The operative 'constitutive elements of the ontological structure of the human being' were so deeply seated, he concluded, that 'they resist, for quite a long time, radical social transformations'.[20] Ćimić's account thus acknowledged religiosity as 'anthropo-psychological fact', a recurring, pervasive human tendency through which religious commitment fulfilled a basic 'need for transcendence'—a scientific truth

[17] Roter (1964) 'Socialistična družba in religija', *Teorija in praksa* 1, no. 6: 809–816, at 810.

[18] Skledar, 'Sociologija religije kao posebna sociologija', 99.

[19] E. Ćimić (1985) 'Marksova kritika religije i/ili ateizma', *Luča: časopis za filozofiju, sociologiju i društveni život* 2, nos. 1–2: 241–245, at 244.

[20] Ćimić, 'Marksova kritika religije i/ili ateizma', 244.

that explained why religion might remain valuable even for members of a society developing along socialist lines.[21]

In some cases explanations grounded in the science of the mind proved even more central to the approach that Yugoslav sociologists adopted, with psychology moved from an initial exposition of premises to an expanded role as a more direct methodological tool. In particular, the work of Štefica Bahtijarević revealed a view of the science-behind-religion with deep connections to personal psychology. While she clearly acknowledged the socially constructed aspects of religion, Bahtijarević insisted that the social production of religion was inextricable from the interior mental and emotional life of the believer, a 'psychological world' that, 'although it is not a complete and genuine picture of total reality, is nevertheless the single true world of each individual'.[22] Elaborating on the generative role of psychological needs and drives to a greater extent than many of her colleagues, Bahtijarević offered a conceptualization in which the individual's religiosity was seen as 'a complex psychic state—a psychological construct' anchored in 'the cognitive, the emotional, and the action-oriented/motivational', while the broader complex phenomenon of religion was presented as one that 'attests to the powerlessness of the person to realize himself as a person, to satisfy his needs in a real fashion, but also ... to his power to master the unknowable and the uncontrollable, to not falter either in space or in time, even carrying his visions beyond their borders—developing in this way a specific form of practice: emotional-illusionist, God-creating'.[23]

Bahtijarević's psychology of religion drew heavily on concepts developed by influential Western theorists, such as Abraham Maslow's well-known hierarchy of human needs. Her interpretation of religiosity and self-identification, for example, relied directly on the concept of attitude articulated by American personality psychologist Gordon W. Allport.[24] She argued that the potent and enduring aspects of religiosity amounted to 'a permanent human disposition, a part of man's personality', and therefore

---

[21] Ćimić, 'Marksova kritika religije i/ili ateizma', 245–245.

[22] Štefica Bahtijarević (1976) 'Religija i crkva u suvremenom svijetu', in *Marksističko poimanje religije i politika SKJ prema crkvi i religiji* (Zagreb: Centar CK SKH [Centralni Komitet Saveza Komunista Hrvatske] za idejno-teorijski rad), pp. 136–172, at 151.

[23] Bahtijarević, 'Religija i crkva u suvremenom svijetu', p. 152.

[24] Bahtijarević (1975) *Religijsko pripadanje u uvjetima sekularizacije društva* (Zagreb: Narodno sveučilište grada Zagreba/Centar za aktualni politički studij), p. 81.

fit the patterns defined by attitude theory.[25] At the same time, however, she cautioned that 'the child is not born religious' and that there was nothing inevitable about the presence of religion in the world: The specific attitude of religiosity was, like all others, derived from and shaped by social experiences and norms, and Bahtijarević therefore balked at Allport's suggestion that religion had, as she put it, 'functional autonomy ... in the sense of original and independent motivation forces and separate needs'.[26] Notably, such a perspective held open the idea that religion *could* disappear, even as Bahtijarević, like so many of her Yugoslav colleagues, dispensed with doctrinally driven assumptions that it *must* disappear in socialist society (or even that it *would* disappear anytime soon). This was an elaboration and refinement of the Marxist science of religion, not a repudiation.

Indeed, Bahtijarević contended that the psychological theorizations she employed, with their emphasis on religion's value as a form of psychic compensation and motivation, confirmed socialist principles. Quoting an assortment of familiar pronouncements from Marx and Engels, she argued in one 1975 study of secularization that they corroborated her own linkage of 'psychic states and experiences' with the critical 'normative-visionary' functions of religion. In the founders' judgments, Bahtijarević said, 'there may be found precisely this aspect of values and its primarily psychological, experiential determination: 'the sigh of the oppressed creature', 'the soul of a heartless world', 'its logic in popular form', 'its enthusiasm', 'its moral sanction', 'its solemn complement', 'spiritual aroma', ... 'the self-consciousness and self-esteem of the man who has either not yet secured possession of himself or has already lost himself again', ... 'a protest against real misery', ... 'the illusory happiness of the people' ... [27] The parallels between Marxist insights and the contemporary science of the mind were, in this view, absolutely clear.

Although psychological ideas were an important recurring element in the Yugoslav sociology of religion, it would stretch the facts too far to conclude that they consistently determined the approaches taken. Indeed, some leading works were remarkable for the extent to which they disre-

---

[25] Bahtijarević (1971) 'Some Characteristics of the Religiosity of Secondary School Attendants', in *Religion et religiosité, athéisme et non croyance dans les sociétés industrielles et urbanisées* [International Conference for the Sociology of Religion, Acts of the 11th Conference, Opatija, Yugoslavia, September 20–24, 1971] (Lille: Édition CISR), pp. 123–144, p. 124.

[26] Bahtijarević, 'Some Characteristics of the Religiosity of Secondary School Attendants', 126, 141.

[27] Bahtijarević, *Religijsko pripadanje u uvjetima sekularizacije društva*, pp. 43–44.

garded concerns about origins and went straight for questions of social functions.[28] Roter's later empirical studies in Slovenia, for example, like his conceptualizations of the sociology of religion more generally, often showed rather little interest in the etiology of religiosity and a far greater concern for its operational dynamics and consequences. To the extent that Roter's interpretations did voice Marxist materialist assertions about religion's function as 'opium', its 'fantastic quality', its potential to provide 'illusory' happiness, and its role as an 'emotional form' of handling outside forces, they were notable for how they tended to destabilize the old orthodoxies and for their claim that other ways of engaging the world—indeed, 'practically every ideology' or 'any other systematized social idea'—might work in precisely the same manner.[29]

Others expressed ambivalence about the importance of the psychological for the workings of religiosity. The early work of Split-based sociologist Srđan Vrcan, for example, viewed religion as a response, 'in a mystified way', to 'real existential human conditions and actual human needs', a treatment that supported Vrcan's call for a deeper recognition of these psychosocial factors and an acknowledgement of 'specific strictly individual aspects ... that cannot be accounted for in purely sociological terms', factors that he insisted should not be ignored or simply dismissed as 'the particularization of ... religion as a collective social fact'.[30] By 1976, however, Vrcan was warning against an 'abstract-anthropological approach' that located the sources of religion in 'the universal characteristics of man in general' or 'the general human situation' without respect to social and political contexts: 'in and of themselves they do not lead, necessarily and inexorably, to God and the church'.[31] Yet, rather than an outright repudiation of psychological factors, this seemed to be an effort to counter some sociological hypotheses of a more fundamental, enduring, perhaps almost

---

[28] E.g., Roter (1988) 'Revitalizacija religije i desekularizacija društva u Sloveniji?' *Sociologija* 30, nos. 2/3: 403–427.

[29] Roter (1976) *Katoliška cerkev in država v Jugoslaviji: 1945–1973: sociološki teoretični vidiki in raziskovalni model* (Ljubljana: Cankarjeva založba), pp. 29–30.

[30] Srđan Vrcan (1971) 'Some Theoretical Implications of the Religiosity as a Mass Phenomenon in a Contemporary Socialist Society', *Internationales Jahrbuch für Religionssoziologie* 7: 150–167, at 163, 165.

[31] S. Vrcan (1976) 'Marksistički pristup religiji', in *Marksističko poimanje religije i politika SKJ prema crkvi i religiji*, pp. 111–135, at 120–121.

insuperable human need for religion.[32] Even with such qualifications, Vrcan continued to acknowledge that humans have 'the need for a definite system of meaning and significance' and that they 'must find an answer to certain critical aspects of human existence', requirements that could result in religiosity but did not compel it.[33] Positions like these kept his work reasonably close to the Yugoslav disciplinary mainstream, with its general affirmation of the psychological.

Probably the most substantial reservations about psychological approaches came in the theorizations advanced by Marko Keršsevan. This influential Slovenian sociologist's early scholarship showed some appreciation of then-fashionable psychoanalytic explanations of religious experience, along with an acceptance of the idea that religion, at least to some extent, served as a mechanism for the satisfaction of basic human needs. Yet it was also remarkable for an insistence that, notwithstanding any indications to the contrary in Keršsevan's own Marxist tradition, reliance on such needs should not be used to explain away religiosity in a reductionist manner.[34] Across the arc of his career, Keršsevan's stance toward any putative psychological science of religion remained uneasy and distanced. While he noted that 'individual popularized psychoanalytic themes and theses' had proved influential for communist conceptualizations of religion, these did not find direct and sustained application in his most important works.[35] Keršsevan's deep skepticism toward a priori assumptions extended to the scientific quest for the sources of religion. 'It does not make sense', he argued along these lines, 'to say that religion is born or originates from human ignorance, powerlessness, social misery, alienation, etc., or from the desire to overcome that condition at least in an illusory way ... If we were to ask in the same way about the origins of science, we could come to the same conclusions: that it was born from the human being's powerlessness in the face of nature and his desire to overcome that condition'.[36] The challenge here to well-worn Marxist orthodoxies was undeniable. Psychology did matter for Keršsevan, but he was insistent that the psycho-

---

[32] See S. Vrcan (1979) 'Nevidljiva i nedoktrinarna religija—zbilja ili čista teorijska konstrukcija?' *Naše teme* 23, nos. 7–8: 1721–1744, at 1737–1738.

[33] S. Vrcan (1986) *Od krize religije k religiji krize: prilog raspravi o religiji u uvjetima suvremene krize* (Zagreb: Školska knjiga), p. 19.

[34] Keršsevan (1967) 'Neka sporna pitanja marksističke teorije religije', *Naše teme* 11, no. 6: 980–993, at 989–990.

[35] Keršsevan, 'Odnos komunistov do religije', p. 82.

[36] Keršsevan (1971) 'Narava in struktura religioznega fenomena', *Problemi: revija za kulturo in družbena vprašanja* 9, no. 102: 56–77, at 63.

logical aspects of religiosity were not the source of religion but instead its result.[37] The domain of the psychological—that is, individual emotional and cognitive experience—thus became a field in which religion operated as just one among many specific forms of social practice.

In the end, the social science of religion in Yugoslavia from the 1960s onward departed significantly from old orthodoxies that saw the disappearance of religion as a goal of scientific practice. While there were still occasional declarations that the full implementation of socialism would eliminate the conditions that sustained religion, many scholars were open to facts that confounded the predictions of Marxist historical materialism. Moreover, they were assertive—adamant, even—about the need to ensure religious freedom and combat anti-religious discrimination, though typically they remained very comfortable, at least in print, with the principle that religious institutions should continue to be excluded from Yugoslav politics: This was a freedom for believers, not churches. Nevertheless, the underlying scientific theories that supported their research remained largely consistent with Marxist ideology. Whether these were the individual-needs-and-drives presuppositions of psychology or the social-construction-of-consciousness premises of sociological theory, such principles could be presented, without much intellectual difficulty, as confirming a socialist understanding of human relations. Marxism may no longer have set the rules of science in a crude, deterministic way, but the new rules were still notable for the extent to which they did not contradict the foundations of Marxist materialism.

Changing Consciousness and the Hegemony of Marxism: The Science of Religion in Socialist Hungary

Nowadays, in Hungary, the demand for religion represents, above all, an emotional demand.

—Hungarian psychologist Mihály Murányi, 1977[38]

Like Yugoslavia, Hungary saw fairly little in the way of empirical research in the psychology of religion. There was, however, plenty of talk about

---

[37] Kerševan, 'Narava in struktura religioznega fenomena', 70–71.
[38] Mihály Murányi (1977) 'A vallásosság struktúrájának és funkcióinak változásai', Világosság 18, no. 1: 21–32, at 27. Unless noted, translations from foreign languages are my own.

the psychological underpinnings of religious belief. This was especially true in the 1960s and 1970s, and even well into the late socialist period, psychological terms and insights held an important place in mainstream academic understandings of religious phenomena. Scientific (or science-styled) explanations of human religious behavior that recognized substantial psychological elements appeared with some regularity in Hungarian sources. The lines between the fields involved—sociology, psychology, philosophy, criticism of religion, and political studies—were blurred at best, and any evaluation must consider that some of the most important social-science conceptualizations of religion during the socialist period came in the sociologically oriented thought of writers who were not, strictly speaking, members of the discipline (though the filter I use here excludes writers like György Lukács who are, notwithstanding their great influence on social theory, better understood as makers of political philosophy and ideology).

Experts disagree sharply about the quality of this sociological work. Emphasizing the subfield's 'evident scientific status and achievements' that surpassed most other communist societies, Croatian sociologist Siniša Zrinščak has concluded that practitioners in Hungary, like their colleagues in Poland and Yugoslavia, 'can be proud of the development of the sociology of religion'.[39] Similarly, Attila Becskeházi and Tibor Kuczi, speaking of the sociological discipline as a whole, outlined an 'emancipatory process' that by the 1960s had resulted in a Hungarian academic culture that 'did not begin from Marxist premises', one that became popular because, with its emphasis on everyday life, it 'signified an effort to understand the true reality of the world, until that time a world mystified by ideology'.[40] Such a reading, if true specifically of the study of religion, would suggest close resemblances to the Yugoslav approach, which was, as shown above, remarkably emancipated.

Some support for such a conclusion comes from Miklós Tomka, a Hungarian sociologist affiliated with Catholic institutions who was a fre-

---

[39] Siniša Zrinščak (2008) 'Sociology of Religion in Croatia 1991–2007', in *The Sociology of Religion in the Former Yugoslav Republics: 15th annual International YSSSR Conference*, ed. Dragoljub Đorđević (Niš: Jugoslovensko udruženje za naučno istraživanje religije), pp. 27–42, at 28–29.

[40] Attila Becskeházi and Tibor Kuczi (1994) 'The Sociology of Reformist Socialism: The Hungarian Model', in Mike Forrest Keen and Janusz Mucha, ed. *Eastern Europe in Transition: The Impact on Sociology* (Westport, Conn.: Greenwood Press), pp. 39–53, pp. 41–42, p. 50, p. 43.

quent contributor to both publications of the socialist academic establishment and, especially later in his career, church-affiliated journals. Writing in May 1989, in what turned out to be the final months of East European communism, Tomka observed that the field in his country represented a 'third model'. Hungarian practice, he maintained, was based neither on a near-monopoly on sociological inquiry by the state (Romania, Czechoslovakia, the Soviet Union) nor on parallel church-sponsored and Marxist sociologies (Poland, East Germany). Rather, researchers avoided the terms of ongoing controversies between communists and their ideological opponents, aiming instead at the straightforward development of empirical findings.[41]

In stark contrast to such approval, French sociologist and East European specialist Patrick Michel in 1988 described a Hungarian practice that pursued a disingenuous tactical variation in its tone and methods but never relinquished its ultimate anti-religious aims. As such, it remained virtually worthless. Only a few of the works published by mainstream scholars were worth reading, Michel suggested, and these writers had contributed little or nothing to the broader sociological understanding of religion or even to a more straightforward comprehension of the workings of religion in Hungary itself. Michel summed up his bleak view of the sociology of religion by recounting the response of émigré sociologist István Kemény, who could 'merely shrug his shoulders and calmly declare that this branch of sociology does not exist in Hungary'.[42]

The sociology of religion under Hungarian socialism thus poses serious problems of evaluation. Was this a shoddy faux science driven by the determinist desiderata of party ideology? Or was it a more honest effort to reach the empirical realities of Hungarian religiosity? There is evidence to support both judgments. Part of the trouble here is the difficulty of determining the meaning within the socialist system, and for the socialist system, of work produced by scholars, such as Miklós Tomka and Zsuzsa Horváth, who positioned themselves as more or less neutral, no doubt part of the 'third model' that Tomka himself described.[43] We should be skeptical of reading vast significance into the willingness of the academic establishment to offer

---

[41] Miklós Tomka (2005) *Church, State, and Society in Eastern Europe* (Washington, DC: Council for Research in Values and Philosophy), p. 168.
[42] Patrick Michel (1988) 'Religion et socialisme réel: l'analyse du sociologue', *Archives de sciences sociales des religions* 65, no. 1: 25–45, at 33.
[43] See Michel, 'Religion et socialisme réel', 36.

periodic presentations that, rather than proceeding from explicitly Marxist premises about the desirability of eliminating religion, sought instead to understand religiosity in more neutral, open-ended terms. And independence could have big costs. Horváth, a specialist in small religious communities who managed to place a straightforward study of Pentecostalists in the leading journal *Világosság* in 1977, was by the last years of communist rule decrying (now in a French journal) a fundamental lack of religious liberty in the country.[44] Her contributions to the social science literature remained marginalized, and she apparently was removed from her position at the Hungarian Academy of Sciences as the result of her ideological waywardness.[45]

Yet with such caveats in mind, it is reasonable to see real import in a variety of facts that indicate a degree of intellectual flexibility. Even as a scholar rendered suspect by his Catholic associations, Tomka built an impressive, frequently cited publishing profile in mainstream venues while avoiding affirmations of party doctrine.[46] And plainly there was some latitude to debate the premises of Marxist ideology. Such openness was evident, for example, in a 1979 piece in *Valóság*, a journal of the party-sponsored Society for the Dissemination of Scientific Knowledge (*Tudományos Ismeretterjesztő Társulat*), where Tomka argued that just as the image of a pre-socialist 'Christian Hungary' was crumbling under the weight of new findings, suppositions about any inevitable 'death of religion' in socialist conditions were likewise proving unsustainable.[47] Directly criticizing prior approaches derived from the 'erroneous impression ... that the achievements attained in the building of socialism should always be (must be) in a functional proportion in accord with a decline in religion', Tomka concluded that the past decade's work in the sociology and psychology of religion in Hungary had forcefully disproved such propositions, with the result that '*factual material* of scientific merit can

---

[44] Zsuzsa Horváth (1977) 'Megteltek mind Szentlélekkel: Az evangéliumi pünkösdi közösség egy gyülekezetéről', *Világosság* 18, 6: 380–386; Horváth (1988) 'Le Ferment de l'église: le mouvement des communautés de base en Hongrie', *Archives des Sciences Sociales des Religions* 65, no. 1: 81–106.

[45] Michel, 'Religion et socialisme réel', 36.

[46] Tomka remained active in the post-socialist period before his death in 2010; a comprehensive bibliography of his early work appears at http://www.phil-inst.hu/intezet/tm.htm.

[47] Tomka (1979) 'A szekularizáció mérlege', *Valóság* 22, no. 7: 60–70, at 60.

be contrasted with the multitude of *assumptions*'.⁴⁸ This was, under the circumstances, a fairly bold challenge to the old, orthodox view. With time, mainstream outlets of academic analysis confronted the persistence of religiosity in increasingly open terms and hosted a rather frank debate over the sources, nature, scope, and implications of Hungarian secularization—or the lack thereof.⁴⁹ Meanwhile, the tone of the Marxist academic approach to religion shifted from one of confident distancing and dismissiveness to a more respectful (if still somewhat patronizing) acceptance of engagement and dialogue.⁵⁰

In the final analysis, the sociology of religion in the Hungarian Marxist mainstream looks somewhat different from the more free-ranging modes of inquiry that typified Yugoslav practice. In the first place, there was simply less in the way of concrete empirical research that made its way into print.⁵¹ That fact complicates any analysis: We must rely more on theoretical and methodological expositions from treatises and essays, with fewer opportunities to look for the underlying science of religion applied in presentations of results from fieldwork. Furthermore, notwithstanding various reformist moves, investigations of religion in the social-scientific mode demonstrated to the very end a greater need to ground theoretical and methodological frameworks in the Marxist classics, to interpret findings with an emphasis on the confirmation of Marxist theory and ideology, and to affirm explicit linkages between the accumulation of scientific knowledge concerning religion and the socialist effort to bring about its 'withering away'. These tendencies surface repeatedly in the work of the primary shapers of the field. Along these lines, a 1987 survey of studies on religion in Hungary by József Lukács called for 'a more radical and concrete response to questions of the social sources and functions of religiosity', cautioning that this had to be done

---

⁴⁸Tomka, 'A szekularizáció mérlege', 60. Revised English version: Tomka (1981) 'A Balance of Secularization in Hungary', *Social Compass* 28, no. 1: 25–42, at 26–27 (emphasis in Hungarian original; quoted passages from uncredited translator).

⁴⁹E.g., Pál Horváth (1987) 'Szekularizáció és vallásosság mai Magyarorszagon', *Társadalmi szemle* 42, no. 1: 29–39.

⁵⁰E.g., Zsolt Papp (1969) 'Ilúzió és valóság: gondolatok a vallásos mozgalmak "eviláságáról"', *Világosság* 10, no. 6: 321–326; János Jóri (1982) *Zu den methodologischen Problemen der marxistischen Religionstheorie* (Pécs: Pécsi Janus Pannonius Tudományegyetem Állam -és Jogtudományi Kara).

⁵¹See József Lukács (1979) 'Harminc esztendő a magyar valláskritika történetében', in J. Lukács, ed. *Történelem, filozófia, vallásosság: tanulmányok* (Budapest: Gondolat), pp. 338–356, p. 351. (originally published in *Világosság* (1975) 16, no. 4: 201–209).

in a context of genuine dialogue with believers and engagement toward common social goals, for only in that way, Lukács argued, could 'the *changing of consciousness* and the true realization of the *hegemony* of Marxism' be assured.[52] As explained above, the Yugoslav sociological mainstream ended up shaking off many of these old habits while remaining, or so it was claimed, reliably Marxist. In Hungary the traditional ways proved somewhat more durable.

That outcome, like much of the social-science framework for the study of religion under Hungarian socialism, attests to wide influence of two leading scholars: Mihály Murányi, the country's most active exponent of an expressly psychological approach to religion during the 1960s and 1970s, and Lukács, specialist in critical studies of religion [*valláskritika*], director of the Institute for Materialistic Religious Research, and editor of the influential journals *Valóság* [Reality, Truth] and *Világosság* [Enlightenment, Clarity], both prominent venues for Marxist thought on religion. Frequently cited in the broader social-scientific and critical literature, Murányi's interpretations were rooted in the conviction that psychological mechanisms offered the key to understanding how the false beliefs and ideas of religion were formed, how they were maintained in the face of the challenges of reality, and how they might be made to disappear. Here he elaborated on the findings of a small group of other socialist scholars concerned with the intersections between religion and the science of the mind, most notably the prominent psychologist László Garai, whose work regularly addressed questions of religiosity.[53]

The guiding concerns of Murányi's inquiry come across clearly in one representative early piece from 1963, where he took up the problem of origins of religiosity and religion's connections with the psyche. 'How is it possible', he asked, 'that in our consciousness there come into existence such imaginings and conceptions that do not correspond to reality? How may these be connected with emotion? What sort of relationship is there

---

[52] J. Lukács (1987) *Vallás és vallásosság a mai Magyarországon: a hazai tudományos kutatások néhány eredménye* (Budapest: Kossuth), p. 109 (emphasis in original).

[53] E.g., László Garai (1969) *Személyiségdinamika és társadalmi lét* (Budapest: Akademiai Kiadó). This text on personality psychology identified three specifically religious modes by which human beings satisfy basic needs in a fictive manner: rituals of sacrifice, rituals of scapegoating, and rituals of penance. Ibid., pp. 202–203. See also L. Garai (1963) 'Csoda és hit I.', *Világosság* 4, no. 1: 22–27; L. Garai (1963) 'Csoda és hit II.', *Világosság* 4, no. 2: 79–83; L. Garai (1962) 'A vallási elidegenedés pszichológiája', *Magyar pszichológiai szemle* 19, no. 2: 213–220.

between faith and a person's needs and aspirations? How does it influence one's behavior and attitudes toward reality as a whole?'[54] The power of science to answer these questions, this psychologist suggested, held great practical significance for the ideological transformation of society, and here he hewed to the Marxist line, treating the persistence of religion as a real hindrance to progress. Concluding that religion was 'the product of vulnerability and ignorance', Murányi declared that 'compromise between science and religion' on the fundamental issues at hand 'is unimaginable, just as it is between the theological and scientific interpretation of the psyche itself' (here employing the word *lélek*, a multivalent word commonly used in Hungarian to indicate soul or spirit in religious parlance and psyche or mind in more scientific terms).[55]

Throughout his career until his death in 1978, Murányi used psychology to probe the sources and functions of religion. His leading work, the wide-ranging 1974 volume *Vallás és illúziók* [Religion and Illusions], traced the workings of religious thoughts and emotions in connection with individual psychological development, social conditioning and the psychology of interpersonal relations, human finitude and the fear of death, and problems of sin, virtue, and free will. Here he continued the line established in his earlier interpretations, blending Marxism with psychology to frame religi on as fundamentally a matter of illusory beliefs, psychic compensation and problem-solving, magical thinking, and false consciousness in the face of overwhelming social and economic alienation.[56]

In his last years, Murányi grappled with the implications of other researchers' empirical findings that documented the durability of religious commitment in the country, results that raised uncomfortable questions about earlier theorizations of secularization and the effects of socialist rule. In keeping with his longstanding emphasis on psychological factors, Murányi concluded that in Hungarian society the persistence of faith in the unreal was the result of inability of 'socialism at its current degree of development' to fulfill satisfactorily the 'present social and individual-psychic needs' of the population using methods that were 'non-religious, thus non-

---

[54] M. Murányi (1963), 'Valóság és illúzió: A vallás kezdetei és a tudat', *Világosság* 4, no. 10: 577–584, at 577.

[55] Murányi, 'Valóság és illúzió', 577.

[56] See M. Murányi (1974) *Vallás és illúziók: a valláspszichológia néhány főbb kérdése* (Budapest: Kossuth Könyvkiadó).

illusory'.[57] The interpretative move here mirrored a larger pattern seen in Hungary and a number of other places in the socialist world (including even Yugoslavia, albeit to a lesser extent): credibility compelled social scientists to amend theory to accommodate empirical facts, but they did not abandon the idea that socialism could, and should, render religion a negligible force.

In their gradual moderation of overt antagonism toward religion, the analyses offered by Lukács likewise illustrate broader Hungarian developments. The early works of this social critic adopted the customary militant tone. Writing in 1961 in *Társadalmi szemle* [Social Review], a prominent party-sponsored theoretical and political journal, Lukács thus portrayed religion as the past and present enemy of scientific and social progress, as shown in the shabby treatment directed at a long list of heroes and martyrs in the cause of empiricism.[58] He rejected what he said were spurious 'modernizing' endeavors by religious thinkers who scrambled to preserve at least some power in the face of scientific advancement by proclaiming two separate 'spheres of competence', a material sphere in which the validity of any assertion was judged scientifically, and a spiritual sphere in which religious truths rightfully held sway. This theological dodge, Lukács insisted, left the main logical problem untouched: 'There are not two truths, just as there are not two worlds, only one. And it is not possible to deduce an adequate image of this unitary world from the dogmas of religion, but rather only from the results of science, without regard to whether or not they contradict the ancient picture of the world, thousands of years old, offered by the Bible'.[59] The 'illustrious place of reason and cognition in theology', Lukács warned, was purely one of appearances: 'bluntly: without *emotional* preconditions (i.e., religious mentality, the feeling of dependency, etc.) and *volitional* motivations (i.e., *striving* for the acquisition of religion out of genuine or assumed interests) ... the *rational* factors of faith still cannot be effective'.[60] There was plenty of the Leninist spirit of 'militant materialism' here.

---

[57] M. Murányi (1977) 'A vallásosság struktúrájának és funkcióinak változásai', *Világosság* 18, no. 1: 21–32, at 22–23.
[58] J. Lukács (1961) 'A tudományos haladás és a mai teológia', *Társadalmi szemle* 16, no. 10: 66–85, at 70–71.
[59] Lukács, 'A tudományos haladás és a mai teológia', 79, 84.
[60] Lukács, 'A tudományos haladás és a mai teológia', 74–75 (emphasis in original).

In later work, Lukács continued to affirm the significance of the emotional in creating and maintaining religiosity. He acknowledged a vital role for individual psychological dynamics, noting that 'ultimately, the *most important subject matter* of religion, in every case, acquires its authentication *emotionally*, gains substance through faith, and within the context of the [social] system becomes undeniable'.[61] The connection to emotions, in turn, brought religiosity squarely into conflict with socialist rationalism. Attempts to formulate a reasoned basis for belief in the supernatural, Lukács said, were doomed to fail, merely revealing tensions and inadequacies in the particular religious system in question. Such refinements and variations were results that 'arise *in frameworks that have been affirmed emotionally; it is not possible to make religion itself rational*'.[62]

To the end of his career, Lukács integrated psychological insights into his broader Marxist interpretation, attributing the greatest causative force to social conditioning but still acknowledging important functions for the psychic make-up of the individual. 'Modern developmental psychology', he concluded in one of his last major works in 1987, has taught us that the interior dimensions of the human personality are not simply a cognitive, intellectual process: 'the external world creates feelings and passions within us, awakens faith and hope or desperation', emotions which are then shaped into consciousness, refined, and governed by a broader system of interpersonal relations, i.e., 'the socially-induced process of the formation of the human character'.[63] Contemporary science, in this view, continued to confirm the fundamental insights of socialist materialism.

Occasionally, however, Hungarian scholars reacted with skepticism toward the inclination to see psychological factors as generative of the illusions of religiosity.[64] The falsity of false consciousness, some insisted, was attributable to actual socio-economic circumstances (which socialism could rectify) not to intrinsic human tendencies (which might be difficult if not impossible to engineer away). As István Pais put it, although the emotional dimensions of religiosity were undeniable, 'there are no specifically religious feelings', for the other-worldly response to feelings of defenselessness and dependency was, in

---

[61] J. Lukács (1979) [1972] *Istenek útjai: a kereszténység előzményeinek tipológiájához*, 3d ed. (Budapest: Kossuth), p. 76 (emphasis in original).

[62] Lukács, *Istenek útjai*, p. 77 (emphasis in original).

[63] Lukács, *Vallás és vallásosság a mai Magyarországon*, p. 155.

[64] E.g., Ákos Károly (1963) 'A vallás mint hamis tudat: pszichológiai és szociológiai tőrténeti szempontok', *Pszichológiai szemle* 20, no. 4: 558–566.

truth, the product of objective social relations, most importantly the class system that was 'the *primary* source of religion and the feelings of fear manifest in the religious form'.[65] As in Yugoslavia, some Marxist interpreters were reluctant to endorse an essentially physiological or neurological science of the mind that risked undermining socialist doctrine on the interpersonal, social creation of consciousness by elevating individual, interior, psychic needs for religion to the status of unshakable universals. A 'hard-wired' psychology of religion implied a future in which faith might be ever blooming, not withering away.

In the main, however, such reservations did not translate into deep resistance, and the social-scientific study of religion by members of the Marxist establishment was marked by tendencies to invoke psychological concepts clearly understood to be consonant with the presuppositions of Marxist ideology. In this respect, practice in Hungary paralleled developments in Yugoslavia; there were likewise significant efforts to realign theory with the unexpected absence of secularization, and a gradual cooling of anti-religious rhetoric. But in the end, Hungarian sociological thought showed considerably less willingness to confront the old political imperative to deploy science against faith in the supernatural. At least among scholars working in the Marxist mainstream, the 'emancipatory process' that Becskeházi and Kuczi describe for the sociological discipline as a whole did not extend deeply into the science of religion.

---

[65] István Pais (1971) 'A vallás érzelmi-függőségi eleme', *Magyar pszichológiai szemle* 28, no. 4: 528–537, at 537 (emphasis in original).

CHAPTER 4

Romanian Spirituality in Ceauşescu's 'Golden Epoch': Social Scientists Reconsider Atheism, Religion, and Ritual Culture

Zsuzsánna Magdó

After Nicolae Ceauşescu became party leader in 1965, top party organs showed a fresh desire to overcome popular religiosity for the sake of atheist convictions. Indeed, during Ceauşescu's short-lived liberalization in the late 1960s and early 1970s, the Central Committee invested unparalleled amounts of resources and professional expertise to transform the spiritual lives of socialist citizens—their moral and aesthetic values, sense of existential meaning, and emotions. Instead of being the 'stepchild' of the regime's ideological programme, in the 1970s the mission of creating atheist beliefs and practices turned into a 'permanent

---

This chapter is based on the dissertation entitled 'The Socialist Sacred: Religion, Atheism, and Mass Culture in Communist Romania, 1948–1989' for which generous funding was provided by the Department of History, University of Illinois, Urbana–Champaign, and the Open Society Institute, New York. I am especially grateful to Victoria Smolkin-Rothrock for sharing her work, which was instrumental in understanding the Romanian case comparatively. I also thank the European Studies Centre at the University of Oxford and the Acton Institute for the Study of Religion and Liberty for providing crucial support that allowed me to participate at the conference from which this edited collection emerged.

Z. Magdó (✉)
Department of History, University of Illinois, Champaign, IL, USA

© The Editor(s) (if applicable) and The Author(s) 2016
P. Betts, S.A. Smith (eds.), *Science, Religion and Communism in Cold War Europe*, St Antony's, DOI 10.1057/978-1-137-54639-5_4

aspect' of building socialism. As a result, atheist education became the subject of repeated party plenums and resolutions; inspired the creation of specialized research laboratories; helped revive psychology, ethnography, and the sociology of religion; and led to the emergence of a new generation of specialized cadres—professional atheists and researchers. Meanwhile, religiosity went from being dismissed as a 'remnant' of 'retrograde societies', in the strict Marxist–Leninist parlance, to being recognized as a 'complex phenomenon' that needed to be discovered empirically. As the individual and experiential components of religiosity came into focus, cadres provided impetus for developing a 'positive atheism'—one that did not simply negate belief but was expected to replace religion's psychological, emotional, and aesthetic functions in daily life. By the late 1970s, these preoccupations coalesced into efforts to produce a socialist spiritual culture. Experts' call for ritual reform culminated after 1976 in the biannual 'Song for Romania' Festival, a vast political–ideological and cultural–artistic event of a decidedly Stalinist bent. Most importantly, by imbuing positive atheism with national values, the regime offered its citizens a so-called Romanian spirituality as a path for fulfilment and self-transcendence.

The importance of religious organizations in Cold War competition and in post-socialist politics has made it seductive to focus on church–state relations in Eastern European socialism and gloss over distinctions between the various elites—leading party members, bureaucrats, scholars, and so on—who were instrumental in shaping official approaches to the religious question. In Romanian scholarship especially, the notion that the regime acted as one has led to the general consensus that the accommodation between the Romanian Communist Party and the Orthodox Church explains how the relationship between religion and communism evolved. This particular bias also sheds light on the following paradox: while the atheism of the Romanian regime is taken to be an old chestnut, until now the ideological arm of the anti-religious endeavour has not been examined. In fact, the question of Marxist–Leninist atheism has received surprisingly little attention in studies of Eastern European socialism overall.

Yet, in the late socialist period, atheist experts transformed the struggle against religion into a struggle for a socialist spiritual life. As Victoria Smolkin-Rothrock argued, what drove the qualitative transformation in atheist thought in Khrushchev's Soviet Union was the recognition that religion was strong not because it constituted a set of intellectual beliefs but because it had spiritual components—aesthetic, emotional, psychological, moral, communal, and ritual—that provided answers to the ultimate questions of human existence.[1] Questions about value, the meaning of life and of being human, or the possibility of transcendence were encoded in Marx's own analysis of the human condition and constituted preoccupations that he shared with religious traditions.[2] Because materialism threatened to empty life, death, and suffering of value, such concerns also preoccupied Soviet ideologists in the 1920s and, in part, explains why socialist regimes cultivated happiness and meaning through festivals or distinctive life-cycle rituals. At the same time, during the late 1950s and early 1960s, a series of events converged to make the Marxist conception of man a topic of virulent debates across the iron curtain. Khrushchev's denunciation of Stalinism, the Second Vatican Council's promotion of Christian humanism, and the maturation of anti-humanist atheism in French thought lent urgency to the long-standing question of whether Marxism was able 'to provide meaningful answers to the problems of human life'.[3] It was in the context of this acute questioning of both 'Western' and 'Eastern' Marxism next to the simultaneous recognition that religion remained relevant to human life that atheism and the spiritual world of socialist citizens became key policy concerns in late socialist states.[4] The subsequent reanimation of the academic study of religion and reliance on professional expertise in atheist policy can also be understood as part of this larger story.[5]

The recomposition of atheism during the liberal phase of Ceaușescu's 'golden epoch' demonstrates that the broader revisionist trends in Marxism

[1] Victoria Smolkin-Rothrock (2010) '"A Sacred Space Is Never Empty": Soviet Atheism, 1954–1971' (PhD Dissertation, University of California at Berkeley).

[2] Benjamin B. Page (1993) 'Introduction' in Benjamin B. Page (ed.) *Marxism and Spirituality: An International Anthology* (Westport, CT: Bergin&Garvey), p. xiii.

[3] The quote is from Miklós Tomka (1982) 'Életfordulók ünneplése a szocialista országokban' in Márta D. Hoffman and Szilágyi Erzsébet (eds.) *Ünnepek a mai magyar társadalomban: az ünnep szociológiai és szociálpszihológiai szempontból* (Budapest: Tömegkomunkációs Kutatóközpont), p. 165. Paul Mojzes (1981) *Christian-Marxist Dialogue in Eastern Europe* (Minneapolis: Augsburg Publishing House). Stefanos Geroulanos (2010) *An Atheism That Is Not Humanist Emerges in French Thought* (Stanford: Stanford University Press). In Romania, debates around the Marxist conception of man resulted, among others, in anthrophilosophical works like C. I. Gulian (1966) *Problematica omului: eseu de antropologie filozofică* (Bucharest: Editura Politică).

[4] By emphasizing these broader trends, I depart from Soviet studies that explain the party-state's turn to spiritual questions after the mid-1950s just in terms of an attempt to depart from Stalinist, mechanistic ways of imagining socialist society. See, for example, Sonja Luehrmann (2011) *Secularism Soviet Style: Teaching Atheism and Religion in the Volga Republic* (Bloomington: Indiana University Press), pp. 165–7.

[5] See Patrick Hyder Patterson's contribution to this volume.

reverberated in late socialist ideology, yet did not cancel out specificities that made Romania 'different' in the socialist camp. Specifically, both the explicit articulation of the atheist project in the national register of a 'Romanian spirituality' and the undertaking of socialist ritual reform in the frame of a mass festival constituted trademarks of an anti-Soviet national Stalinist regime. As Vladimir Tismăneanu observed namely, the Ceaușescu regime emerged as such in response to precisely what Romanian communists lacked since the underground years and for much of the early socialist period: national credentials.[6] Thus, while Romanian atheists echoed revisionist trends in the socialist bloc, Marxist–Leninist atheism under Ceaușescu evolved rather differently in two respects. In contrast to socialist states that underwent de-Stalinization, the rhetoric of spirituality in Romania did not, in the end, entail the regime's move away from collective principles in the interest of private, individual happiness.[7] Also, the Ceaușescu regime in the late 1980s did not accept religion in the spirit of the ideological pluralism promoted by Mikhail Gorbachev; instead, it embarked on the destruction of urban churches and continued to issue resolutions as late as 1988 calling for the strengthening of the atheist drive in a relentless pursuit of social homogenization.[8]

From Gheorghiu-Dej to Ceaușescu

Socialist states claimed Marxism–Leninism as a scientific mantle and spent considerable effort to establish 'religion' as the other of everything 'socialist', modern, and rational; in ideological terms, after all, the bright communist future was predicated upon the eradication of religion. Because Soviet models provided the initial guidelines in Eastern Europe, under party leader Gheorghe Gheorghiu-Dej (1948–1965) the Romanian

[6] Vladimir Tismăneanu (2003) *Stalinism for All Seasons: A Political History of Romanian Communism* (Berkeley: University of California Press), pp. 26–27. For Romanian specificities compare with Patrick Hyder Patterson's contribution to this volume.

[7] 'Introduction' (2011) in Marina Balina and Evgeny Dobrenko (eds.) *Petrified Utopia: Happiness Soviet Style* (London: Anthem Press), p. xxi. Paulina Bren (2010) *The Greengrocer and His TV: The Culture of Communism after the 1968 Prague Spring* (Ithaca: Cornell University Press).

[8] Christopher Williams (2009) 'The Fight for Religious Freedom and Pluralism in post-Soviet Russia' in George McKay et al. (eds.) *Subcultures and New Religious Movements in Russia and East-Central Europe* (Bern: Peter Lang), p. 232. ANIC, inv. 3291, File 29/1988, f. 2–18.

regime was quick to adopt Bolshevik ritual culture and a host of measures in order to restrict religious life 'within church walls'. The Society for the Dissemination of Science and Culture (SDSC), an institution modelled on the late-Stalinist Znanie Society, was similarly called to arms beginning 1949; next to its publications, its nationwide lectures of political, cultural, and scientific enlightenment were expected to banish faith from the hearts and minds of every citizen.

However, as regime differentiation set in after Stalin's death in 1953, Soviet models underwent increasing domestication. This process was unmistakable, first, in the party-state's relations with Romanian Orthodoxy and, second, in the regime's retreat from scientific-atheist education. Interwar religious nationalism had propelled the institutional expansion and financial growth of the Orthodox Church, a process that was not entirely interrupted by the communist take-over. While the 1948 unification of the Greek Catholic Church with the Orthodox Church was inspired by Stalin and applied elsewhere in Eastern Europe, in Romania this reunification showed a meeting of interests between the party-state and the interwar aspirations of Orthodox nationalists. The church remained heavily regulated throughout the entire socialist period and yet, the privileged status of Romanian Orthodoxy was evident even during the repressions of the late 1940s and early 1950s: no bishops were imprisoned; the religious and economic life of monasteries revived; and, above all, the church was the only religious organization in the socialist camp that canonized saints (1955) and published significant religious and liturgical works.[9]

Because of their tenuous national credentials, Romanian party elites resisted Nikita Khrushchev's de-Stalinization efforts after 1956; but in the early 1960s Soviet plans to reduce Romania to an agricultural hinterland for the more industrialized socialist countries aggravated interparty relations further, as a result of which the Dej regime issued a declaration of independence from Moscow in April 1964. As the Romanian party elite embarked on de-Sovietization instead of de-Stalinization, foreign and domestic policy priorities shifted and the official stance towards religion relaxed further. The external relations of the Orthodox Church and the higher clergy proved particularly instrumental in repairing diplomatic relations with the west and in obtaining funds for socialist industrialization from outside of the Soviet camp. Meanwhile, party-state attempts

---

[9] Lucian Leuștean (2009) *Orthodoxy and the Cold War: Religion and Political Power in Romania, 1947–1965* (Basingstoke: Palgrave Macmillan), pp. 93–5, 163 and 204.

to reconcile with Romanian citizens both on the religious and national front prepared the way for the symbolic re-appreciation of cultic buildings. These began to be incorporated into the circuits of socialist tourism and patriotic education based on their aesthetic and historic value. The retreat from the ideological battle against religion culminated finally in 1963 when the Soviet-inspired SDSC was reduced to small department within the Ministry of Culture—an institutional erosion that had prohibitive effects on science popularization, the promotion of alternative socialist rituals, and the professionalization of atheist work in the late Dej era.

A fundamental question regarding religion and ideology in late socialist Romania is why the regime became invested in atheism at the height of the thaw in the late 1960s when it had practically abandoned this enterprise a couple of years earlier. Following in Dej's footsteps, Ceaușescu amplified some of his predecessor's liberal policies. Thus, the re-classification of churches as historical monuments, their state-funded restoration or the circulation of religious art through various exhibits constituted efforts to establish the independence of Romanian socialism from Moscow in front of both domestic and international audiences.[10] Religion was, after all, central to the ideological competitions of the Cold War not only across the iron curtain but also within the respective camps.[11]

However, the road towards an 'autochthonous' socialism also required the re-invention of Marxist–Leninist ideals and the institution of ideological control. Indeed, during the Ceaușescu era's first campaigns in atheist education (1966 and 1969), it became clear that the relaxation in antireligious measures along with other aspects of the thaw—the amelioration of postal and press censorship, the opening of borders to western tourists, or the 1964 amnesty of political prisoners, including priests—animated religious life across the country and produced confusion among the lower

---

[10] In 1960–1977, 79 churches were restored from state funds. They retained their religious functions while also serving as religious-historical museums and tourist destinations. ANIC, inv. 2898, File 27/1977, f. 14–6. Although monasteries in Romania were closed in great numbers between 1959 and 1960, the symbolic career of churches in late socialist Romania stands in contrast with the fate of ecclesiastical buildings in the Soviet Union, which were closed *en masse*, converted to anti-religious museums, demolished, or, at best, declared architectural monuments. See contributions by Catriona Kelly and Igor J. Polianski in this volume. Romanian Orthodox religious art was circulated through international exhibits in the United Kingdom (1965–1966), West Germany (1971), and Japan (1977).

[11] Philip E. Muehlenbeck, ed. (2012) *Religion and the Cold War: A Global Perspective* (Nashville: Vanderbilt University Press).

party ranks. The dynamism of the religious landscape at the time of the regime's relative openness to 'creative Marxism' soon led to a consensus that Marxist–Leninist atheism needed revision because it did not give an accurate picture of popular religiosity, explain its persistence in socialism, or provide directions for how to disenchant society.[12] But the renewed commitment to atheist work was also fuelled by the Central Committee's anxiety over maintaining ideological control. Having sided with China in the Sino-Soviet split and condemned the Warsaw pact invasion of Czechoslovakia (1968), Ceaușescu was at the height of his isolation in the socialist camp and the possibility of Soviet invasion seemed real enough to warrant calls for the re-establishment of ideological unity around party and leader.[13]

Accordingly, discussions on atheism and religion at the top party level achieved their peak in 1969–1973 in terms of both frequency and verve. Top organs issued repeated instructions to clarify the ideological line on religion. Yet, despite the Central Committee's oft-repeated claim to ideological leadership, the task of addressing the foundational problems of atheism fell on the shoulders of a new category of specialists: professional atheists and scholars.

For most who joined this sphere of ideological work, the task of discovering religion empirically and reforming atheist work both in terms of theory and method was far from trivial. There were two important obstacles along the path. Interwar sociology, ethnology, and psychology produced some valuable works on popular religiosity based on state-of-the-art methods for the time: surveys, participant observation, and systematic collection of literary and musical folklore.[14] In Greater Romania, this interest in religious mentalities was animated by the political and cultural Orthodoxism of the era. In fact, numerous social scientists ascribed the Orthodox Church and the peasantry, at this time overwhelmingly portrayed to be mystical-religious, a significant role in the preservation of 'Romanian spirituality' in the modern nation-state.[15] In 1948, however,

---

[12] On the short-lived openness to Marxist revisionism, see Tismăneanu, *Stalinism for All Seasons*, p. 31.

[13] Adam Burakowski (2011) *Dictatura lui Nicolae Ceaușescu, 1965–1989* (Bucharest: Polirom), pp. 129–131.

[14] These studies emerged primarily in the framework of the Cluj School of Experimental Psychology and the Bucharest School of Sociology.

[15] Antonio Momoc (2013) 'Sociology and Theology: Building the Romanian Cultural Nation', *European Journal of Science and Theology*, 9, no. 4, 101–9.

these fields were denounced as bourgeois pseudoscience, which precipitated a nearly 30-year hiatus in the scientific study of religion. Second, to become aware of recent trends in Marxist–Leninist atheism, experts also had to overcome the setbacks that de-Sovietization in the 1960s imposed on their theoretical understandings and instruments in propaganda work. For atheists in the 1970s, all of this meant then that they needed to construct a field for the scientific study of religion and atheism, train a new generation of specialists, revive discussions on research methods, and conduct actual investigations in order to improve policy. The impulse to confront these challenges was strong. Leading Romanian atheists were convinced namely that they lagged behind their more 'modern' Soviet and East European atheists had already become involved in the post-Stalinist revision of Marxism-Leninism and were also more in tune with new scientific perspectives on religion and secularization that emerged across the Cold War divide. Leading Romanian atheists were therefore convinced that they lagged behind their more "modern" colleagues.[16]

And yet, while the Ceașescu regime enlisted their expertise to advance socialism, Romanian atheists soon found their efforts increasingly circumscribed. The party-state's commitment to reformism, half-hearted already during the thaw, waned by the late 1970s. Meanwhile, the knowledge atheists produced about religion became routinely appropriated for the re-Stalinization of Romanian society that commenced around the same time.

'What Do We Put in Its Place?' Atheism and Romanian Spirituality

Before religious belief became a central party concern in the late 1960s, Romanian cadres had operated with the understanding that atheism was predicated above all on the opposition between science and religion. Formed in the scientific enlightenment programme of the early Dej era, proponents of this view had exalted the natural sciences for the ability to disprove religious cosmologies and regarded Marxism–Leninism, given its stature as a social science, to be uniquely suited for revealing the twinned, cognitive and sociohistorical, roots of religion. When considering these conventions, it is perhaps not surprising that in the early 1970s some atheists took the party's call to improve their work as a ritualistic cue to reiterate previous ideological formulas. Indeed, throughout the Ceaușescu era,

---

[16] Leading atheists analysed the state of their work in a comparative perspective at an Agitprop meeting on 31 January 1969. ANIC, inv. 2898, File 5/1969, f. 15–46.

textbook authors taught the military, in particular, that ending the life-cycle of religion hinged ultimately on scientific enlightenment.[17]

For the vast majority of professional atheists, however, the bustling religious life of the 1960s proved that the approaches of scientific enlightenment were inherently problematic and downright untenable. Atheism required insight from other fields of knowledge, primarily sociology, philosophy, and psychology, and an answer to the question of why religiosity continued to persist, even revived, in the inhospitable conditions of socialism. Indeed, echoing earlier remarks made in discussions with Agitprop in January 1969, atheists and other experts added further arguments in the print media for why enlightenment workers failed to reach their intended audience and simultaneously provided suggestions for the future direction of atheist work. As atheists *publicly* circumscribed the party line on religion and started to outline 'scientific-atheism' as an academic discipline, many reiterated claims that religion was in crisis on the global scale and that the advent of an atheist society was therefore inevitable.[18] Such confidence was deeply embedded in much broader narratives about secularization, according to which modernity inadvertently rendered faith antiquated and irrelevant. Popular periodicals, scientific journals, and other forms of media reproduced such pronouncements with verve throughout the Ceaușescu era. And yet, while atheists never fundamentally questioned the fateful march towards a disenchanted world, their expectations about the straightforwardness and rapidity of this process became increasingly tempered.

In fact, as they revised atheist theory and their understanding of religion, Romanian atheists drew considerably on the broader philosophical revisions of Marxism that were underway in Eastern Europe since the mid-1950s. Historian Pavel Kolář observed that the events of 1956 shattered the idea of a quick and straightforward march towards the communist future.[19] History slowed down and became more indeterminate for Romanian atheists, much like to prominent Marxists and communist parties across Eastern Europe. In line with the broader post-Stalinist ideo-

---

[17] For example, Simion Asandei (1985) *Omul, știința și religia* (Bucharest: Editura Militară), pp. 25–7.

[18] Petre Datculescu (1976) 'Dezvoltarea atitudinilor materialist-științifice ale tineretului', *Caiet Documentar* no. 5 (Bucharest: Centrul pentru Studierea Problemelor Tineretului), 8–9.

[19] Pavel Kolář (2012) 'The Party as a New Utopia: Reshaping Communist Identity after Stalinism', *Social History*, 37, no. 4, 404, 422–4.

logical shifts, Romanian atheists also 'discovered' that socialist societies generated their own contradictions and possibilities for alienation, reasons for which they increasingly came to argue that, contrary to previous pronouncements, religion was not a form of estrangement particular to past societies. Religion had profound roots in socialism too and overcoming it would be a long endeavour.[20]

As part of this process of ideological revision, prominent atheists also pointed out that the source of the remaining utopian confidence in secularization was that their colleagues failed to appreciate the distinction between abstract understandings of religion and the actual religiosity of believers on the ground. Haralamb Culea, a sociologist at the Academy of Socio-Political Sciences, noted that atheists expected watershed changes in believers' mentality because they understood religion exclusively as an epiphenomenon of the macrosocial sphere. But faith, Culea observed, was mediated by the entire, 'concrete' human personality. Religiosity persisted in socialist modernity precisely because its 'last bastion' and deepest roots lay in a sphere 'relatively autonomous' from the social—'man's internal life'.[21] Indeed, atheists increasingly began to contend was that what made religion resilient in socialist society had less to do with cosmological explanations, its embeddedness in sociocultural life, or its social origins. The strength of 'religiosity as religiosity' lay in its *spiritual* functions—that is, in religion's ability to give meaning to death, suffering, and life through its promise of salvation.

The critique of previous conceptions of religion had two important implications: first, for how atheists reconsidered the trajectory of secularization and, second, for how they reimagined atheism as an ideal. In a series of works intended for the local propagandist and the common citizen, Culea observed that religious life and modern society revealed a 'bushy dialectics' (*dialectică stufoasă*): the involution of religion proceeded

---

[20] On the theoretical discovery of the possibility of alienation in socialist societies see James H. Satterwhite (1992) *Varieties of Marxist Humanism: Philosophical Revision in Postwar Eastern Europe* (Pittsburgh: University of Pittsburgh Press), p. 183. Petru Berar (May 1972) 'Educația științifico-ateistă a maselor: cerință a umanismului socialist', *Era socialistă* LII, no. 5, 25–6.

[21] Haralamb Culea (5 January 1973) 'Implicații ale analizei sociologice a fenomenului religiozității după o investigație concretă', *Viitorul Social*, 2, no. 1, 201. Haralamb Culea (1973) 'Explicarea sociologică a fenomenului religios' in Petru Berar (ed.) *În sprijinul educației materialist-științifice a tineretului: Planuri tematice și bibliografice* (Bucharest: Editura Politică), p. 28.

in stages along a 'sinuous, jagged road', occasionally giving way to the recrudescence of faith.[22] The piecemeal character of this process meant that distinctions had to be drawn between certain stages. The spread of secularization, desacralization, irreligiosity, and even scientific rationalism in society constituted only 'preliminary phases', Culea argued, when the 'demolition' of religion was the important task at hand. But the move towards an atheist society raised the problem of 'reconstruction'. In order to decisively uproot faith, atheists needed to offer a belief system that provided 'positive' existentially meaningful answers to the key questions of human life and that had a salvation theory its own. Culea explained:

> The task is not only to answer the question 'What do we put in its [religion's] place?' [...] Instead of faith in heavenly immortality, there is the need for an optimistic belief in the earthly immortality of humanity. [...] Marxist humanism assures the spiritual premises of this model, and communism secures its material, social foundations. But these conditions are not sufficient. It is also imperative to erect on a mass scale a Promethean human type, who is capable of experiencing fully (in thought and feeling) the happiness of a life based on creative action in the service of humanity, the sole humanist answer to the problem of death.[23]

By presenting this 'earthly' model of immortality, Culea distinguished 'integral atheism', as he termed it, from lay mentality or scientific-atheism proper. In effect, what he suggested was that atheism's and, by implication, socialism's success ultimately hinged on whether they could acquire a spiritual–emotional dimension to address the problems of human existence in a meaningful and 'optimistic' way. For this purpose, developing an atheist art and culture were key, he argued, for these areas in social activity provided individuals with feelings of ecstasy, crucial for anchoring an atheist worldview in the human personality and for making atheist belief fully liveable.[24]

Arguments that held up atheism as an existential model, as opposed to a simple rejection of belief in supernatural cosmologies and religious organizations, indicate that Romanian ideologists in the 1970s made the transition from a negative to a positive atheism. However, the question

---

[22] Haralamb Culea (1975) *Religia sub semnul cercetării filozofice-sociologice: Dezbateri ideologice* (Bucharest: Editura Politică) pp. 7 and 60.

[23] Culea, *Religia sub semnul cercetării*, p. 70.

[24] Culea, *Religia sub semnul cercetării*, pp. 71–2.

still remained: what made their ideas distinctly Romanian? Resolving this issue required atheists to reconstruct the genealogy of their thought in a twinned national and Marxist–Leninist frame.

On the surface, this task entailed obscuring the continued influence of Soviet atheism in keeping with the regime's claims to independence from Moscow. The issues raised by Romanian atheists in the 1970s overlapped extensively with the kinds of questions Soviet atheists began to ask in the Khrushchev period, which suggests that familiarity with both the debates and works published in the metropole existed.[25] Indeed, specialized bibliographies largely marked for the 'internal use' of the party academy provided extensive lists on Soviet scholarship and major thinkers in Marxist–Leninist atheism.[26] Such works, however, were rarely mentioned in Romanian publications much less translated. Indeed, the only exceptions to this rule throughout the late socialist period were Petru Berar's 1976 anthology on religious psychology, which included selections from the opus of Soviet scholars such as Dmitri Ugrinovici, Vladimir Pavliuk, and Vladislav Serdakov, and a handful of translations from the works of key atheists, most notably Yaroslavski and Iuri Frantzev.[27]

To cast their endeavour into a national mould, atheists also had to re-evaluate prominent arguments from the interwar period that had identified religion as a cornerstone of Romanian civilization and spirituality. As Herderian notions of the *Volksgeist*—'the spirit of the people'—gained appeal in the Romanian national movement, the religiosity of the peasantry became a matter of debate particularly in folklore. Between the two world wars, however, the Romanian soul's penchant for the metaphysical was no longer a question but a matter of consensus, especially among intellectuals adhering to the cultural and political Orthodoxism hegemonic at the time.[28] As historian Sorin Antohi argued, the pre-socialist imagery of 'an autochthonous Geist', at peak in the interwar era, involved a 'vertical escape' from the stigmatizing realm of Eurocentric symbolic geographies in that it placed national being 'in a protective relationship

---

[25] Smolkin-Rothrock, '"A Sacred Space Is Never Empty"'.

[26] Petru Berar, ed. (1981) *Fenomenul religios în istoria umanității: cercetare bibliografică* (Bucharest: Biblioteca Academiei Ştefan Gheorghiu).

[27] Petru Berar, ed. (1976) *Psihologia religiei* (Bucharest: Editura Ştiinţifică şi Enciclopedică).

[28] Keith Hitchins (1994) *Rumania, 1866–1947* (Oxford: Clarendon Press), pp. 298–319.

with a divine or (in the secular versions of this way of thinking) a transcendental principle'.[29]

The essentializing links that interwar thinkers like Lucian Blaga, Mircea Eliade, and Constantin Noica established between national essence and the peasant appealed to the party, which from the mid-1960s promoted the Romanian peasantry and its culture (folklore) as 'progressive' revolutionary forces in history. In part, this was the reason why such writers underwent partial rehabilitation and their works were published in the early Ceauşescu period.[30] But from the perspective of the ideological establishment, these intellectuals had also denounced materialism as alien to 'the Romanian soul'.[31] As a result, nationalizing atheism in the 1970s imposed the need to affirm the paganism, or better yet, the secularity of peasant mentality and spiritual life.

To posit the essential laity of the nation, intellectuals undertook concerted efforts over the course of the decade, above all around those classics of folk literature that had provided the foundation for arguments about national specificity and religiosity in the first place. Thus, commenting on the ballad of *Meşterul Manole*, the literary historian Liviu Rusu dismissed interwar interpretations and identified instead peasant 'spiritual values' such as egalitarianism, love of humankind, and charity as part of a quintessentially lay ethical system.[32] In the epic poem *Miorița*, the philosopher Pavel Apostol discovered a 'human ideal' for whom the problem of death strengthened 'trust in the unlimited ability of man to overcome suffering and defeat'.[33] This peasant morality and 'profane attitude' towards the meaning of life, the philosopher Alexandru Tănase concluded, was not sufficient to assure 'the lay orientation' of Romanian society. Yet, the peasant spiritual perspective had an invaluable 'ethical and aesthetic potential' for socialist humanism.[34] To return to the question of Romanian atheism

---

[29] Sorin Antohi (2002) 'Romania and the Balkans: From Geocultural Bovarism to Ethnic Ontology', *Transit—Europäische Revue*, no. 21, http://www.iwm.at/read-listen-watch/transit-online/romania-and-the-balkans, date accessed 24 May 2015.

[30] Katherine Verdery (1995) *National Ideology under Socialism*: Identity and Cultural Politics in Ceauşescu's Romania (Berkeley: University of California Press), p. 125.

[31] Al. Posescu (1967) 'Tendințe și poziții ateiste și materialiste în filozofie în perioada interbelică' in N. Gogoneață, Simion Ghiță, Radu Pantazi et al. (eds.) *Curente și orientări în istoria filozofiei româneşti* (Bucharest: Editura Academiei), pp. 169–224.

[32] Liviu Rusu (1967) *Viziunea lumii în poezia noastră populară* (Bucharest: Editura Politică și Literară), p. 34.

[33] Pavel Apostol (1970) *Trei meditații asupra culturii* (Bucharest: Dacia), pp. 190–2.

[34] Alexandru Tănase (1973) *Cultură și religie* (Bucharest: Editura Politică), p. 137.

then, what such arguments effectively proposed was that the Romanian peasant was the ancestor of a quintessentially national Promethean man.

By the late 1970s, ideological aspirations to articulate an existentially meaningful atheism and the discourse on national authenticity converged to incorporate 'Romanian spirituality' into the political vocabulary of the Ceaușescu regime. A staple term of pre-socialist mystical and religious nationalism, in its recent iteration 'Romanian spirituality' constituted a new semantic constellation that subsumed the ideal of self-transcendent man alongside constitutive socialist values such as happiness, work, and culturedness. In line with the regime's return to Stalinism, it simultaneously referenced the subordination of the individual to the collective principles of the socialist nation, which was in turn placed in a vertical relationship with other Stalinist transcendental principles: the party and, increasingly, Ceaușescu. Articles during the last decade of the 'golden epoch' used 'Romanian spirituality' consistently in this sense.[35]

Much like its predecessor, socialist Romanian spirituality also reflected a symbolic geography that inverted dominant imaginaries of centre and periphery. As a term, it was inflected namely by two additional discourses that gained currency after 1974: protochronism and the anti-colonial rhetoric of the nonaligned movement. Protochronism, Katherine Verdery writes, was a 'rescucitation of interwar indigenist arguments about national essence'; it stressed the sovereignty, creative independence, and temporal priority of Romanian cultural creations in relation to western and Soviet forms while emphasizing indigenous accumulation of value and Romanian contributions to the 'world stock' of cultural capital and historical development.[36] As the Ceaușescu regime became involved in the nonaligned movement, protochronists absorbed the movement's anti-colonial language and posited the humanism of a historically anti-imperial Romanian nation to be essentially superior to both the western and the Soviet variant. This position was best summed up in the words of the key protochronist Dan Zamfirescu who expressed his belief that the Romanian people, 'thirsty for justice, universal in their capacity for sympathy and for centuries devoted to the supreme value of humanism', were ideally suited to work towards a 'new universal mentality' which could defeat 'the demons that ruled [...] the destiny of the human species in the twentieth century'.[37]

---

[35] For example, Constantin Răducu (March 1981) 'Permanențe ale spiritualității românești', *Contemporanul*, 27, 2.

[36] Verdery, pp. 168, 174–180.

[37] See, for example, Dan Zamfirescu (1979) *Via Magna* (Bucharest: Editura Eminescu), p. 259.

Studying Believers

Besides relying on experts for key ideological revisions, the regime's commitment to reform atheist work during the 1970s was also evident in that higher education and research on religion commenced at various institutions across the country. For the future generation of intellectuals, courses on scientific-atheism opened at university departments of scientific socialism, philosophy and sociology. To train cadres while providing ideological control at the highest level, the Academy of Socio-Political Sciences set up its own laboratory of scientific-atheism.[38] Since the need to explain shortcomings and devise new methodologies for atheist education hinged on investigations of popular religiosity, research required systematic coordination. As a result, research groups were formed first under the aegis of the Council for Socialist Education and Culture and the Centre for Research on Youth Problems.[39] From 1976, however, the Central Committee assigned the task of coordinating research to the Academy in the framework of 'priority themes' set in the 5-year plans.[40]

The revival of the academic study of religion in the 1970s was part of the growing reliance on social science expertise, sociology in particular. After some timid steps under Dej in the early 1960s, the Ceaușescu regime rehabilitated, among others, the name of the Bucharest School in sociology, social psychology, and ethnology from the interwar years. Beyond rooting the party in 'progressive' national heritage, the School's international fame served an important role in the regime's efforts to reconstitute ties with western social science. In fact, most sociologists who rose to prominence in the late socialist period, including those studying religion, reaped the benefits of the party-state's foreign policy interests, being permitted to embark on extended research trips to the USA, West Germany, and France.[41] In addition to having access to Soviet and Eastern European scholarship on religion, Romanian social scientists also became familiar with empirical research and seminal theoretical works produced after 1945 on both sides of the iron curtain. All of this suggests that, beyond the veneer of ideological discourse, the ideas and methods developed to study religion and atheism in late socialist

---

[38] ANIC, inv. 2574, File 1/1973, f. 13. ANIC, Fond C.C. al P.C.R., inv. 3293, File 13/1986, f. 10v.

[39] Culea, 'Implicații ale analizei sociologice', 202.

[40] ANIC, inv. 3125, File 10/1976, f. 182.

[41] I give three examples here: Achim Mihu (born in 1931), Ilie Bădescu (born in 1948), and Septimiu Chelcea (born in 1940).

Romania were of broader transnational inspiration and can hardly be understood adaptations of Marxist–Leninist scholarship. Indeed, as historian Călin Cotoi observed, in the sociology of the 1970s 'the gambit of scientificity was played between historical materialism, Western structural-functionalism, empirical research, and fieldwork, but also interwar models of national science'.[42]

Above all, however, the unprecedented attention to the study of religiosity emanated more broadly from official views regarding sociology's centrality to socialist modernization.[43] The party-state's expectations were that empirical analysis of popular religiosity could describe the gap between Marxist–Leninist ideology and social reality and simultaneously provide policy prescriptions on how to achieve convergence between the two. Ideological parameters certainly imposed limitations on scholarship. Yet, the amount of material sociologists and their research groups gathered around the country in the 1970s and 1980s produced the most extensive map of religiosity in Romania thus far. Some of the findings confirmed previous suspicions about the nature of religion in socialism but also prompted atheists to re-assess understandings of the typology of believers and the direction of atheist propaganda.

Among others, sociological investigations devoted a lot of attention to the emotional elements of religious experience arguing that the strength of modern faith was rooted in its affective and spiritual dimensions.[44] Researchers most often emphasized the capability of religious organizations to respond to 'real' needs in an individual's life by providing moral guidance and emotional relief from suffering during times of challenge.[45] The thrust of such arguments was naturally that atheist education and party-state institutions overall failed to fulfil this 'therapeutic' function. As Petre Datculescu suggested, this omission carried particularly high stakes in the case of the young and the ill who converted to sectarian belief because of traumatic life-events such as lengthy diseases, suffering from family or romantic relationships, the death of a loved one, and even in the

---

[42] Călin Cotoi (2011) 'Sociology and Ethnology in Romania: the Avatars of Social Sciences in Socialist Romania' in Ulf Brunnbauer, Claudia Kraft and Martin Schulze Wessel (eds.) *Sociology and Ethnography in East-Central and South-East Europe: Scientific Self-Description in State Socialist Countries* (Munich: Oldenbourg Verlag), p. 135.

[43] Cotoi, 'Sociology and Ethnology', p. 140.

[44] Ion Drăgan, Ion Mihail Popescu, and Ilie Bădescu (5 June 1982) 'Factori sociali ai fenomenului religios și cerințele educației materialist-științifice', *Era Socialistă* 11, no. 11, 44.

[45] Petre Datculescu, 'Dezvoltarea atitudinilor materialist-științifice', pp. 14–5.

case of personal failures at school or in the work environment.[46] Indeed, confirming atheist fears, interview subjects inadvertently reported positive feelings after conversion, which ranged from the sense of being guided and protected to 'spiritual contentment', 'moral stability', and even 'indescribable happiness'.[47] Importantly, Datculescu noted, religion's compensatory function appeared to work as young atheists exhibited 'significantly' higher levels of tension, anxiety, and depression than believers.[48] Atheists likewise observed that all churches embraced the socially marginal elements—the Romany, unmarried women, alcoholics, the unemployed, poor workers, and peasants—providing economic aid, moral guidance, and community support to those who benefited less from the blessings of socialism.[49] Far from being expressions of mere cynicism of religious charity and social welfare, these observations showed the abiding concern that religious organizations were turning the weaknesses of socialism to their own advantage. By extension, they also underscored that atheism needed to fulfil the 'pastoral' functions in an individual's everyday life—functions that traditionally belonged to religion.

As findings confirmed that the battle against religion was going to be a protracted affair, atheists increasingly concluded that the new task in their endeavour was to strengthen atheism spiritually and emotionally and thus humanize socialist life. Given the compensatory functions of 'religious myths, representations, and practices', Datculescu argued, atheist education required above all the cultivation of 'inter-human relations characterized by solidarity, mutual respect, friendliness, and mutual aid in difficult situations or moments in personal life'.[50] Indeed, Aurel Olaru and Gheorghe Dumitru added, to address suffering and negative emotions, atheist policy needed to perfect 'psycho-social relations' among citizens in all areas of socialist life, whether in public institutions, mass organizations, the work environment, schools, places of leisure and

---

[46] Petre Datculescu (1980) *Educația materialist-științifică a tineretului: cercetări psihosociale* (Bucharest: Editura Politică), p. 114.

[47] Datculescu, *Educația materialist-științifică*, pp. 115 and 125.

[48] Datculescu, *Educația materialist-științifică*, pp. 136–7.

[49] Ion Mihail Popescu (1981) *Aspecte ale fenomenului religios și probleme ale educației ateiste a tineretului de pe teritoriul județului Caraș-Severin* (Bucharest: Centrul de Sociologie, Universitatea București), pp. 111–2, 121, and 150.

[50] Petre Datculescu (1979) 'Probleme actuale ale formării tineretului în spiritul concepției materialist-dialectice' in Fred Mahler (ed.) *Dezvoltarea conștiinței socialiste a tinerei generații* (Bucharest: Editura Politică), p. 200, p. 218.

living, or the more private domain of the family.[51] In practical terms, this task of humanizing social relations was far from straightforward. First, sociological data motivated atheists to revise the categories by which a person was classified as a believer or, conversely, an atheist. Second, experts also had to return to old questions about the place of atheism in mass culture, especially in connection to socialist rituals. Inextricably linked, these two issues marked the ways in which the Ceaușescu regime would envision socialist spiritual culture in its late years.

The objective of providing alternatives to religious holidays and life-cycle rites had re-emerged in 1966, during the first atheist campaign of the Ceaușescu regime but beyond instructions to intensify such efforts at the local level, the ritual question remained a low priority for the top party ranks. It was what sociological findings revealed about the relationship between religious belief and the practice of rites that propelled the problem of socialist ritual reform to the centre of party concerns. Throughout the 1970s, the material gathered on the ground indicated that levels of observance for both life-cycle and seasonal rituals were 'strikingly high' even in areas that researchers classified as being of 'low religious intensity'. Thus, the Centre of Sociology at the University of Bucharest reported that in Gorj county 66.8 % of subjects turned to the church for baptisms, weddings, and funerals.[52] Researchers also devoted considerable care to assess the popularity of various ceremonies, parish feast days, and pilgrimages in comparison to the socialist festivals that local organs set up for such occasions. Popescu noted for instance that, gathering from five counties, tens of thousands of believers celebrated the Dormition of the Mother of God at the 600 year old monastery of Tismana. Meanwhile, taking place 3 km away, the local socialist feast degenerated into 'a common fair'.[53] What sociological findings revealed then was that religious rituals continued to pose a challenge to the atheist project—a challenge that its various enlightenment measures, 'cultured leisure' programmes, red-letter days, and rites of passage had not yet adequately addressed.

What compounded the question of ritual reform was that experts identified profound contradictions between ritual observance and,

[51] Aurel Olaru and Gheorghe Dumitru (20 February 1982) 'Cercetarea sociologică a formelor și mijloacelor educației materialist-științifice', *Era Socialistă* LXII, no. 4, 27.

[52] Ion Mihail Popescu and Ilie Bădescu (1980) *Factori economico-sociali și culturali ai fenomenului religios. Raport de cercetare (Județul Gorj)* (Bucharest: Centrul de Sociologie), p. 103.

[53] Popescu and Ilie Bădescu, *Factori economico-sociali*, pp. 49–50.

conversely, people's actual beliefs and motivations to practice. Based on national data gathered in 1971–1972, Haralamb Culea remarked, for instance, that whereas 76.9 % of subjects practiced rituals, only 5.7 % actually held religious convictions.[54] Years later, researchers from the Centre for Sociology at the University of Bucharest similarly observed that numerous young people participated at the consecration of churches not because of any religious motivation but out of curiosity to see the bishop's apparel or see how a ritual was conducted.[55]

By the late 1970s, such contradictions between ritual practice, attitude towards religious organizations, and belief warranted separate sections in the Centre's reports and demonstrated that discussions about the typology of belief had undergone a radical shift.[56] Experts embraced new criteria to assess degrees of conviction on a spectrum that went from active religiosity to atheism, the middle section being occupied by a type that researchers varyingly described as 'transitional', 'indifferent', or 'non-atheist'. A crucial trait of persons who belonged to this category was the disconnect between belief and religious practice: while they may have identified themselves as believers and practiced rituals, their commitment to religious worldviews and church dogmas was either not strong or not present at all.[57]

Since findings indicated that this demographic category far outweighed either the convinced believer or the atheist, sociologists presented two main arguments to explain why 'indifferent' believers were attracted to religious rituals. According to researchers at the Centre for Sociology in Bucharest, one of the answers was that as religious organizations modernized, they increasingly focused the centre of their activity on the existential and spiritual domains of human life. Instead of insisting on supernatural cosmologies foreign to the essence of the believer and the essence of nature, religious organizations acquired new functions in response to modern socio-economic and spiritual developments, the argument went. Churches transformed themselves into 'meeting places' for youth or into places of 'retreat and meditation, without purely religious meaning' for

---

[54] Haralamb Culea (1973) 'Religie, mentalitate laică, ateism', *Forum: Științe sociale* 5 (1973), 167.

[55] Popescu, *Aspecte ale fenomenului religios*, p. 65.

[56] See Ion Mihail Popescu (1978) *Factori economico-sociali și culturali ai fenomenului religios. Probleme ale educației ateiste (Județul Tulcea)* (Bucharest: Centrul de Sociologie), pp. 72–5.

[57] Popescu, *Aspecte ale fenomenului religios*, pp. 110 and 187.

citizens facing difficult situations, economic setbacks, or moral troubles.[58] More importantly, however, the Centre's national findings revealed that 'tradition' was the most widespread motive for religious behaviour. Citizens declared that religious ceremonies were necessary in human life. But such views, Popescu warned were not rooted in faith. 'Indifferent believers' reinterpreted religious rituals as part of 'cultural tradition' and customs.[59]

The discovery of 'indifference' as a category on the Romanian religious landscape was particularly troubling. On the surface, the socialist press was quick to claim the declining numbers of convinced believers was proof of atheist success. But in reports presented to the restricted audience of party leaders, researchers were far less confident. The Centre of Sociology warned top organs that the population exhibiting such 'religiosity of transition' was just as likely to embrace religious conviction as a lay mentality.[60] The process of secularization was not linear, these sociologists reminded, and as a result, atheist work had to focus primarily on this category of believer.

The conclusion was that if spiritual needs and the attachment to cultural traditions explained the religious behaviour of most believers, then atheists had to provide attractive lay ceremonies. For Datculescu, these needed to be infused with socialist humanism and had to be 'well-grounded from a historical, ethnographic, folkloric, psychological and sociological perspective'.[61] Popescu also imagined 'sumptuous lay ceremonies' rooted in folklore. In practical terms, he added, this meant that atheists had 'to uncover the lay pip of certain religious celebrations'. After all, 'Christian holidays (Easter, Annunciation, Whitsunday, Dormition of the Mother of God, Christmas, and Epiphany, as well as religious celebrations tied to church patrons) constituted either lay celebrations tied to memorable events in people's real lives—primarily harvest cycles—or pagan festivities', which church fathers and theologians subsequently endowed with new meanings. Atheists were to imitate such efforts, Popescu suggested, while naturally emphasizing the lay sense of such holidays.

One implication of such arguments was that in order to return to the 'original meanings' of such celebrations, research in folklore was

---

[58] Popescu and Bădescu, *Factori economico-sociali*, pp. 44–5.
[59] Popescu, *Aspecte ale fenomenului religios*, pp. 181–3.
[60] Popescu, *Aspecte ale fenomenului religios*, p. 187.
[61] Petre Datculescu, 'Probleme actuale', p. 216.

imperative and had to encompass not only festival culture but also customs around life-cycle rituals.[62] Yet, in order for atheist propaganda to be successful with 'the indifferent' believer in particular, the new folkloristic ceremonies also necessitated 'sumptuous' spaces. Every locality needed houses of baptism and marriage, funeral homes, and lay cemeteries that surpassed those of local churches. The urgency for a suitable material culture was particularly high in the case of funereal rites. According to the findings of the Centre for Sociology at the University of Bucharest, most cemeteries in the country had not been nationalized and, as a result, were church-owned. Without a proper sepulchral culture that provided rituals and a material environment, Popescu argued, a socialist country could not give meaning to individual life. If it could not 'love its dead and respect their memory', socialism 'was like a sanctuary without an altar, or a heroic army without a flag'.[63]

Conclusion: Ritual Reform and the 'Song for Romania' Festival

Over the course of the 1970s, sociological research around the country, most of which was carried out under the aegis of the Academy of Socio-Political Sciences, provided unprecedented information about popular religiosity in Romania. However, this religious map alongside the various theoretical revisions in atheism bore ambivalent results for propaganda work. In line with the policy function of the social sciences, the reports of the Centre of Sociology in Bucharest lobbied for the creation of special institutions for atheist education and research by proposing, first, the creation of the National Commission for the Problems of Religion and Atheism and, second, the establishment of a National Institute for Research on Religion and Atheism at the Academy. Among others, sociologists also suggested the creation of atheist popular and scholarly journals and a national census of the religious population, the first since 1930.[64] But top party organs continuously rejected such proposals. Indeed, the only new institution established at the turn of the decade was a small

---

[62] Petre Bartoş and Mihai Merfea (5 September 1982) 'O cercetare sociologică privind educaţia materialist-ştiinţifică', *Era Socialistă* LXII, no. 17, 43.

[63] Popescu, *Aspecte ale fenomenului religios*, p. 208.

[64] ANIC, inv. 2898, File 5/1969, f. 16–7. Popescu, *Factori economico-sociali (judeţul Tulcea)*, pp. 76–7, 81.

interdisciplinary laboratory at the Party Academy 'Ştefan Gheorghiu'. Alongside the Centre for the Study of Youth Problems and a similar laboratory at the Academy of Socio-Political Sciences, such institutions provided the framework for limited research during the 1980s.[65] Nationwide investigations became uncommon, however, as the Party increasingly marginalized sociology and psychology as policy-relevant fields for socialist construction. As Ceauşescu's rapid industrialization programme gradually prepared the way for a full-scale economic collapse, state funding became increasingly unavailable and investigative research on 'social reality' politically undesirable.[66] But the limiting and eventual supplanting of semi-official expertise by the late 1970s reflected, most of all, the regime's unapologetic return to the Stalinist ethos and its subsequent intolerance of any centres of scientific knowledge that could detract from the party's power and its leading role in society.[67]

The re-Stalinization drive built on sociological findings, ushering in two significant changes in atheist propaganda. On the one hand, the religiosity among the avant garde of socialist construction—party members and the youth—prompted leaders to instate an ideological hard line. Thus, the Agitprop's report for the Central Committee's July conference in 1979 noted that in the previous 4 years (1974–1978) the party excluded over 2200 citizens for converting to 'sects'.[68] At the meeting, Ceauşescu himself demanded the removal of pedagogues from the education system unless they had clear 'materialist convictions'. He also declared that the religiosity of party members and of the communist youth was unacceptable in stark contrast to his previous pronouncement in 1970 when he expressed a preference for the religious cadre who was disciplined in work to the one 'who does not go to church but introduces disorder, indiscipline, [and] chaos in society'.[69]

More importantly, however, the question of ritual reform—that is the task of providing socialist folkloristic ceremonies for holidays and

[65] ANIC, inv. 3293, File 33/1987, f. 43–7.
[66] Academic sociology was banned in the late 1970s. Cotoi, 'Sociology and Ethnology', p. 143. Psychology's demise in 1982 became linked with the so-called Transcendentalist Meditation Affair. Adrian Neculau (2004) 'Afacerea "Meditaţia Transcendentală"—o răfuială cu psihologia şi psihologii' in Adrian Neculau (ed.) *Viaţa cotidiană în comunism* (Bucharest: Polirom), pp. 264–88.
[67] Tismăneanu, p. 30.
[68] ANIC, inv. 3129, File 54/1979, f. 8–11.
[69] Ibid., f. 37. ANIC, inv. 2574, File 16/1970, f. 46–54.

milestone moments in individual life—moved to the centre of the regime's concerns. Informed by empirical investigations over the course of the decade, the Agitprop's report for the July 1979 conference of the Central Committee posited the problem of rituals as an urgent matter.[70] As the laicization of folk culture became a new objective in atheist work, the need for appropriate repertoires of seasonal and life-cycle rituals put its mark on ethnological research. Thus, commissioned to develop the first Ethnographic Atlas of Romania, the Institute of Ethnology and Dialectology in Bucharest conducted research between 1972 and 1982 in over 530 villages on all aspects of village life and material culture, including seasonal celebrations and life-cycle rituals.[71] The regime's support for the publication of the Atlas waned but the ethnographic material gathered came to inform the repertoires that the Institute was obliged to compile for the numerous song and dance ensembles participating in the biannual Song for Romania Festival.[72]

Created in 1976, the Festival was a mass competition that, among others, staged local folklore traditions in order to produce a revolutionary national culture and thus provide a realm for the becoming of a self-transcendent Promethean man, distinctly Romanian and atheist. The festival's national-Stalinist dimension was unmistakable in its militantly mobilizational character and in its celebration of collective—socialist and national—identity under a leader over and above the private and the personal domain of the individual citizen. Indeed, programmatic articles about the festival hailed Ceaușescu for developing the 'past spirituality' of the people in line with the 'needs of the future'.[73] When the Central Committee gathered once more to evaluate atheist education in February 1986, Ceaușescu himself militated that the festival become a framework not only for the staging of laicized seasonal and life-cycle rituals inspired by folklore but a medium for their large-scale implementation.[74] Song for Romania was a venue for social and cultural homogenization with a 'lasting social impact', anthropologist Vintilă

---

[70] ANIC, inv. 3129, File 54/1979, f. 36v–7.
[71] Interview with Dr. Ion Cuceu, Director of the Romanian Academy's Institute 'Archive of Folklore', Cluj-Napoca, 7 April 2011.
[72] ANDJHR, inv. 663, File 83/1981–1987, f. 302–15.
[73] Suzana Gâdea (Mai 1981) 'Partidul—promotor al spiritualității românești', *Contemporanul* 8, 2.
[74] ANIC, inv. 3293, File 13/1986, f. 6–9.

Mihăilescu writes. 'Folk culture (and its experts) never recovered after this vast social experiment'.[75]

Vying with Poland, today's Romania counts as one of the most religious post-socialist countries.[76] This religious scene makes it tempting to consider it as yet another case where Marxist–Leninist atheism 'failed', albeit for historically specific reasons—that is, the problematic national credentials of the Romanian communists and their consequent reliance on the Orthodox Church throughout their stay in power. The belated manner in which the Romanian party-state decided to put atheism at the centre of its ideological programme and the extent to which it nevertheless circumscribed the work of experts can certainly be understood to mean that, overall, Romanian party elites did not have 'their heart' in the atheist project, at least in comparison to some of their socialist neighbours. And yet, the conceptual and methodological transformations of atheism in the 1970s outlined in this chapter demonstrate that the Central Committee shifted both the means and the ends of its anti-religious endeavours. Despite being supplanted during the re-Stalinization drive, social science expertise was central to this process; indeed, the particular intersection between the atheist reach for spirituality, ritual reform, and the ethno-folkloristic mass culture embodied in the Song for Romania Festival cannot be understood otherwise.

The evolution of atheist work during the 'golden epoch' also qualifies the totalitarian aspects of the Romanian party-state. Adopted in 1966, Ceaușescu's pro-natalist policies are often cited to distinguish Romania in the socialist bloc by indicating the particularly invasive ways in which the late socialist regime 'appropriated the private realm' of the family and the individual.[77] And yet, the fact that the Central Committee did not instruct for the implementation of new life-cycle rituals until 1986 or undertake the creation of the necessary material culture for baptisms, weddings, and funerals in any centralized manner provide examples for how it did not

---

[75] Vintilă Mihăilescu (2008) 'A New Festival for the New Man: The Socialist Market of Folk Experts during the Singing Romania National Festival' in Vintilă Mihăilescu, Ilia Iliev and Slobodan Naumović (eds.) *Studying Peoples in the People's Democracies II: Socialist Era Anthropology* (Berlin: LIT Verlag), p. 55.

[76] Irena Borowik, Branko Ančić and Radosław Tyrała (2013) 'Central and Eastern Europe' in Stephen Bullivant and Michael Ruse (eds.) *The Oxford Handbook of Atheism* (Oxford: Oxford University Press), p. 630.

[77] Gail Kligman (1998) *The Politics of Duplicity: Controlling Reproduction in Ceaușescu's Romania* (Berkeley: University of California Press), p. 34.

subordinate the private sphere to collective principles. Marking the important moments of human life remained the domain of religious organizations and, in this sense, the sociologist Ion Mihail Popescu was right. For all the popular support that Ceaușescu's personality cult and virulent nationalism could muster up, in some crucial ways, Romanian advanced socialism did not provide meaning for the individual.

PART II

Science, Religion and the Paranormal

CHAPTER 5

Inculcating Materialist Minds: Scientific Propaganda and Anti-Religion in the USSR During the Cold War

James T. Andrews

There was a fragmented nature to anti-religious propaganda and scientific education among the Soviet masses in the years between the 1917 Revolution and the end of the Second World War. Yet in the post-Second World War years, especially after 1947, a dramatic turn occurred with the centralization of scientific materialist propaganda following the creation of the *Znanie* (Knowledge) Society in the USSR and its collaboration with officials from the Soviet Academy of Sciences. After 1947, as a consequence of the liquidation of the Union of Militant Atheists, in addition to scientific propaganda, the *Znanie* Society, was tasked with the job of propagating atheistic views. As the Cold War and the conflict with the 'capitalist & religious West' intensified, spreading anti-religious, materialist views became critical to Stalin and his regime. It became one of the essential and defining signifiers of the ideological ethos of this Marxist state as it remade itself on the world stage after 1945.

In the past, historians tended to neglect the role of religion and culture in the Cold War historiography, though over the past decade, new Cold War studies have begun to emphasize the importance of both religion

J.T. Andrews (✉)
Modern Russian and Comparative Eurasian History,
Iowa State University, Ames, IA, USA

© The Editor(s) (if applicable) and The Author(s) 2016
P. Betts, S.A. Smith (eds.), *Science, Religion and Communism in Cold War Europe*, St Antony's, DOI 10.1057/978-1-137-54639-5_5

and anti-religion in the analysis of the conflict between the USA and USSR.[1] Religion and atheism certainly provided the Cold War contestants with paradigmatic conceptualizations of their belief and value systems.[2] However, as will be argued in this article, religious institutions and believers were able to respond (even adapt) to the new modernized, scientific cultures of the post-Second World War era. Religious organizations could even appropriate science to try to undermine irreligious Marxist, communist regimes.[3]

This chapter will analyze how the battle between scientific atheism and religious belief was waged from above domestically by the Soviet regime in the era of the Cold War. It will begin with an overview of the relationship between anti-religious propaganda and scientific enlightenment in the period of 1918–1946, leading into the immediate post-Second World War era. That section will highlight the diversified nature of anti-religious propaganda, and the tenacious sustainability of religious belief in Soviet Russia prior to the Second World War. The next section will then begin with an analysis of the *Znanie* Society's role in this development during the early Cold War, document various well-known scientists and propagandists involved in its formation, as well as discuss its eventual fracturing from within. When anti-religious activism was linked to scientific enlightenment it tapped into the public's thirst for scientific knowledge; however, when it focused myopically on anti-religious dogma it often fell short of its lofty goals.

At times, both the intelligentsia and general public responded with mixed perceptions of the significance of the spread of Soviet scientific and technological achievements in the post-Sputnik era (post-1957). However, especially in Khrushchev's times, space exploration, large-scale technologies and astronomical conceptions of the universe were a vibrant (if not new) focus of science popularization that indeed captured the popular imagination. Furthermore, the Soviet public (like its Western

---

[1] Dianne Kirby (ed.) (2003) *Religion and the Cold War* (London: Palgrave Macmillan). For an understanding of how Cold War propagandists were influenced by either religious or anti-religious beliefs, see David S. Fogelsong (1999) 'Roots of Liberation: American Images of the Future of Russia in the Era of the Cold War, 1948–1953', *The International History Review*, 21(1), 57–59.

[2] Andrew Preston (2012) 'The Religious Cold War', in *Religion and the Cold War—A Global Perspective*, ed. Philip E. Muehlenbeck (Nashville: Vanderbilt University Press).

[3] Carl J. Friedrich and Zbigniew K. Brzezinski (1966) *Totalitarian Dictatorship and Autocracy*, 2nd ed. (New York: Praeger), pp. 314–315. Elizabeth Shakman Hurd (2008) *The Politics of Secularism in International Relations* (Princeton: Princeton University Press).

counterparts) exhibited pride in its country's technological achievements—especially because educated Russians, unlike earlier pre-Second World War citizens (due to their technical educational levels), had an even better understanding of the scientific relevance of these events globally.[4] The chapter will end by additionally assessing concerns that anti-religious activists had about the effectiveness of more ideological, anti-religious propaganda in the Khrushchev and Brezhnev eras with a more educated, scientifically inclined populace. In the end, as the USSR approached its twilight, this more technologically savvy public may have been the root cause of anti-religion's failures—maybe more so than the resurgence of a spiritual awakening among part of the Russian populace.

TEACHING THE MASSES SCIENCE: THE FRAGMENTED NATURE OF SCIENTIFIC ANTI-RELIGIOUS PROPAGANDA IN PUBLIC, 1917–1946

After the Russian Civil War subsided, and the New Economic Policy era was ushered into the 1920s, the Bolshevik state created the setting for support of a diversified, yet fragmented, scientific anti-religious campaign. Unlike the post-Second World War era, when scientific propaganda (and anti-religious activity) was more centralized, the Soviet state in the 1920s, one could argue, cultivated both pre- and post-revolutionary intellectuals, editors, scientists, and propagandists to inculcate a scientific worldview among the populace. This campaign could thus be located or situated in diverse venues such as governmental institutions, voluntary associations, scientific organizations, and even publishing houses (some that had survived from the pre-revolutionary era). Furthermore, it heavily stressed, at least initially prior to Stalin's times, scientific education as a driving force in this secularization component of its overall cultural revolution.

To begin, a variety of Soviet governmental institutions took part in widespread anti-religious activity just after the October Revolution of 1917, highlighting the use of numerous venues to inculcate scientific,

---

[4] In this chapter, I suggest that by the 1960s citizens of the West and East were engaging science and technology in similar ways irrespective of ideological differences. For an interesting look at how Soviet economic relations could also be seen as more practical, seeking global trade partnerships like their Western counterparts, see Oscar Sanchez-Sibony (2014) *Red Globalization: The Political Economy of the Soviet Cold War from Stalin to Khrushchev* (Cambridge: Cambridge University Press).

anti-religious propaganda. During the Russian Civil War era, the task of initiating an anti-religious campaign actually came under the auspices of the Commissariat of Justice, which was commissioned to organize specifically scientific lectures, posters, and leaflets that debunked religious superstitions. This Commissariat published a journal entitled *Revoliutsiia i tserkov'* (Revolution and the Church), and linked its early work in this area with Anatole Lunacharskii's Commissariat of Enlightenment (who, by the way, would write occasional articles for these journals and pamphlets).[5] While the first official Communist Party pronouncement on scientific-anti-religious propaganda was officially delineated at the Eighth Party Congress in March 1919, it was not until the Russian Civil War had subsided by 1922 that a more organized scientific-enlightenment and anti-religious campaign was spawned in diverse arenas and manifestations.[6]

From the beginning of the 1920s, the Communist Party placed more emphasis on the nexus between anti-religious activity and a broader scientific-enlightenment campaign. The Central Committee was even concerned with the methods of a variety of its own institutions such as the *Komsomol* (Communist Youth League) that initially carried out very coarse anti-religious festivals that mocked peasant religiosity. The Central Committee thus, in 1923, issued a very staunch directive on moderating anti-religious activity and focusing on science education: 'We must take all steps to avoid giving offense to religious feelings, shifting the weight of the work to a scientific explanation of the origin of religious holidays, especially of *paska* (Easter)'.[7] The Central Committee issued a series of directives (both earlier in February 1922 and again in January 1923) to its Agitational-Propaganda Department (Agitprop), headed by an individual who had been involved in anti-religious propaganda himself,

---

[5] G. Vorontsov (1959) *O propaganda ateizma* (Moscow). Also see Joan Delaney (1971) 'The Origins of Soviet Antireligious Organizations', in *Aspects of Religion in the Soviet Union 1917–1967*, ed. Richard H. Marshall, Jr. (Chicago: The University of Chicago Press).

[6] *Kommunisticheskaia partiia sovetskogo soiuza v rezolutsiiakh i resheniiakh s"ezdov, konferentsii i plenumov TsK*, 8th ed. (1970) (Moscow), 2: pp. 37–38.

[7] *Kommunisticheskaia partiia i sovetskoe pravitelstvo o religii i tserkvi* (1959) (Moscow), 76. With these staunch directives by the Central Committee aside, the Communist Party as a whole still continued to have residual internal disagreements about anti-religious methods, goals, and tactics: namely, whether religion should fade away as a byproduct of broader secularization and scientific campaigns, or whether secularization could be more quickly advanced through anti-religious propaganda and administrative methods.

Ia. Iakovlev, to shift to long-term secularizing goals, programs stressing natural-scientific knowledge, and the spread of technical information.[8]

The Communist Party and its Agitprop Department continually, at least in the provincial areas, confronted more problems on this front than success. Even with the more moderate, scientific-educational method of anti-religious propaganda, ethnographic reports of anthropologists (particularly those such as V. G. Bogoraz-Tan) attest to the fact that folk religion was positively flourishing in Russia in the 1920s and early 1930s. A few Russian historians, such as A. Angarov, have also noted a rise in Christian sectarian influences in the provinces and countryside during the 1920s and interwar era in general.[9] Furthermore, interestingly enough, even the reports of agitational-propaganda activists themselves in regional areas (sent back to central authorities) attested to a continual frustration that science and anti-religious lectures did not attract, but instead alienated, rural people who did not want to give up their religious rituals.[10]

In an attempt to strengthen its society-propaganda activities, the Communist Party tried to coordinate its work with more disparate forces such as voluntary (non-Party) anti-religious groups, scientists, as well as editors of journals. These non-Party groups had proliferated since the revolution, each with different goals in mind for anti-religious propaganda. By 1922, popular scientific journals that had significant anti-religious themes expanded particularly in urban areas. They included *Nauka i religiia* (Science and Religion), edited by the publicist Mikhail Gorev, which highlighted the role of scientific-enlightenment in anti-religious activism. By early 1923, E. Iaroslavskii started a weekly newspaper entitled *Bezbozhnik* (The Godless). The editorial board also emphasized long-term scientific education as a method to stamp out religion, targeting urban workers, and professional and administrative social categories of Soviet readers. In 1924,

---

[8] Iakovlev, in turn, sent these directives out to regional and provincial agitational-propaganda departments throughout Soviet Russia. See 'Tsirkuliarno, fevralia 1922, vsem obkomam, oblbiuro i gubkomom RKP, o postanovke antireligioznoi propagandy, zam. zaved. agitprop otdel TsK RKP (b), Ia. Iakovlev', in RGASPI—Russian State Archive of Socio-Political History (formerly 1991–1999, Rossiiskii tsentr khraneniia i izucheniia dokumentov noveishei istorii—RTsKhIDNI, earlier the Central Committee's Communist Party Archive of the Institute of Marxism-Leninism), f. 17, op. 60, d. 146, l. 50.

[9] Vladimir G. Bogoraz-Tan (1928) *Khristanstvo v svete etnografii* (Moscow). Also see the work of Angarov on Christian sectarian movements in provincial Russia in the 1920s: A. Angarov (1929) *Klassovaia bor'ba v sovetskoi derevne* (Moscow: Izd-vo Kommunisticheskoi Akademii).

[10] RGASPI, f. 17, op. 60, d. 405, l. 162.

anti-religious activists formed the *Obshchestvo druzei gazety Bezbozhnik* (Society of the Friends of the Newspaper 'The Godless'), the most important voluntary association of opponents of religion in the former Soviet Union. In June 1925, the organization became the 'League of the Godless'. By 1929, during the Stalinist Cultural Revolution, the organization changed its name once again to the 'Militant League of the Godless'.[11]

In the mid-to-late 1920s, and into the era of the Cultural Revolution (1928–1931), debates also began between radicalized younger Soviet scientists and Tsarist-era trained pedagogues, scientists, and editors. The debates involved whether scientific-educational propaganda should be laden with a clear ideological message.[12] One of the main proponents of this perspective was the biologist and propagandist Boris M. Zavadovskii, who taught at the Ia. M. Sverdlov Communist University in Moscow between 1920 and 1932. A dedicated Marxist thinker, he had also been the director and founder of the K. A. Timiriazev Institute in Moscow, which had a museum dedicated in the 1930s to the history of biology, agriculture and science education. Zavadovskii wrote supplementary pamphlets for the publishing house of the newspaper *Bezbozhnik*, and attacked religious explanations of the evolution of plant and animal life using a scientific materialist perspective.[13] Zavadovskii, and some of his associates at the Sverdlov Communist University, helped the State Publishing House (*Gosizdat*) create a variety of more politicized popular-science and anti-religious series during this period as well.[14] These debates during and just after the Stalinist Cultural Revolution may have been a precursor for the late 1940s era regarding both the potential centralization of anti-religious propaganda and the resurgent focus on ideological vigilance.

---

[11] For a look at the ineffectiveness of early Soviet anti-religious propaganda, see Daniel Peris (1998) *Storming the Heavens: The Soviet League of the Militant Godless* (Ithaca: Cornell University Press).

[12] William Husband has argued convincingly that anti-religious propaganda, even prior to Stalin's Cultural Revolution, was not an end in itself for the Communist Party. William Husband (2000) *Godless Communists: Atheism and Society in Soviet Russia, 1917–1932* (Dekalb: Northern Illinois University Press).

[13] B. M. Zavadovskii (1929) *Sushchestvuet li dusha?* (Does There Exist a Soul?) (Moscow: Akts. Izd. O-vo Bezbozhnik). These short pamphlets, published by the Militant League of the Godless' publishing house, were often written by scientists housed in newer Marxist institutions like the Institute of Red Professors in Moscow.

[14] See, for example, the popular Gosizdat Series *Nauka dlia vsekh* (*Science for All*) as examples of these popular scientific works and genres used by anti-religious activists during this era.

However, regardless of this imminent ideological fervor and precursor for times to come, the Communist Party and anti-religious activists were still committed in the 1930s to a tactic that focused on the spread of scientific educational material, adding an emphasis on technical education for the masses—especially (in the 1930s) shifting its emphasis to the urban sector due to both its frustrations in the countryside, as well as the influx of migrants into the industrial sector. One institution entitled VARNITSO (All Union Society of the Workers of Science and Technology for the Promotion of Socialist Construction) was particularly aggressive in aiding anti-religious activists (as well as Agitprop itself) in furthering science and technical propaganda in the urban sector.[15] VARNITSO in the 1930s helped the Communist Party launch popular press campaigns and condemned non-Communist educators and scientists who did not take part in these broader technical and anti-religious propaganda campaigns.[16]

Given that many new unskilled migrants to the cities were from the countryside, VARNITSO and Agitprop activists continually complained in their reports to central authorities about 'peasant religiosity' that was now being transplanted into the urban environment. Peasants who migrated to large cities, such as Moscow and Leningrad, not only retained their religious beliefs but also attempted to get married in those churches that remained opened in the cities. This certainly changed during the Great Terror in the late 1930s, when Orthodox clergy were persecuted at alarming rates. Timothy Colton estimated, for example, that in Moscow in the early 1930s there were still hundreds of operating churches, but that diminished tenfold after the Terror's height in 1938.[17] It is interesting to note that according to the long-suppressed 1937 Soviet Census, almost 60 % of the adult population in major Soviet Cities in European Russia still claimed to be religious believers. Question number five of the census was labeled 'religion', and this was

---

[15] For an excellent overview analysis of the organization and its relationship with scientists, activists, and the Academy of Sciences, see I. A. Tugarinov (1989) 'VARNITSO i akademiia nauk SSSR, 1927–1937', *Voprosy istorii estestvoznaniia i tekhniki*, no. 4: 46–54.

[16] Agitprop had close ties with VARNITSO during this period, and A. I. Krinitskii, head of Agitprop, urged VARNITSO members to produce scientific and technical pamphlets for workers in factories. Krinitksii was particularly interested in material that would not only help overcome their meagre scientific skills, but he also wanted to instill in them a reverence for science and technology over superstition. For archival holdings referencing VARNITSO's inter-related popular-enlightenment and anti-religious work see Rossiiskii gosudarstvennyi arkhiv ekonomiki (RGAE), f. 4394, op. 1, d. 41, ll. 1–26.

[17] Timothy J. Colton (1995) *Moscow: Governing the Socialist Metropolis* (Cambridge: Harvard University Press), pp. 260–264.

unique among pre-Second World War Soviet censuses to include a question on religiosity.[18] What is even more fascinating is that only 40 % of Soviet respondents to the 1937 census identified themselves as 'non-believers'.[19]

Religion was certainly a major component of the Second World War, as propaganda, legitimacy and the power of ideas was paramount between the combatants and their allies.[20] By the time the Nazis invaded the former Soviet Union in June 1941, while Stalin was still hostile to Orthodoxy and the established church, he realized tactically he would temporarily need the help of Metropolitan Sergei (and other Orthodox believers) to defeat the Germans. In 1942, interestingly enough, Metropolitan Sergei offered to raise money for tank production, and after Stalin accepted his help a thaw in church–state relations developed even more. In 1943, Stalin allowed Sergei to be elected Patriarch and once again increase the number of clergy and functioning churches on a limited level.[21] Even after the war, in 1945–1946, a few hundred formerly closed monasteries and churches, especially in major cities, were temporarily re-opened. Also, by 1947, the League of the Militant Godless, the cradle of anti-religious scientific propaganda in the USSR, was disbanded. Yet with all the temporary concessions in the period 1941–1946, a period of renewed ideological restraints, especially starting with the era of the *Zhdanovshchina* (1946–1948, named after Leningrad Party Chief Andrei Zhdanov, and Politburo aide to Stalin), ushered in another difficult period for the church,[22] and a revitalized, but more centralized scientific, anti-religious campaign orchestrated from Moscow.[23]

---

[18] *Vsesoiuznaia perepis' naseleniia v 1937 g. Perepisi naseleniia. Al'bom nagliadnykh posobii* (1938) (Moscow), pp. 25–26.

[19] Felix Corley (1994) 'Believers Responses to the 1937 and 1938 Soviet Censuses', *Religion, State and Society*, 22(4).

[20] Steven Merritt Miner (2003) *Stalin's Holy War: Religion, Nationalism, and Alliance Politics, 1941–1945* (Chapel Hill: University of North Carolina Press).

[21] M. I. Odinstov (ed.) (1995) *Religioznoe organizatsii v SSSR nakanune i v gody Velikoi Otechestvennoi voiny, 1941–1945* (Moskva: Nauka).

[22] For an analysis of how this renewed tension between Soviet state and church affected Soviet foreign policy and relations to other states, see W. C. Fletcher (1973) *Religion and Soviet Foreign Policy, 1945–1970* (Oxford: Oxford University Press). Anna Dickinson has argued that after the Second World War, the strengthening of relations with the Russian Orthodox Church outside Russia was primarily a means of strengthening the domestic church in order to facilitate its assertion of monopoly control in the liberated areas of western Russia and Ukraine, where the Uniate Church was strong. Anna Dickinson (2003) 'Domestic and Foreign Policy Considerations and the Origins of Post-War Soviet Church-State Relations, 1941–1946', in *Religion and the Cold War*, ed. Dianne Kirby, pp. 23–36.

[23] Dmitrii Pospielovsky (1984) *The Russian Church Under the Soviet Regime, 1917–1982* (Crestwood, NY: St. Vladimir's Seminar Press).

The Znanie (Knowledge) Society and the Centralization of Scientific and Anti-Religious Propaganda in the Cold War Era

At the beginning of the Cold War in 1947, with the aid and orchestration of the Communist Party, prominent Soviet scholars (such as the physicist Sergei Vavilov), and the Academy of Sciences, helped form a massive new institution dedicated to conducting anti-religious propaganda, scientific-materialist popularization, and the promotion of Marxist ideology—The All-Union Society for the Dissemination of Political and Scientific Knowledge (or the *Znanie* Society). Vavilov, a physicist at the Academy and well-known science-popularizer, became its first President from 1947 to 1951. Ultimately the Society was a key component in the post-Second World War Soviet regime's effort to centralize all atheistic and scientific propaganda as it renewed and slowly revitalized its assault on religious belief that it had previously conducted in the 1930s.

While the Society played a role in propagating and popularizing knowledge from a wide arena of subject matter—including the social and natural sciences—a primary goal of the All-Union Society was devoted to inculcating an 'atheist scientific-materialist' worldview among the Soviet populace. At its very inception in 1947, after the liquidation of the Union of Militant Atheists by the Soviet government, the Society had transferred to it the primary function of propagating atheistic values among a broad audience throughout the former USSR. It therefore began a wide publicist campaign of editing journals that combined anti-religious themes with scientific propaganda as well. These journals included *Nauka i zhizn'* (Science and Life), as well as its main journal on this topic entitled *Nauka i religiia* (Science and Religion). The latter featured articles by the Society's Presidents and key activists in the struggle against religion. Later in the Khrushchev and Brezhnev eras the Society would struggle internally as it attempted to juggle its two main functional components, emphasizing either science education or more dogmatic anti-religious propaganda.

The organizational structure of the Society was similar to the Communist Party and Soviet government in that it had a central governing body in Moscow and then various societies in the respective republics of the former Soviet Union. While it was technically labeled a 'voluntary society' it was closely supervised from above by the Communist Party and had ties also to both Agitprop and the leadership of the Soviet Academy of Sciences in

Moscow.[24] The fact that it had numerous republican, regional, and local chapters, however, allowed it to cast a wide net among the populace both through its publications and public events.[25]

At its height, the Society claimed to have nearly 130,000 primary organizations all over the former Soviet Union with over 2.5 million Soviet citizens as members or participants in its organization. While it claimed to draw in participants from all segments of Soviet society, no less than 1 of every 15 members had a higher academic (or university) degree, and therefore many came from the social category (or stratification) labeled the Soviet 'intelligentsia'.

Under its supervision was the famed Polytechnical Museum in Moscow and the Central Polytechnical Library. In the post-Second World War era, up to the Gorbachev era starting in 1985, the Polytechnical Museum, on average, had hundreds of thousands of visitors per year and had a vast collection stretching over 60 exhibit halls in downtown Moscow. The *Znanie* Society also ran approximately 32 planetaria in various cities in the former USSR, as well as the oldest, the Moscow Planetarium itself.[26] At these planetaria, it would also sell pamphlets in its stores not only on scientific and astronomical themes but also anti-religious propaganda that included a series of brochures such as *Nauchnyi ateizm* (Scientific Atheism).[27]

The first President of the *Znanie* Society, Sergei I. Vavilov (the brother of the famed geneticist Nikolai Vavilov) had been an important member of the Soviet Academy of Sciences. He had been one of the six directors of the various sections, and thus a founding member, of the Institute for the History of the Natural Sciences and Technology in Moscow starting in 1932. He worked alongside Nikolai Bukharin, who was the Institute's first director in March 1932; and Bukharin was flexible at the time regarding the directors' interests. As a collective, the directors of this Institute in the

---

[24] I. P. Tsamerian and M. I. Shkhnovich (1964) 'Ateizm v SSSR', *Kratkii nauchno-ateisticheskii slovar'* (Moscow), pp. 39–46.

[25] For an understanding of the vast administrative and institutional structure of the *Znanie* Society from the central to republican level, see Gosudarstvennyi arkhiv Rossiiskoi federatsii (GARF), f. 9547, op. 1, d. 1048, ll. 3–6.

[26] The Moscow Planetarium became a critical site for anti-religious activists to hold public science-enlightenment events and displays. GARF, f. 9547, op. 1, d. 1429.

[27] For an overview of data on the Society, though probably inflated slightly, see Iu. K. Fishevskii (1972) 'Znanie, vsesoiuznoe obshchestvo', in *Bol'shaia sovetskaia entsiklopedia*, 3rd ed., vol. IX, pp. 555–556. For more updated statistics of the Society in the late Soviet era, see N. Basov (1989) 'Statistics of Our Activities', in *Nauka i zhizn'*, no. 5: 2–11.

1930s made plans to create a scientific museum and had a very ambitious publishing program, including encyclopedias in the classics of science and technology. Vavilov was also critical in the late 1930s in assembling a large collection of scientific and popular-scientific books for the Institute. It is thus not surprising that someone like Vavilov would be tapped by the regime to head this institution—he was a respected academician/physicist, had ample directorial experience, and finally was interested in museum work and popular publishing. All these traits were essential given the various needs of the *Znanie* Society. It is interesting to note though that unlike earlier scientist-activists in the 1930s, such as Boris Zavadovskii, Vavilov was not necessarily a dedicated Marxist thinker, yet interested all the same in scientific enlightenment and atheistic propaganda. His interests thus as a scientific publicist were critical in the regime's mind, and his past experience seemed to have temporarily converged with the broader ideological orientation of the Society in the post-Second World War era.[28]

Vavilov, unlike some other members of the Soviet scientific intelligentsia, could uniquely navigate the line between scientific expertise, directorial administrative necessity, and propagandist tendencies—variegated experience so needed in such an ideological, institutionalized position. Indeed, he held a powerful political-scientific position as President of the Soviet Academy of Sciences from 1945 till his death in 1951. He had also been Head of the Lebedev Institute of Physics in Moscow since 1934 and trained doctoral students in physical optics and luminescence. His most famous doctoral student, with whom he collaborated, P. Cherenkov, earned a Nobel Prize in Physics in 1958.[29]

Yuri Vavilov, Sergei's nephew (and Nikolai's son), has argued in his reminiscences that his uncle struggled accepting the position of both President of the Academy of Sciences (on Stalin's request) and the *Znanie* Society given the fate of his brother at the hands of the regime (as well as the highly politicized pressures on scientists during high Stalinism). Yet Yuri (and other scientists of the time) argued that Sergei

---

[28] V. D. Esakov (1988) 'N. I. Bukharin i akademiia nauk', *Priroda*, 9: 94–96.
[29] P. A. Cherenkov began his physics research, under his supervisor S. I. Vavilov, on the action of radiation on liquids that would later become known as the Cherenkov effect, for which in 1958 he received the Nobel Prize in Physics alongside Igor Tamm and I. M. Frank. Loren R. Graham (1993) *Science in Russia and the Soviet Union* (Cambridge: Cambridge University Press), pp. 208–209. Tamm himself made contributions to quantum field theory that made him well known in world scientific circles. See I. E. Tamm (1934) 'Exchange Forces between Neutrons and Protons and Fermi Theory', *Nature*, 133: 981.

accepted the powerful position of Academy President with the altruistic goal of improving the working conditions of scientists in this period, while *Znanie*, on the other hand, offered him a venue to continue his popularization goals from the pre-war era.[30] Leon Orbeli, one of his colleagues, while agreeing that Stalin had pressured Vavilov tremendously during his time as administrator and publicist, also believed that Vavilov was instrumental in advancing and protecting pure scientific research in Russia. Sergei was concerned that laboratories were better supplied with equipment and instruments, pushed the state to increase salaries of researchers, and pressed to expand the number of scientific institutes of the Academy of Sciences.[31] In his own diary, Sergei Vavilov reflected pensively on his meetings with Stalin in July of 1949 during his Presidency. He felt much pressure after these meetings, as Stalin wanted not just more research productivity out of scientific scholars in the Academy, but Vavilov hinted that Stalin wanted him to expand dramatically his role in supporting scientific education and anti-religious propaganda.[32]

Ultimately, he would therefore be highly valued, both by Stalin's regime and his colleagues, as President of the *Znanie* Society, for his publicist and propagandist skills, especially as a popularizer of science. He had been, at one point, Chief Editor of the Great Soviet Encyclopedia, so he knew how to guide his staff of writers (and some of his students) in assembling pamphlets on scientific-enlightenment. He had skillfully written short biographies for public consumption on scientific luminaries from the West and East: Lucretius, Galileo, and Russia's own Renaissance man, M. Lomonosov. Indeed, under his stewardship, the *Znanie* Society's scientific popular press expanded its publications to target the mass reader interested in the lives of individual scientists, as well as broader scientific themes ranging from astronomy to geology.[33] He was thus influential in

---

[30] For his nephew's views on why S. I. Vavilov was willing to be the President of the Soviet Academy of Sciences, even though Stalin had a hand in his brother's arrest and subsequent death in prison, see Yuri Vavilov (2002) 'Different Fates of the Vavilov Brothers', *Chelovek*, no. 4: 144–145.

[31] For his colleagues' reminiscences of Vavilov, and his correspondence with them concerning the fate of scientists in the late Stalin era, see Arkhiv Rossiiskoi Akademii Nauk (ARAN), f. 596, op. 3, d. 440, ll. 1–10.

[32] Sergei Vavilov (2004) 'Diaries and Reminiscences (1939–1951)', *Voprosy istorii estestvoznaniia i tekhniki (VIET)*, no. 2: 33–35. Also see, S. I. Vavilov (1999) *Ocherki i vospominaniya: epokha i lichnost', fiziki (1891–1951)* (Moskva: Nauka).

[33] Graham, *Science in Russia*, pp. 140–141. While historians of science, such as Loren Graham, have briefly touched on Sergei Vavilov's life, his collective biography has either been

creating a nexus between the Academy of Sciences, science popularizers, and anti-religious propagandists in this period.³⁴

SCIENTIFIC-ENLIGHTENMENT, THE SPUTNIK ERA, AND THE FRACTURING OF ANTI-RELIGIOUS DOGMA IN KHRUSHCHEV'S AND BREZHNEV'S TIMES

After Vavilov's era as President, and subsequently Stalin's death in 1953, *Znanie* began in Khrushchev's times to focus its efforts on popular journals that juxtaposed scientific themes against religious and superstitious vestiges in Soviet society. Besides its principal popular science magazine *Nauka i zhizn'* (Science and Life), it expanded the circulation of anti-religious periodicals such as *Nauka i religiia* (Science and Religion). It made a more concerted effort in its anti-religious publications to target the republics and their readers as well. Its most directed foray into this area came initially with the publication and wide circulation of its first Ukrainian anti-religious journal entitled *Liudina i svit* (Man and the World). The Society also published a series of brochures for mass circulation, such as *Nauchnyi ateizm* (Scientific Atheism).

During the late 1950s, the Communist Party, however, insisted that *Znanie*, in addition to scientific education (the strength of Vavilov and his team of publicists), expand the ideological struggle with religious values—particularly as the anti-religious campaign became virulent again under Khrushchev. By the time the first issue of *Nauka i religiia* came off the presses in 1959, Khrushchev's Central Committee called for an increase in atheistic propaganda in society and asked *Znanie* leaders to also increase and expand their public lectures. The Communist Party was seriously interested in renewing atheistic education at many levels at this time, especially at the university level where courses on scientific education were reintroduced into the curriculum.³⁵

---

overshadowed by the tragic life story of his tormented brother Nikolai, or has simply not focused enough on his administrative, scientific-enlightenment, and publicist activities.

³⁴ For an analysis of Sergei Vavilov and his cohort of Stalin-era physicists, see Alexei Kozhevnikov (2004) *Stalin's Great Science: The Life and Times of Soviet Physicists* (London: Imperial College University Press).

³⁵ For an good overview of the Communist Party's interest in courses on science and atheism in higher schools and at the university level, see Michael Froggatt (2006) 'Renouncing Dogma, Teaching Utopia: Science in Schools under Khrushchev', in *The Dilemmas of De-Stalinization: Negotiating Cultural and Social Change in the Khrushchev Era*, ed. Polly Jones (New York:

In the 1960s, *Znanie* then expanded a network of People's Universities that were dedicated to not just scientific and technical education, primarily in rural or provincial areas, but also were foci for the Society's 'houses of atheism'. While scientists who worked for the organization were certainly not keen on doing work that solely discredited religion, *Znanie* was, however, involved indirectly in aiding the party in documenting how religious belief was conducted in outlying areas. Using this information, and its own operative sources, between 1958 and 1964, when Khrushchev was ousted from power, the Communist Party closed some 7000 (out of 18,500 registered) places of worship, and the Orthodox Church was affected the most in this period, especially in outlying areas.[36]

Yet simultaneously with these efforts on the part of the regime to focus relentlessly on anti-religious propaganda, the Soviet government also continued to foster a climate that encouraged mass-publicity of science and technology as a long-term means toward supporting secular norms. While the Communist Party had certainly supported this strategy of mass scientific-enlightenment as far back as the 1920s, with a more scientifically educated public in the Khrushchev era, it could now propagandize the virtues of Soviet technological feats without focusing on remedial scientific education.[37] Furthermore, a new emphasis on space exploration, astronomy, and the utilitarian aspects of large-scale technologies inherently touched a chord with the post-Stalinist Soviet public. Thus when *Znanie* and its publicists tapped into the public's thirst for scientific information on contemporary issues (particularly with popular journals such as *Nauka i zhizn'*), as well as steered away from anti-religious dogma, it was more successful in its long-term goals of secularization.[38]

Routledge), pp. 250–267. For an earlier work, David E. Powell (1975) *Antireligious Propaganda in the Soviet Union: A Study of Mass Persuasion* (Cambridge: MIT Press).

[36] John Anderson (1994) *Religion, State, and Politics in the Soviet Union and Its Successor States* (Cambridge: Cambridge University Press).

[37] For an understanding of how the scientific-technical revolution during the Khrushchev and Brezhnev eras was affecting the working classes in the USSR, see O. N. Ianitskii (1972) *Urbanizatsiia, nauchno-tekhnicheskaia revoliutsiia i rabochii klass* (Moskva: Nauka).

[38] In many respects, *Znanie* was influential when it carried on the previous scientific-enlightenment goals that its first prominent President Sergei Vavilov set in motion in the late 1940s and early 1950s. The historian Michael Froggatt argues convincingly that *Znanie*'s key journals such as *Nauka i zhizn'* were indeed popular among Soviet readers when they were able to tap into current technological trends (i.e. space exploration) that sparked interest in a new generation. Michael Froggatt (2006) 'Science in Propaganda and Popular Culture Under Khrushchev 1953–1964' (Ph.D. Dissertation, Oxford University).

In a way, the long-term battle between ideologically driven anti-religious propaganda and scientific education as a means of achieving a more secular society continued to naturally play out even during Khrushchev's reinvigorated anti-religious campaign. While ideologically keen on reinforcing atheistic tenants, the Khrushchev regime was equally interested in spreading technical information as a signifier of modernization and progress. The Soviet government, outside institutions such as *Znanie*, really began to focus more on its technological achievements vis-à-vis the West, especially once *Sputnik 1* was successfully launched in 1957. In fact, the country witnessed an array of publicity on Soviet aeronautical, cosmonautic, and technological feats, and a host of newspapers and journals were literally filled with laudatory articles on Soviet rocketry, the history of Russian spaceflight, and cosmonautic achievements in the Cold War era. Articles on Soviet feats in outer space appeared regularly from 1957 through 1962 in both newspapers such as *Pravda* (The Truth) and *Izvestiia* (The News), as well as literary and popular cultural journals such as *Literaturnaia gazeta* (Literary gazette).[39] The press was as engrossed with the feats of space dogs such as Laika as it was in covering the flights of cosmonauts such as Yuri Gagarin.[40] Furthermore, celebrations and mass rallies (particularly in capital cities) became an important site for the Soviet public to become involved in the spectacle of technological display.[41]

During the early Cold War in Khrushchev's times, Soviet planetariums (many organized by *Znanie*) hosted popular lectures on outer space, short stories for adults and children were written with exaggerated platitudes by Soviet writers, and composers created popular songs to be sung at schools celebrating *Sputnik*.[42] This propaganda from above, however, was also met at times with sincere popular adulation and engagement from below, especially during parades such as in April 1961 when many from the Soviet baby

---

[39] See the litany of articles in late 1957 on this topic in the Soviet press, such as 'O nabliudenii iskusstvennykh sputnikov zemli', *Pravda*, 8 November, 1957.

[40] Amy Nelson (2010) 'Cold War Celebrity and Courageous Canine Scouts: The Life and Times of Soviet Space Dogs', in *Into the Cosmos: Space Exploration and Soviet Culture*, ed. James T. Andrews and Asif Siddiqi (Pittsburgh: University of Pittsburgh).

[41] For the relationship between big science, technology and public culture, see Paul R. Josephson (1995) 'Projects of the Century in Soviet History: Large Scale Technologies from Lenin to Gorbachev', *Technology and Culture*, July.

[42] For the relationship among space technology, the cosmos, popular culture, and scientific atheistic propaganda, see S. Ostrovskii (1958) 'Pesenka o sputnike', *Kul'turno-prosvetitel'naia rabota*, I: 30–33.

boomer generation marched through Red Square to celebrate Gagarin's feat circumnavigating the globe.[43] Furthermore, space travel touched a chord with schoolchildren as astronomy clubs and young inventor competitions in the late 1950s and early 1960s forged a bond between popular youth culture and official forms of ideology. Young rocket and airplane modelmakers all over the USSR (especially those in the Komsomol and Pioneers) were encouraged to compete in the V. M. Komarov and Iu. A. Gagarin prizes. Furthermore, at pioneer gatherings, on occasion, these famous cosmonauts like Gagarin would visit inventor competitions or youth rallies.[44]

In 1959, 2 years after *Sputnik I's* launch, the Moscow Planetarium came under the jurisdiction of the *Znanie* Society, and the Khrushchev regime had hoped that the planetarium could effectively combine scientific-enlightenment with atheist instruction. However, the planetarium's new staff, much like the Society itself, was bifurcated, so to speak, with those radical critics who believed scientific lectures lacked atheistic militancy versus those who believed their popular astronomical and physics lectures were effective long-term in creating scientifically educated citizens in Russia. Many of the popular science lecturers also did not want to exploit the public enthusiasm for astronomy by imbuing the narrative with atheistic propaganda.[45] One of *Znanie*'s, and the Soviet regime's, biggest problems during Khrushchev's times, was to get the Soviet scientific intelligentsia aligned to agitate against religion when they were actively involved in scientific popularization itself.

As the public's enthusiasm for Soviet technological progress widened, even some representatives of the Khrushchev era artistic intelligentsia enthusiastically engaged scientific and technical progress. Vladimir Tubin, a famous art critic, argued in a 1961 treatise that the artistic community should embrace scientific and technical advances.[46] Furthermore, Ernst

---

[43] See Loren R. Graham (2006) *Moscow Stories* (Bloomington: Indiana University Press) pp. 18–20. Graham has argued convincingly that Sputnik's generation enthusiastically embraced these new technological feats as expressed in heartfelt public support at outdoor rallies and celebrations in the 1960s.

[44] S. Soloveichik (1 June 1962) 'Eto Vam, schastlivye!' *Komsomol'skaia Pravda*, 4–5. For popular gatherings at meeting places such as the Pioneer Palace in Moscow, see Susan E. Reid (2005) 'Khrushchev in Wonderland: The Pioneer Palace in Moscow's Lenin Hills, 1962', *The Carl Beck Papers in Russian and East European Studies*, no. 1606 (Pittsburgh: The University of Pittsburgh Press), pp. 27–28.

[45] GARF, f. 9547, op. 1, d. 1324, ll. 26–29.

[46] V. Turbin (1961) *Tovarishch vremia i tovarishch' iskusstvo* (Moscow: Iskusstvo), pp. 51–53.

Neizvestnyi, the well-known Khrushchev era modernist sculptor, believed the scientific-technological revolution demanded new artistic expression mimicking those developments.[47] Neither Turbin nor Neizvestnyi, however, advocated that art should be a venue to combat religion, no matter how much they engaged technological progress. Yet with all this enthusiasm for science amidst the Soviet intelligentsia, however, the Khrushchev and Brezhnev regimes conversely came under fire by prominent Russian intellectuals who were concerned at the regime's obsession with technological pursuits and shortsighted science enlightenment. The writer Il'ia Ehrenburg, among others, in the pages of *Literaturnaia gazeta* (Literary Newspaper or Magazine) in the early 1960s, began to question publicly the regime's obsession with science and technological feats and the exorbitant funds expended in these arenas to the detriment of the arts and humanities. Ehrenburg was concerned that these technical achievements had possibly not only obscured, but maybe doomed the arts and literature in the modern secular age: 'Leaders of the Communist Party, economists, engineers, scientific agronomists boast they have quickly raised the standard of living of the Soviet population. Astrophysicists sound out their way to the moon. Yet people will judge our work according to our daily, human existence ... and the relationship we have to one another comrades'.[48]

Ehrenburg's words may have been an expression of the concern the intelligentsia had at this time for the regime's celebratory focus on Soviet technology. All the same, these warnings continued in popular journals and newspapers. One interesting debate became known as the *Liriki-fiziki* (lyrical poets vs. physicists), while others dealt with concern over the expansive funding of big science.[49] These debates, also featured in Communist Youth League presses during the Khrushchev thaw, focused on how the arts could survive in an age where scientific propaganda and technological feats reigned supreme.[50] Though they did not discuss necessarily

---

[47] E. Neizvestnyi (1962) 'Otkryvat' novoe!' *Iskusstvo*, no. 10: 10–11.
[48] Il'ia Ehrenburg (1 January 1960) 'O lune, o zemle, o serdtse', *Literaturnaia gazeta*, 3–4.
[49] B. Slutskii (1991) 'O drugikh i o sebe', *Biblioteka Ogonek* (Moscow), no. 40. Many of the battles between the 'physicists and lyricists' raged in the pages of the newspaper *Komsomol'skaia Pravda* between 1959 and 1960.
[50] Polly Jones (ed.) (2006) *The Dilemmas of De-Stalinization: Negotiating Cultural and Social Change in the Khrushchev Era* (New York: Routledge). Also see Victoria Smolkin-Rothrock (2011) 'Cosmic Enlightenment: Scientific Atheism and the Soviet Conquest of Space', in James T. Andrews and Asif Siddiqi, eds. *Into the Cosmos: Space Exploration and Soviet Culture* (Pittsburgh: The University of Pittsburgh Press), pp. 159–194.

the specific issue of religion versus science, they did question the regime's search for a technological panacea to society's ills under socialism.

Znanie therefore seemed to be very successful when it engaged scientists, artists, and intellectuals in these broader issues but not very influential when it supported anti-religion and technological progress as ends in themselves. By the Brezhnev era and beyond, this led to the *Znanie* leadership itself, the Communist regime's epicenter of anti-religious and scientific propaganda, to reflect on its own misperceptions, miscalculations, and concerns on how religious activists responded to propaganda during the era of the Cold War. In a stark self-criticism in 1972, one of its own top members, writing in the journal *Nauka i religiia* (Science and Religion), acknowledged problems with its pamphlets and popular lectures on anti-religious themes. V. Maslin argued that while Soviet propagandists knew their scientific material well (in diverse disciplines), they had been unable to convince audiences of their broader world-view. Furthermore, Maslin believed anti-religious activists, while successful in propagating scientific and technical information, could not easily counter the citizens' fervor for spiritual awakening. Maslin believed the essential problem was that *Znanie*'s activists lacked the ability to visualize how religious views were reconcilable with science and contemporary social conditions, instead of seeing them constantly as diametrically opposed.[51] In the journal *Nauka i zhizn'* (Science and Life), another key member of *Znanie*, Iu. Fishevskii, echoed Maslin's concerns, though focused on another related issue. Fishevskii reminded members (and readers) that what is most important for the Society was not to focus necessarily on eliminating religion (and viewing it as diametrically opposed to science) but instead to harp on the progressive educational component of science. Fishevskii, in frank terms, argued that science enlightenment had been part of the mission of the Russian intelligentsia going back to the pre-revolutionary era, and thus it was also in the best traditions of 'indigenous Russian science & scientists' to enlighten the public on contemporary scientific trends first and foremost and think about anti-religion as a byproduct of this development.

Indeed it was within the pages of *Znanie*'s anti-religious journals themselves (such as *Science and Religion* and *Science and Life*) that self-criticism

---

[51] V. Maslin (1972) 'Vysokoe prizvanie Sovetskoi intelligentsii', *Nauka i religiia*, no. 6: 2–11. Also see the very frank discussion of Brezhnev era science propaganda and the spread of atheistic values by Iu. Fishevskii (1972) in *Nauka i zhizn'*, no. 6, 16–21.

of ideological inflexibility began to seep through the cracks. By the Gorbachev era, even the last President of the *Znanie* Society, Academician N. G. Basov, began to reflect on ideological heavy-handedness, both past and present, and hinted indirectly at many myopically constructed fault lines that might have ostracized religious believers. Basov argued that there was much to be praised in the way propagandists lectured on anti-religious themes in the post-Khrushchev era, but criticized the practical effectiveness of ideological work if it could not keep up with the informational and 'scientific-technical' needs of a more modernized citizenry. He argued that there had to be more revised strategies in the late twentieth century for scientific self-education. In the era of a scientific and informational revolution, even propagandists must assimilate new forms of science, employing computer technology and better ways to address the scientific needs of its citizens if it wanted to simultaneously promote notions of the virtues of secularism and anti-religion.[52]

Epilogue

Basov's words, uttered in the Gorbachev era, may have resonated with many other socialist activists and atheists as they struggled to reconceptualize not only whether science and religion where diametrically opposed but how 'ideological myopia' may have hindered the regime's ability to capture the hearts, minds, and 'souls' of Soviet citizens in its competition with the West. Gorbachev's momentous reforms gave much hope to religious movements. By 1988, the drafting of laws on freedom of conscience, alongside the recognition by the state of more religious communities, was part of a greater pledge by the Soviet government to work cooperatively with church leaders together to improve Soviet life—not necessarily in opposition to one another.

By the end of the Gorbachev era, the Supreme Soviet of the USSR passed a new law 'On Freedom of Conscience and Religious Organizations'. It was remarkable in that it finally granted to citizens and religious organizations the same types of rights and status enjoyed in Western democratic countries. While the law generally adhered to a more modern notion of the separation of church and state, and prohibited the state financing of 'religious organizations', what is most fascinating is that in one of its stipulations it

---

[52] N. G. Basov (1989) *Nauka i zhizn*, no. 5: 2–7.

prohibited finally the 'state from financing atheistic propaganda'.[53] It is this seemingly minor, if not submerged, article of the law that in retrospect was most significant; especially because the Soviet state, since 1947, had spent so much money, energy, and resources on centralized scientific anti-religious propaganda (and the funding of its premier institution *Znanie* in this arena). Yet it may have legally acknowledged what its own activists (going back to Brezhnev's times) had been concerned with all along: namely, that ideologically charged anti-religious activism itself had been part of the tactical problem and may not have even been necessary in that form given the success of scientific enlightenment campaigns.

By the twilight of the Soviet Union in the Gorbachev era, it may have been slow modernizing trends, a more scientifically inclined public, and more engagement in technological progress and professional opportunities (not years of militant atheism) that created a secularized citizenry (even with the opening of religious institutions and expression again in the 1980s). Soviet sociologists in the Gorbachev era noted a significant growth in the sector of the scientific and technical intelligentsia, with even more urban Soviet citizens participating in the broader technical sector of the economy.[54] Historians, such as Moshe Lewin, who were interested in how remnants of Russian peasant religiosity were slowly receding in urban sectors of the former Soviet Union during the Cold War, have suggested that urbanization and scientific education could have been leading slowly toward a more secular, technical-oriented, and less religious society. This more long-term sociological trend may have been a byproduct of socio-economic and progressive educational factors outside of the realm of anti-religious propaganda.[55]

---

[53] For an overview of the repercussions of the law on freedom of conscience see Anderson, *Religion, State and Politics*. The legality of religious conditions themselves, as Anderson points out, were complex in this period of the collapse of the USSR. Soviet republics were passing their own laws, and differences and contentions amongst believers were also rising to the surface.

[54] See V. F. Sbytov (1986) 'Shtrikhi k portretu sovetskoi nauchno-technicheskoi intelligenti', in *Sotsiologicheskie issledovaniia* 3. Sociologists have documented a rapid growth in the 1970s and 1980s in the USSR of the scientific and technical intelligentsia: by the beginning of the Gorbachev era, scientists and engineers working in research and development numbered 1.5 million, and this was flanked by a sizable cadre of auxiliary technicians. Also see, G. Iagodin (1986) 'Vyshee obrazovanie: sostoianie i perspektivy perestroika', *Kommunist*, 16.

[55] Moshe Lewin (1991) *The Gorbachev Phenomenon: A Historical Interpretation* (Berkeley: University of California Press), pp. 48–50.

At times, in the pre-Second World War era, scientific education was a tool for educators and propagandists alike to use in assimilating citizens into a more scientific, urban, and secularized world using diverse means and tactics. Yet in the ideologically charged Cold War era, science, technology and a Marxist worldview had to compete with a Western polity that was itself struggling to craft its own synthetic worldview, especially considering how religion would integrate into their own secular forms of governance under the Eisenhower Administration in the 1950s and beyond.[56] American politicians saw communism as threatening the spiritual foundations of western life, while Soviet socialists alternatively may have seen the Cold War as a war of the scientific godless against the spiritual god-fearing. Yet this reductive dichotomy became an ever more complex battle, as Eisenhower's administration responded to Sputnik's launch with more investment in technological education, rethinking his more simplistic notions of American spiritualism and governance, while Soviet activists alternatively realized that less combative scientific enlightenment may have been the more practical means to secular goals.[57]

---

[56] Recent American historiography on the relationship among science, politics, and religion has also underscored the ongoing influence foreign affairs had on spiritual development. T. Jeremy Gun (2008) *Spiritual Weapons: The Cold War and the Forging of an American Religion* (New York: Praeger).

[57] After Sputnik's launch, Eisenhower launched a national campaign in the USA that funded not only space exploration but also a major strengthening of science and higher education. He both created NASA as a civilian space agency and signed a landmark science education law. Yankek Mieczkowski (2013) *Eisenhower's Sputnik Moment: The Race for Space and World Prestige* (Ithaca: Cornell University Press).

CHAPTER 6

Tsiolkovskii and the Invention of 'Russian Cosmism': Science, Mysticism, and the Conquest of Nature at the Birth of Soviet Space Exploration

Asif Siddiqi

As cosmonauts rocketed into the cosmos in the 1960s, the Soviet Communist Party and its mouthpieces produced inspirational rhetoric that repeatedly reinforced the link between its achievements in space and the foundational requirement that made the space program possible, socialism. Most famously, in an oft-quoted aphorism, Nikita Khrushchev was said to have noted that 'socialism—this is the steadfast launch pad from which the Soviet Union successfully sent into space its powerful modern space ships'.[1] This connection to socialism was frequently articulated as the outcome of historical forces dating back to the disjuncture of 1917, an acknowledgement of the Bolsheviks' instrumental alignment of their state project with the power of science

---

[1] Anonymous (1962a) 'Startovaia ploshchadka—sotsializm', *Sovetskii voin* 17, 1–4. The quotation was originally published as part of a lengthy joint statement of the Central Committee of the Communist Party of the Soviet Union, the Presidium of the USSR Supreme Soviet, and the Council of Ministers in the aftermath of the joint space mission of *Vostok-3* and *Vostok-4* in August 1962.

A. Siddiqi (✉)
Department of History, Fordham University, New York, NY, USA

© The Editor(s) (if applicable) and The Author(s) 2016
P. Betts, S.A. Smith (eds.), *Science, Religion and Communism in Cold War Europe*, St Antony's, DOI 10.1057/978-1-137-54639-5_6

and technology. The space program of the 1960s was seen as both an *outcome* of the long and close relationship between socialism and science and technology and an *expression* of that association. The triangulation among socialism, science, and space served as a scaffolding for a multiplicity of meanings and symbols to be ascribed to Soviet cosmic achievements and all were deployed in the service of Cold War rivalry with the USA.

The discourse of socialism, science, and space needed key figures to humanize the Soviet surge into the cosmos. They included a holy trinity of actors, Konstantin Tsiolkovskii, the early twentieth century theorist of space travel, Sergei Korolev, the designer who was the architect behind Sputnik, and Iurii Gagarin, the young and handsome fighter pilot who became the world's first space voyager in 1961. Each, especially as deceased martyrs for the cosmic cause, served as a discursive node around the massive upswell of enthusiasm for space exploration that gripped the Soviet populace in the 1950s and 1960s. For each, there were statues, conferences, memorials, and thousands of books and articles. And each, in their constructed biographies, seemed to embody the kind of rationalism and materialism that fit well with Bolshevik aspirations (rooted in Marxist rhetoric) for a unity between the scientific and the social under Soviet socialism.

While the biographies of Korolev and Gagarin provided relatively uncomplicated canvases for Soviet authorities to imprint their expectations, Tsiolkovskii presented more of a challenge. For one thing, Tsiolkovskii had produced most of his important works during the Tsarist era. In addition, he had never joined the Communist Party although had benefited from its largesse in the last years of his life. But these 'inconsistent' aspects of his career were accommodated or at least de-emphasized given his worthy scientific accomplishments, which the Soviet media claimed had gone unrewarded until the Bol'sheviks came to power in 1917. What appeared in the Soviet media was a linear story of genius: after a long period of self-education in the late nineteenth century, in 1903, Tsiolkovskii, the village school teacher, identified the basic mathematical equations required to make sense of the practical possibility of space travel. Most critically, he did this before contemporaries in Western Europe and the USA. By the 1960s, long after his death (in 1935), Tsiolkovskii's scientific contributions to the advancement of Soviet cosmonautics had been entrenched and then continuously asserted by a growing industry of public fealty orchestrated by official organs of the Soviet state. This campaign eventually became a deluge as Tsiolkovskii became the center of cultlike devotion that structured his scientific claims around priority, prescience, and progress.[2]

Yet, if the Soviet space program was the apotheosis of a kind of post-Enlightenment project that merged instrumental views about science and technology with ideas about social engineering ('the Soviet man'), it has become clear in the past two decades that other sensibilities, some seemingly at odds with Bolshevik (and later, Stalinist) exhortations on the materialist context of modern science and technology were somehow responsible for the Russian interest in the cosmos. There is now quite a large body of literature in English—much of it popular—suggesting that the inspiration for Soviet space achievements had little to do with Bolshevik interests in modern science and technology but rather rooted in a uniquely Russian philosophical tradition dating from the late nineteenth and early twentieth centuries called 'Cosmism' or 'Russian Cosmism'. Many connect these ideas to the writings of the well-known Russian Orthodox Christian philosopher Nikolai Fedorovich Fedorov (1829–1903). We learn in many of these writings that Fedorov's 'cosmist' ideas about human destiny in outer space and of the quest to achieve 'victory over death' profoundly influenced Tsiolkovskii and other Russian and Soviet actors in the early twentieth century, thus producing a kind of space *mentalité* that provided the fire to Soviet popular enthusiasms for the cosmos.[3] These works all suggest—some explicitly, others implicitly—that the roots of the Soviet space program of the 1960s lie *not* in the Bolshevik materialist and

---

[2] For useful biographies in Russian, see Sergei Samoilovich (1969) *Grazhdanin vselennoi (cherty zhizni i deiatelnosti Konstantina Eduardovicha Tsiolkovskogo)* (Kaluga: GMIK Named After K. E. Tsiolkovskii); Valerii Demin (2005) *Tsiolkovskii* (Moscow: Molodaia gvardiia). For a short treatment, focusing primarily on Tsiolkovskii's pedagogy and fiction, see James T. Andrews (2009) *Red Cosmos: K. E. Tsiolkovskii, Grandfather of Soviet Rocketry* (College Station, TX: Texas A&M University Press).

[3] Albert A. Harrison (2013) 'Russian and American Cosmism: Religion, National Psyche, and Spaceflight', *Astropolitics*, 11, 25–44; Vladimir Lytkin, Ben Finney, and Liudmilla Alepko (1995) 'Tsiolkovsky, Russian Cosmism and Extraterrestrial Intelligence', *Quarterly Journal of the Royal Astronomical Society*, 36, 369–76; R. Diordievic (1999) 'Russian Cosmism (with the Selective Bibliography) and Its Uprising Effect on the Development of Space Research', *Serbian Astronomical Journal*, 159, 105–09; Michael Holquist (1985–1986) 'The Philosophical Bases of Soviet Space Exploration', *The Key Reporter*, Winter, 2–4; Vladimir V. Lytkin (1998) 'Tsiolkovsky's Inspiration', *Ad Astra*, November/December, 34–39; Nader Elhefnawy (2007) 'Resurrecting Nikolai Fedorov', *The Space Review*, May 21, http://www.thespacereview.com/article/873/1. See also Chap. 9 ('The Scientific Cosmists') of George M. Young (2012) *The Russian Cosmists: The Esoteric Futurism of Nikolai Fedorov and His Followers* (Oxford: Oxford University Press), pp. 145–76.

teleological conception of science and progress but rather in rather more murky epistemologies rooted in mystical dreaming.

There is much to recommend such a line of thinking: it is indeed true that Fedorov's ideas, published in the first decade of the twentieth century, called for migration of humanity into outer space. Fedorov's calling, genealogically part of a long and rich Russian sensibility of mystical, philosophical, and sometimes occultish thought, was indirectly echoed by a number of early twentieth century Russian intelligentsia. Tsiolkovskii (who might have met Fedorov as a teenager) later sprinkled his work with Fedorovesque musings about the rationale for human migration into space. A large part of Tsiolkovskii's writings were undoubtedly mystical and religious in nature, and they are inextricably linked to his more mathematical musings. Yet, much of the recent fascination with 'Russian Cosmism' and its influence over the Soviet space program also misses the mark. Andrew Thomas, for example, ascribes Soviet space exploration largely to Fedorovian philosophy.[4] A recent documentary by George Carey focusing on Fedorov and Tsiolkovskii helpfully synthesizes the argument: 'it was … Nikolai Fedorov, who set Russians on course for Yuri [Gagarin's] moment of glory'.[5] Busy with analyzing the content of Fedorov's and Tsiolkovskii's works—peculiar and esoteric to modern ears—these pronouncements weave a grip on the popular imagination at the cost of historical specificity.[6]

---

[4] Andrew Thomas (2010) 'Kul'tura Kosmosa: The Russian Popular Culture of Space Exploration', M.A. thesis, De Montfort University, Leicester, UK.

[5] George Carey (2011) 'Why Russia Won the Space Race', *The Telegraph*, April 8, http://www.telegraph.co.uk/culture/tvandradio/8437995/George-Carey-Why-Russia-won-the-space-race.html. The documentary in question is *Knocking on Heaven's Door* (2011), BBC Storyville documentary series. I should note that I was thanked in the credits of the documentary for assistance to Mr. Carey.

[6] The best writing in Europe or the USA on 'cosmism' and 'Russian Cosmism' is the work of Michael Hagemeister. Hagemeister (1997) 'Russian Cosmism in the 1920s and Today' in Bernice Glatzer Rosenthal (ed.) *The Occult in Russian and Soviet Culture* (Ithaca, NY: Cornell University Press); Hagemeister (2011) 'The Conquest of Space and the Bliss of the Atoms: Konstantin Tsiolkovskii' in Eva Maurer, Julia Richers, Monica Rüthers, and Carmen Scheide (eds) *Soviet Space Culture: Cosmic Enthusiasm in Socialist Societies* (Basingstoke: Palgrave Macmillan); Hagemeister (2012) 'Konstantin Tsiolkovskii and the Occult Roots of Soviet Space Travel' in Birgit Menzel, Michael Hagemeister, and Bernice Glatzer Rosenthal (eds) *The New Age of Russia: Occult and Esoteric Dimensions* (Munich: Verlag Otto Sanger), pp. 135–49.

The purpose of my essay is *historically* to contextualize the relative importance of the two dominant narratives of the roots of the Soviet space program—one derived from socialism's supposed social setting as a launching pad for modern science and technology and the other rooted in the metaphysical musings of 'Russian Cosmist' thought, but to do so with an eye to disentangling how history itself has colored our understanding of the relative importance and relationship between the two sensibilities. By doing so, I hope to position the philosophical foundations of Soviet space activities as part of a broader conversation about the meaning of *modernity* in the Russian context, through its instrumental invocation of the 'mastery over nature', and its unsavory eugenicist worldview. I argue that these impulses, the Bolshevik cult of science and technology and the mystic ideas of 'Russian Cosmism', each representing particular rationales for the conquest of space, were responses to fundamentally *modern* questions opened in the early twentieth century by the emergence of discrete scientific disciplines distinct from natural philosophy.

The essay is divided into two parts. In the first section, I highlight the emergence of a culture-wide fascination with desirable futures enabled by scientific and technology in Russia in the late nineteenth century and its alignment with Bolshevik ideology after 1917 and then briefly track its evolution and manifestation across the Soviet era through to the end of the twentieth century. My focus here will be on how popular enthusiasm for space exploration in the 1920s was eventually subordinated to positivist state discourses that understood modern science and technology as a measure of 'progress' under socialism. Such messages inspired many of the main actors in the Soviet space program while also preparing a generation of postwar Soviet citizens to the fact that space exploration by the Soviet state was not only inevitable but absolutely necessary. The key symbol at the center of this discourse was the memory of Konstantin Tsiolkovskii. In the second and longer part, drawing from the writings of Fedorov and Tsiolkovskii, I summarize the basic contours and substance of mystical and Cosmist thought about space exploration, especially as it was embedded in particular historical moments in Russia in the twentieth century. I then conclude with some thoughts on how these disparate strands might be understood as a part of larger Russian narrative about humanity's relationship with the natural world and one possible embodiment of an aspiration for modernity in twentieth century Russia.

The Cult of Science and Technology

Popular Russian fascination with the instrumental role of science and technology emerged in the late nineteenth century as part of an intersection several phenomena: the appearance of discrete scientific disciplines in Russian universities, the acceleration of industrialization (and consequent popular interest in the aesthetics and mechanics of machines), the rise of private publishers interested in science popularization, and a growing bourgeois and educated demographic.[7] The tenor of popular engagement with late nineteenth century science was also catalyzed by the predilection of established scientific luminaries—such as the chemist Dmitrii Mendeleev (1834–1907) and the botanist Kliment Timiriazev (1843–1920)—to join editorial boards of several major publishing houses specializing in popular science. Inspired by the example of prominent scientists, Russian intelligentsia increasingly embraced public science as a vocation, producing a spate of new textbooks for the coming generation of young Russians.[8] The works of foreign writers such as Jules Verne, Camille Flammarion, and H. G. Wells flooded the Russian publishing market and encouraged native Russian writers to take up the cause of scientific fiction. By the first decades of the twentieth century, there were dozens of popular science journals flooding the market. One of the largest private publishers of popular science, Soikin, produced a fleet of magazines with such descriptive names as *Priroda i liudi* (Nature and People), *Mir prikliuchenii* (World of Adventure), and *Vestnik znanii* (Journal of Knowledge) and issued more than 80 million books that made a deep impression on the generation that came of age at the turn of the century.[9] Now firmly separated from natural philosophy, modern science appeared as a discrete category of knowledge, marked and admired by many for its explicit stance against 'superstition' and 'backwardness'.[10]

---

[7] Asif A. Siddiqi (2010) *The Red Rockets' Glare: Spaceflight and the Soviet Imagination, 1857–1957* (New York: Cambridge University Press), pp. 16–42.

[8] E. A. Lazarevich (1984) *S vekom naravne: populiarizatsiia nauki v Rossii: kniga, gazeta, zhurnal* (Moscow: Kniga). For more on the emergence of 'public science' as a distinct category, see also James T. Andrews (2003) *Science for the Masses: The Bolshevik State, Public Science, and the Popular Imagination in Soviet Russia, 1917–1934* (College Station, TX: Texas A&M University Press).

[9] A. Admiralskii and S. Belov (1970) *Rytsar knigi: ocherki zhizni i deiatelnosti P. P. Soikina* (Leningrad: Lenizdat).

[10] Loren R. Graham (1994) *Science in Russia and the Soviet Union: A Short History* (Cambridge, MA: Cambridge University Press); Alexander Vucinich (1963) *Science in*

If these early strands of technological fascination were inchoate and sporadic before the Revolution, they acquired a quasi-utopian tenor after: the Revolution propelled technological visions to move from the wisp of dreams to the arena of possibility. Helped by the hyperbolic claims of early Bolshevik ideologues, after 1917, fascination with science and technology acquired a millenarian tone. The richest expressions of this meeting between utopia, science, and possibility occurred during the years of the New Economic Policy (NEP) when the country moved through rapid economic recovery matched by what Sheila Fitzpatrick has called 'an upsurge of optimism among the Bolshevik leaders'.[11] With economic stability more in evidence to the average urban resident, people from all walks of life conjured up possible futures in new and surprising ways. We find a superb chronicle of these anticipations, some of them utopian in language and meaning, in the work of the late Richard Stites who invited readers into a world of mad excitement as actors from all walks of life invoked, debated over, wrote about, and confronted the future.[12] From ritual to religion, mannerisms to machines, and art to architecture, idealized visions of the future pervaded Soviet society at all levels. These were not all suffused with an optimistic glow; there was a deep ambivalence and anxiety to these anticipations, especially in relation to the social consequences of modern science and technology, but the reactions show a grappling with the fundamental shifts in the structure of knowledge evident at the turn of century as 'science' became both a category and a lever for social improvement.[13]

As I have shown elsewhere, ideas about cosmic travel constituted a highly visible aspect of the obsession with science and technology in the 1920s and early 1930s. Its manifestations were multitude: in the hundreds of

Russian Culture: A History to 1860 (Stanford, CA: Stanford University Press); Alexander Vucinich (1970) *Science in Russian Culture, 1861–1917* (Stanford, CA: Stanford University Press); Michael Gordin (2004) *A Well-ordered Thing: Dmitrii Mendeleev and the Shadow of the Periodic Table* (New York: Basic Books).

[11] Sheila Fitzpatrick (1994) *The Russian Revolution*, 2nd ed. (Oxford, UK: Oxford University Press), p. 113.

[12] Richard Stites (1989) *Revolutionary Dreams: Utopian Vision and Experimental Life in the Russian Revolution* (New York: Oxford University Press).

[13] Anindita Banerjee explores the nature of ambivalence about science and technology in her recent book: Banerjee (2012) *We Modern People: Science Fiction and the Making of Russian Modernity* (Middleton: Wesleyan University Press). See also Julia Vaingurt (2013) *Wonderlands of the Avant-Garde: Technology and the Arts in Russia of the 1920s* (Chicago: Northwestern University Press).

articles and books in the Soviet media, driven by key popularizers such as Vladimir Riumin, Iakov Perel'man, and Nikolai Rynin; in the innovative cinema of *Aelita* (1924) and *Kosmicheskii reis* (Space Voyage, 1935), the latter of which featured two spaceships named after Iosif Stalin and Kliment Voroshilov; in the activities of burgeoning amateur enthusiast societies based on Moscow and Leningrad; in the heavily publicized and attended exhibitions on space travel in Kiev (in 1925) and Moscow (in 1927); in art, architecture, poetry, and literature; and in the networks of thousands of disparate activists throughout the country who wrote to each other and exchanged information about space topics.[14] The one single sustained commonality in all of this activity was the figure of Konstantin Tsiolkovskii, who, although largely ignored by the Bolsheviks, was elevated to saintly status by faithful correspondents. One of the organizers of the Moscow cosmic exhibit wrote breathlessly to Tsiolkovskii that united in their zealous belief in the power of 'invention' and 'inventors', they saw Tsiolkovskii as a 'prophet' of the new era of science and technology, 'superior even to [Thomas] Edison', the most recognizable American icon of the age.[15]

By the late 1930s, the Bolshevik cult of science and technology had transformed itself from one extreme—idealized visions of the future—into another—a fetish for large-scale infrastructural projects designed to literally submit nature to Soviet power.[16] The Moscow Metro, the industrial factory city of Magnitogorsk, the White Sea-Baltic Sea Canal—these were the living artifacts of Soviet civilization, built by the labor of the new Soviet man but given life by the markers of modern technology—electricity, steel furnaces, diesel tractors, and the assembly line. Beyond their functionality, these projects also represented powerful symbols of national identity; it was assumed and then reiterated in public rhetoric that Soviet power was synonymous with the ability to remake the social, physical, and cultural landscape of the Soviet Union. This link among large-scale technologies, the natural landscape, and national identity were not unique to the Soviet context, but what gave them a particular frisson was the way in which socialism seemed to uniquely energize and embody a populist relationship between nation and technology: socialism made possible the proper social

---

[14] Siddiqi, *Red Rockets' Glare*, pp. 74–113.

[15] Efofbi (Ol'ga Kholoptseva) to Tsiolkovskii (3 December 1928), Archive of the Russian Academy of Sciences (ARAN), f. 555, op. 3, d. 199, ll. 5–6.

[16] Paul R. Josephson (1995) '"Projects of the Century" in Soviet History: Large-Scale Technologies from Lenin to Gorbachev', *Technology and Culture* 36, no. 3, 519–59.

organization to foster world class science and technology, which improved the lives of the working class, thus serving as a measure of national prowess on the international stage. This was nowhere more evident than during the early years of the Cold War. In contrasting the 'basic economic law of socialism' with that of capitalism, Stalin himself described the socialist order in 1952 as 'the maximum satisfaction of the constantly rising material and cultural needs of society through the continuous expansion and improvement of socialist production on the basis of better technologies'.[17]

The Cold War added fuel to the Stalinist project to use science and technology to conquer nature. Few could ignore that science and technology had helped to fundamentally alter the nature of war, and now, with the arrival of the Cold War as a motivating framework for national imperatives, it promised to transform the nature of 'peace'. Under Stalin and then Khrushchev, the Soviet populace learned (and many undoubtedly believed), the future of science and technology was firmly tied to the future of the Soviet Union. In the 1950s, new popular science journals joined already existing publications such as *Tekhnika-molodezhi* (Technology for Youth), *Znanie-sila* (Knowledge is Power), and *Nauka i zhizn'* (Science and Life), while the market was flooded with biographies of great Russian scientists written for popular consumption. *Nauka i zhizn'* alone had monthly print runs into millions, comparable to lifestyle magazines such as *Ogonek* (Spark).[18]

Enthusiasm for space, much like the fascination with nuclear energy, was a significant component of this intersection of postwar scientific and technological optimism with Cold War exigencies.[19] As with the mass interest in the 1920s, Tsiolkovskii was a central figure in this burst of writing on cosmic topics. Soon after the end of the war, a few influential space enthusiasts actively canvassed to elevate Tsiolkovskii's name as a native-born talent within the pantheon of Soviet science. The Soviet Academy of Sciences, for the first time, recognized the value of a self-educated genius to advance the standing of 'Soviet' (as opposed to 'Russian') science

---

[17] I. Stalin (1952) 'Ekonomicheskie problemy sotsializma v SSSR', *Pravda*, October 4.

[18] Siddiqi, *Red Rockets' Glare*, pp. 301–13; Mark Kuchment (1990) 'Bridging Two Cultures: The Emergence of Scientific Prose', in Loren R. Graham (ed.) *Science and the Soviet Social Order* (Cambridge, MA: Harvard University Press), pp. 325–40.

[19] Paul R. Josephson (1996) 'Atomic-Powered Communism: Nuclear Culture in the Postwar USSR', *Slavic Review* 55, no. 2, 297–324; Paul R. Josephson (1990) 'Rockets, Reactors, and Soviet Culture' in *Science and the Soviet Social Order*, pp. 168–91; Matthias Schwartz (2011) 'A Dream Come True: Close Encounters with Outer Space in Soviet Popular Scientific Journals of the 1950s and 1960s' in *Soviet Space Culture*, pp. 232–50.

and fostered what would become an industry of activity around the late Tsiolkovskii. Numerous celebrations of the late theoretician's birthday after World War II helped to reanimate the possibility of space exploration among the broader public. Commentators in the postwar era took pains to emphasize repeatedly that Tsiolkovskii had conceived his formulae on rocket and space motion before others in the USA and Germany. Arcane and rather lukewarm quotes from Tsiolkovskii on his generally ambivalent feelings toward the Bolsheviks were embellished and traded in publication. Writers repeatedly linked Tsiolkovskii to the Bolshevik cause, both in spirit and practice. Both sought their answers in the materialist conception of the world, it was said, and both had achieved success.[20]

By their own admission, the positivist science and technology-identified cosmic discourse influenced a generation of space activists and enthusiasts and future managers, engineers, and cosmonauts. For example, in the 1920s and 1930s, the founding designer of the Soviet space program and the architect behind Sputnik, Sergei Korolev, read the works of science popularizer Iakov Perelman (and in fact corresponded with him in the 1930s).[21] Valentin Glushko, the man who designed the engines that powered Gagarin into orbit, recalled in the 1980s that Perelman's book, *Interplanetary Travels*, sent him 'in the correct direction in his hobby of spaceflight'.[22] Future cosmonauts Georgii Grechko, Konstantin Feoktistov, and Boris Egorov, all admitted their future careers were driven to a large degree by their encounter with the works of Perelman and Rynin. Other cosmonaut heroes who gained fame at the height of the Cold War recall reading of the grand future of space travel as young boys and girls, seeing in the technophilic colorful images of *Ogonek* and *Krokodil*, possibilities for both the Soviet Union and themselves.[23]

The launch of Sputnik in 1957 cemented the link among socialism, science, and space. In the very first communiqué issued after the satellite's launch, *Pravda* noted that 'artificial earth satellites will pave the

---

[20] For Tsiolkovskii's re-appropriation by the Soviet state in the postwar era, see Siddiqi, *Red Rockets' Glare*, pp. 294–301.

[21] Some of this correspondence has been published. See for example, M. V. Keldysh (ed.) (1990) *Tvorcheskoe nasledie akademika Sergeia Pavlovicha Koroleva: izbrannye trudy i dokumenty* (Moscow: Nauka), pp. 52–53, 79–80.

[22] Valentin Glushko, introduction to Grigorii Mishkevich (1986) *Doktor zanimatelnykh nauk* (Moscow: Znanie).

[23] Many of these reminiscences are interspersed through cosmonaut biographies and autobiographies. See for example, L. Lebedev, B. Lukianov, and A. Romanov (1971) *Syny goluboi planety* (Moscow: Politizdat).

way to interplanetary travel, and ... our contemporaries will witness how the freed and conscientious labour of the people of the new socialist society makes the most daring dreams of mankind a reality'.[24] An unprecedented run of space achievements in the 1960s, year after year, helped to reinforce the international image of the Soviet Union as a nation, not of dreary collective farms and obsolete technology but one at the vanguard of a new modern and dynamic future. The congruent nature of Khrushchev's Thaw and the first early burst of cosmic enthusiasm was not coincidental, as the former gave the (discursive) space for the latter to flourish. Both were characterized by optimism for a future aligned with the original (and still unrealized) past dreams of the Bolshevik Revolution.[25]

By the late 1960s, the Soviet space program had its own trinity of dead heroes—Tsiolkovskii, Korolev, and Gagarin—who loomed over the cosmic discourse. Now the Soviet future in the cosmos had a stable and fixed past from which to access its dreams for the future. Anticipations of the future were frequently articulated in terms of what the trinity did not live to see, thus seamlessly mapping the past into the future in a grand narrative of Soviet space history. In a rare public lecture in 1970 given by one of the architects of the Soviet space program, Mikhail Tikhonravov, the future was still framed in terms of the power of science to remake society. His talk, tellingly titled 'K. E. Tsiolkovskii and the Future' was ostensibly about Tsiolkovskii's visions for what Soviet socialism might do in space. Yet, the speech, even as it described Tsiolkovskii's humanistic vision of human expansion and settlement in outer space, was punctuated with esoteric technical terms and mathematical equations, as if to emphasize the materialist and scientific contours of space travel. Here, according to Tikhonravov, only the cold hard language of modern science (enabled by socialism) could draw a single progressive line from the Soviet past to the Soviet future.[26]

---

[24] Anonymous (1957) 'Soobshchenie TASS', *Pravda*, 5 October.
[25] Much has been written about this period of popular fascination with space. For two recent edited volumes, see James T. Andrews and Asif A. Siddiqi (eds) (2011) *Into the Cosmos: Space Exploration and Soviet Culture* (Pittsburgh, PA: University of Pittsburgh Press); Maurer, Richers, Rüthers, and Scheide (eds) (2011) *Soviet Space Culture*.
[26] M. K. Tikhonravov (1970) 'K. E. Tsiolkovskii i budushchee' in A. D. Ursal et al. (eds) (1974) *Idei Tsiolkovskogo i problemy kosmonavtiki* (Moscow: Mashinostroenie). I elaborate in more detail on the connection between the past and the future in Soviet popular visions of space exploration in: Siddiqi (2012) 'From Cosmic Enthusiasm to Nostalgia for the Future' in *Soviet Space Culture*, pp. 283–306.

Into the Mystic

The question remained: why *did* Tsiolkovskii advocate for human expansion into outer space? In Tikhonravov's talk in 1970, there is only an enigmatic footnote: 'Justification of the need for occupation of cosmic space is given in some unpublished manuscripts of K. E. Tsiolkovskii'.[27] This is partly true. Tsiolkovskii did provide a justification for human migration into space but it was not unpublished. In fact, he published quite a bit of his rationales and thoughts on the issue in the 1910s and 1920s. These copious philosophical musings about the place of humanity in the cosmos represented an odd juxtaposition to his more mathematical works and drew on sensibilities rooted in Orthodox Christianity, mysticism, and the occult. By the time of the Soviet collapse and into the 1990s, this set of articulations about the true destiny of humanity in the cosmos—by then integrated into a larger Russian philosophical worldview named 'Russian Cosmism' linked with the nineteenth century philosopher Nikolai Fedorov—all but eclipsed other possible rationales as the historical root for Russian fascination with the cosmos. So strong was its appeal that one author even imagined that the entire Bolshevik project was Fedorovian in nature.[28] Such connections, made largely on the basis of literary echoes of Fedorov in the pronouncements of Bolsheviks unfortunately reveal a kind of ahistoricism, unmoored from specificity, as if Fedorov's ideas were at once everywhere and with everyone to access. Recovering Fedorov and Tsiolkovskii as historical actors, exploring their original philosophical contributions in their particular moments, and reconstructing their influences leads to a more complex and chaotic story.

Fedorov never used the term 'cosmist' or 'Russian cosmist' in his writings, but the text that he left—difficult to follow, often contradictory, and drawing from Eastern and Western philosophical traditions, theosophy, pan-Slavism, and Russian Orthodox thinking—was in the service of one goal: human migration off the planet Earth.[29] What one Western writer has called 'one of the most profound, comprehensive, and original ideas in the history of Russian speculation' was, at its core, a striking oeuvre of philosophical thought about the evolution of humanity and the universe,

---

[27] Tikhonravov (1970) 'K. E. Tsiolkovskii i budushchee'.

[28] Dmitry Shlapentokh (1996) 'Bolshevism as Fedorovian Regime', *Cahiers du monde russe: Russie, Empire russe, Union soviétique, États indépendants* 37, no. 4, 429–65.

[29] Michael Hagemeister (1989) *Nikolaj Fedorov: Studien zu Leben, Werk und Wirkung* (Munich: Sagner); George M. Young, Jr. (1979) *Nikolai F. Fedorov: An Introduction* (Belmont, MA: Nordland); Young, Jr. (2012) *The Russian Cosmists*.

and the inviolable relationship between the two.[30] Fedorov first formulated this philosophy while a librarian at the Rumiantsev Library (later, the Lenin Library) in Moscow, and although he never published anything during his lifetime, he was said to have communicated his ideas to many. He began preparation of a comprehensive set of his thoughts soon after the turn of the century but died suddenly in 1903. Three years later, they were prepared and issued in a small run of 480 copies by two of his disciples. A second volume was issued in 1913. Neither was for sale but simply distributed to interested intellectuals in a very small community of like-minded thinkers. In this work, grandly entitled *Filosofiia obshchego dela* (The Philosophy of the Common Task), Fedorov outlined his basic doctrine: that the 'common task' of all of humanity is to resurrect the dead. This mission stemmed from a distinctly theocratic view of the universe in which he saw Christianity as primarily a religion of resurrection; he believed that humanity's moral task was to emulate Christ and make bodily resurrection possible. Mass resurrection, he argued, would finally eliminate the artificial boundaries among the 'brotherhood' of humanity, that is, between previous and current generations. In other words, none of the ills of society could be solved without devising a solution to the inevitability of death. Using all of the resources at its disposal, including science and technology, humanity, Fedorov believed, should engage in a quest to reassemble the corporeal particles lost in the 'disintegration' of human death, leading to an ideal utopian setting ('as it ought to be'), where there would be no birth and no death, only the progressive reanimation of the deceased millions from history.

Two aspects of Fedorov's 'philosophy of the common task' concerned voyages into the cosmos. First, to achieve his ultimate goal of 'liberation from death', Fedorov called for subordinating humanity's natural environment, which for him included not only the Earth but also the entire universe, to its will. As if to emphasize his point, Fedorov began one key section of his book ('Regulation of Nature') with the words: 'Nature—This is our enemy'.[31] Indeed, his entire corpus of writings is punctuated with calls for humanity to come together in unison so as to fight, conquer, and submit nature to its own domination, an ideology that found common resonance among many technocratic visionaries of the late nineteenth and early twentieth centuries. In the Soviet context, this stance

---

[30] Young (1979) *Nikolai F. Fedorov*, p. 7.
[31] N. F. Fedorov (1995) 'Regulation of Nature' in A. G. Gacheva and S. G. Semenova (eds) *N. F. Fedorov: Sobranie sochinenii v chetyrekh tomakh*, t. 2 (Moscow: Progress), p. 239.

took on a particularly state-identified sheen, as buoyed by Bolshevik claims of remaking the social universe, many Soviet intelligentsia consequently committed to remaking the natural one. The famous Russian geochemist Vladimir Vernadskii, for example, shared similar views (although he probably never heard of Fedorov), and headed the government's Commission for the Study of the Natural Productive Forces (KEPS), a body whose goals encompassed such transformative projects as harnessing solar and electromagnetic forces for the good of Russian society.[32]

Beyond subordinating nature to human exploitation, Fedorov believed that humans from Earth would have to travel into the cosmos—to the Moon, the planets, and stars—to recover the disintegrated particles of deceased human beings that are spread throughout the universe. Once the bodies of the deceased were reconstituted (in forms that might not resemble humans), the resurrected would then settle throughout the universe. In his *Philosophy of the Common Task*, Fedorov saw this as a duty, one that was imposed upon *each and every single* human being, living and dead. Citing the most advanced aeronautical technology of the day, he explained that:

> The balloon, hovering over the ground, would foster courage and ingenuity ... it would serve, so to speak, as an invitation for all minds [to create] an open path to heavenly space. The duty of resurrection requires such an opening, as without possession of heavenly space the co-existence of all generations is impossible, while on the other hand, without resurrection it is impossible to achieve full possession of heavenly space.[33]

There was an explicit Malthusian rationale at work in his ideas:

> The need for transfer [from one world to another] is beyond doubt for those who have contemplated all the problems associated with the birth of an ideal society in which social vices and evils would be abolished ... To refuse [to go to] heavenly space would be to refuse to solve the economic problems predicted by Malthus, [and] in general, to reject the moral existence of humanity.[34]

---

[32] Kendall E. Bailes (1990) *Science and Russian Culture in an Age of Revolutions: V. I. Vernadsky and His Scientific School, 1863-1945* (Bloomington, Ind.: Indiana University Press); G. P. Aksenov (1993) 'O nauchnom odinochestve Vernadskogo', *Voprosy filosofii* no. 6, 74–87.

[33] N. F. Fedorov (1995) 'Part IV: What is our Goal?' in A. G. Gacheva and S. G. Semenova (eds) *Sobranie sochinenii v chetyrekh tomakh*, t. 1 (Moscow: Progress), p. 255.

[34] Fedorov, 'What is our Goal?', p. 256.

As a prelude to migration off the Earth, Fedorov also proposed two other technical projects: the 'regulation of atmospheric phenomena' and 'control over the movement of the Earth' as a kind of giant spaceship. All these, he predicted would be possible with the use modern science and technology.

During his lifetime, Fedorov's unusual ideas—at least his more religious Christian-identified ones—were known and admired by some very famous exemplars of late nineteenth-century Russian intelligentsia, for example, Tolstoy and Dostoyevsky, but by and large his legacy was rather limited. The few who took up the case, the so-called *Fedorovtsy*, were small in number and gained some visibility only in the 1920s. By then, most were captivated not by an engagement of the spiritual and Christian dimensions of Fedorov's plan—although they were committed to those—but more by how science and technology could make Fedorov's ideas come true. The most active of these *Fedorovtsy* was Aleksandr Gorskii (1886–1943), a rather obscure Moscow-based poet and musicologist who became obsessed with the idea of conquering death. Gorskii published a small print run book about Fedorov in the late 1920s—probably the only book about Fedorov published in that decade—but mostly he served as the center of a small network of like-minded thinkers, many of whom had a shared interest in theosophy and the occult—and indeed had connections with well-known occultists and mystics both in Russia and abroad.[35] They were not terribly successful in disseminating and sharing information about Fedorov, and although there were others in the NEP cultural era who also shared similar goals of immortality, none ever explicitly cited Fedorov as a direct influence.[36] Besides the small group of *Fedorovtsy*, the 'Anarchist-Biocosmists' were the most active group whose claims echoed those by Fedorov. Organized around 1920, the group's explicit goal was to make immortality a reality in the Soviet era. Upon Lenin's death, the group, essentially an artists' collective, announced that they would initiate

[35] They included Vasilii Chekrygin (1897–1922), Nikolai Setnitskii (1888–1937), Aleksei Brusilov (1853–1926), Olga Forsh (1873–1961), and Iuliia Danzas (1879–1942). Hagemeister provides a detailed survey of the *Fedorovtsy* in his Hagemeister (1989) *Nikolaj Fedorov*, pp. 343–62. Besides Gorskii, the most well-known Fedorov devotee in the 1920s was probably Valerii Muravev (1885–1930) about whom little is known. See V. G. Makarov (2002) 'Murav'ev V. N.: Ochelovechennoe vremia', *Voprosy filosofii* no. 4, 100–28; V. G. Makarov (2003) '"Otnosias sochuvstvenno k sushchestvuiushchemu stroiu …": sledsvennoe delo 1929 g. filosofa-kosmista V. N. Muraeva', *Otechestvennye arkhivy*, no. 1.

[36] Nikolai Krementsov (2014) *Revolutionary Experiments: The Quest for Immortality in Bolshevik Science and Fiction* (New York: Oxford University Press).

a program to resurrect Lenin.[37] For a short while, they published two journals, *Biokosmist* and *Bessmertie* (Immortality) with contributions mostly in the form of poems—paeans to fighting death. The group, also known as the Biocosmist-Immortalists, didn't last very long and left a light imprint on cosmic discourse in the 1920s, recovered only by historians recently.[38]

One major Soviet intellectual figure who did speak of Fedorov on occasion in the 1920s was Maksim Gorkii who saw in the emergence of Soviet power a new ability of society to master and 'regulate nature' in the service of social 'progress'. His animus toward death no doubt also made Fedorov's ideas more attractive, although Gorkii distanced himself from the possibility that immortality was actually possible. In his memoirs, we learn that Gorkii considered Fedorov 'remarkable' and an 'original thinker' and what Gorkii remembered about Fedorov was at least once noticed by a highly placed Bolshevik. In a short piece to note the 100th birthday of Fedorov, *Izvestiia* reported that M. I. Kalinin, the nominal head of the Soviet state, was told of a Fedorov aphorism by Gorkii himself, that 'Freedom without command over nature—this is the same as the liberation of peasants without land'.[39] This rare invocation of Fedorov's ideas in the public eye notwithstanding, the 'philosophy of the common task' did not widely circulate among Soviet intelligentsia in the 1920s and 1930s. Fedorov's original works were few in number, and although there are some sporadic references to Fedorov in the Soviet press in the 1920s, no major work ever seriously considered his ideas during the NEP era at a time when many mystical, occultish, and metaphysical ideas were routinely tolerated in public discourse. Undoubtedly there were a number of prominent Soviet intelligentsia—for example, the geochemist Vladimir Vernadskii—who had similar ideas about the relationship between humanity and nature, or whose ideas one might consider metaphysical or mystical nature, but it is difficult if not impossible to attribute any of this to the *Philosophy of the Common Task*. Certainly, there is no evidence to suggest that *any* of the active cosmic travel activists of the 1920s, such as Fridrikh Tsander, Mikhail Lapirov-Skoblo, Moris Leiteizen, Iakov Perelman,

---

[37] A. Sviatogor, N. Lebedev, and V. Zikeev (1924) 'Golos anarkhistov', *Izvestiia*, January 27; A. Sviatogor, P. Ivanitskii, V. Zikeev, and E. Grozin (1924) 'Deklaratsiia kreatoriia rossiiskikh i moskovskikh anarkhistov-biokosmistov', *Izvestiia*, January 4.

[38] For the Biocosmist manifesto, see A. Sviator (2000) 'Biokosmizm: biokosmicheskaia poetika' in S. B. Dzhimbinov (ed.) *Literaturnye manifesty ot simvolizma do nashikh dnei* (Moscow: XXI vek-soglasie), pp. 305–14.

[39] A. Gornostaev (1928) 'N. F. Fedorov', *Izvestiia*, December 28.

Nikolai Rynin, Grigorii Kramarov, and Vladimir Vetchinkin ever took Fedorov's ideas as sources of inspiration.[40]

Fedorov was not forgotten in the 1940s and 1950s—his name and contributions could be found in encyclopedias, for example—but by and large, there were none who critically re-evaluated his contributions and legacy. The study of Fedorov was infused with life only in the late 1970s, when a few working philosopher-academics, led by a graduate of the Gorkii Literary Institute, Svetlana Semenova, began publishing articles in the Soviet media advocating a singular place for Fedorov in the longer tradition of modern Russian philosophy. For their project, they appropriated the neologism 'Russian Cosmism', which had been circulating since about 1970 to describe Fedorovian ideas about ensuring the future of humanity through controlling nature, resurrection, and migration into the cosmos.[41] Inspired by the work of Semenova (who prepared a rather hastily edited selection of Fedorov's writings in Moscow in 1982), others joined to comment on the 'Russian Cosmist' tradition.[42] Seeing in it resonances of the work of many other prominent Russian intellectuals of the *fin-de-siècle* era, including Vladimir Vernadskii, Vladimir Solovev, Aleksandr Chizhevskii, and, most strikingly, Konstantin Tsiolkovskii, the founding theorist of Soviet space travel, the literature on 'Russian Cosmism' ballooned in the 1990s into a full-fledged intellectual movement that has now intertwined with both ultra-nationalist Russocentric strands in post-socialist Russia *and* growing transhumanist movements that seek to imagine human life beyond its current human form. These modern day Russian Cosmists see the older turn-of-the-century thinkers as part of a lineage of likeminded individuals who were confronting a suite of humanistic problems. As part of this constructed history, what was rather disparate and atomized in individual writing has been structured into a more cohesive

---

[40] For the space activists, see Asif A. Siddiqi (2008) 'Imagining the Cosmos: Utopians, Mystics, and the Popular Culture of Spaceflight in Revolutionary Russia' in Michael Gordin, Karl Hall, and Alexei B. Kojevnikov (eds) *Osiris*, 2nd Series, Vol. 23 (*Intelligentsia Science: The Russian Century, 1860–1960*) (Chicago: University of Chicago Press), pp. 260–88.

[41] Probably the first use the term 'Russian cosmism' in a publication can be found in R. A. Gal'tsev (1970) 'V. I. Vernadskii' in F. V. Konstantinov (ed.) *Filosofskaia entsiklopediia*, t. 5 (Moscow: Sovetskaia entsiklopediia), p. 624. See V. P. Rimskii and L. P. Filonenko, 'Sudba termina "Russkii kosmizm"', paper presented at 2012 Tsiolkovskii Readings, http://readings.gmik.ru/lecture/2012-SUDBA-TERMINA-RUSSKIY-KOSMIZM.

[42] A. V. Gulyga and S. G. Semenova (eds) (1982) *N. F. Fedorov: Sochineniia* (Moscow: Mysl).

philosophical tradition. Resonance and echoes of each other writings are deployed as evidence of an active ideological movement in the early twentieth century.[43]

Besides Fedorov, the central actor in this constructed genealogical tradition is Konstantin Tsiolkovskii whose mystical pronouncements provide rich evidence of the so-called spiritual and mystical roots of Soviet space exploration. Fedorov's ideas of restructuring humanity and the cosmos, and especially the central role of science and technology in this transformation, anticipated Tsiolkovskii's writings, which are also suffused with the Promethean urge to remake everything that surrounds us. There is also a key piece of evidence that connects Fedorov with Tsiolkovskii: the two may have met. While in Moscow as a teenager, from 1873 to 1876, Tsiolkovskii had seen Fedorov, who worked as a librarian at the Chertkovskii Public Library. In an autobiography published after his death (in 1939)—and repeated many times over the decades—Tsiolkovskii noted that 'in the Chertkovskii Library I *met* one of its employees ... [who] turned out to be the well-known ascetic Fedorov ... [he] donated his salary to the poor. I now realize that he wanted to give me his pension [too]. But in this he did not succeed: I was too shy'.[44] Over the years, many have embellished these brief observations, and without any evidence, created an entire narrative of the supposed friendship between these two men; it was noted that during their many meetings, Fedorov inculcated the young Tsiolkovskii with the idea of space travel.[45] Others have suggested that Fedorov influenced the young

---

[43] There is a vast body of literature on the imagined history and current concerns of Russian Cosmism. For general overviews from the early 1990s, see L. V. Fesenkova (ed.) (1990) *Russkii kosmizm i sovremennost* (Moscow: IFAN); Svetlana Semenova (1992) 'Russkii kosmizm', *Svobodnaia mysl'* no. 17, 81–97; S. G. Semenova and A. G. Gacheva (eds.) (1993) *Russkii kosmizm: antologiia filosofskoi mysli* (Moscow: Pedagogika-Press); O. D. Kurakina (1993) *Russkii kosmizm kak sotsiokulturnyi fenomenon* (Moscow: Moskovskii fiziko-tekhn. in-t); V. N. Demin and V. P. Seleznev (1993) *K zvezdam bystree sveta: russkii kosmizm vchera, segodnia, zavtra* (Moscow: Akademiia kosmonavtiki im. K. E. Tsiolkovskogo).

[44] K. E. Tsiolkovskii (1939) 'Cherty iz moei zhizni' in N. A. Islentev (ed.) *K. E. Tsiolkovskii* (Moscow: Aeroflot), pp. 15–25 (see page 27 for the quote).

[45] The legend that Fedorov pointed Tsiolkovskii in the direction of space travel probably originated from scientist Viktor Shklovskii in Shklovskii (1971) '"K" in "Kosmonavtika ot A do Ia"', *Literaturnaia gazeta*, April 7. See also V. E. Lvov (1977) *Zagadochnyi starik: povesti* (Leningrad: Sov. pisatel).

man to take up a kind of 'cosmist' philosophy.⁴⁶ The evidence in fact suggests no such thing. An inspection of Tsiolkovskii's original handwritten biography shows clearly that publications added details that were absent in Tsiolkovskii's own writings; in the original, Tsiolkovskii merely noted that he had only *seen* Fedorov, possibly once, and implied strongly that they never spoke a word.⁴⁷ In all his copious writings, both published and private, Tsiolkovskii mentioned Fedorov only that one instance, and it remains unclear how or when Tsiolkovskii was exposed to the works of Fedorov.⁴⁸ Rumor and legend have grown around this meeting, but the circumstantial evidence suggests nothing of substance was exchanged.

After his brief stay in Moscow, Tsiolkovskii moved around several times, from provincial town to town, ending up in Borovsk (in 1880) and then finally Kaluga (in 1892). It was then that he began to publish modest articles in popular science journals about a variety of topics, all loosely grouped around aeronautics. His passion was to design a passenger airship. As he became more prolific, and influenced by Jules Verne, he published science fiction dealing with space travel. Then, finally in 1903, he issued his most important mathematical work, substantiating the possibility of space travel using liquid propellant rockets. Followed by supplements in 1911–1912 and finally, in 1914, this set of writings on the mechanics of launching objects into space and the physics of human spaceflight constitute the first fully coherent mathematical intervention anywhere in the world into what until that moment had been entirely a discourse of speculation and myth. In the second part of his series, he noted very clearly his most important influence: 'the first seeds of the idea [of using rockets for space exploration] were cast by the famous fantasy writer Jules Verne [who] awakened in my mind [thoughts] in this direction'.⁴⁹

⁴⁶ Konstantin Altaiskii (1966) 'Moskovskaia iunost Tsiolkovskogo', *Moskva*, no. 9, 176–92. See also all the sources cited in Refs. 3, 4, and 5.
⁴⁷ In his original pencil-written notes, Tsiolkovskii notes that in the library he '*noticed* one of [the library] employees' rather than he '*met* one of the [library] employees'. See K. E. Tsiolkovskii, 'Cherty iz moei zhizni' (1934–1935), ARAN, f. 555, op. 2, d. 14, ll. 1–29ob (See especially. ll. 12ob–13).
⁴⁸ One of Tsiolkovskii's earliest biographers (who also knew him well), B. N. Vorobev, wrote in 1940 that according to the testimony of scientist's family, Tsiolkovskii learned of the philosophical works and personal life of Fedorov from magazine articles in his adulthood and that there is no evidence to suggest that the two ever spoke to each other. See B. N. Vorobev (1940) *Tsiolkovskii* (Moscow: Molodaia gvardiia), pp. 29–30.
⁴⁹ K. Tsiolkovskii (1911) 'Issledovanie mirovykh prostranstv reaktivnymi priborami', *Vestnik vozdukhoplavaniia* no. 19, 16–21. Quote on p. 16.

Through this period Tsiolkovskii read voraciously, not only the standard and latest contributions of European modern science but also, for the first time, more esoteric works, what even in those days might be considered mystical in content, such as the theosophist writings of Madame Helena Blavatsky, which, as Maria Carlson has shown, were not difficult to obtain in Russia.[50] By the mid-1910s, after publishing his mathematical writings, he began to articulate a semi-coherent philosophy about the justification for space travel, one that reflected his reading interests, a hodgepodge of Christian eschatology, theosophy, and various Western and Russian mystics. As with his more mathematical considerations which were also self-published in Kaluga, he began to publish his 'philosophical' ideas in small booklets, the first being *Nirvana* (1914), followed by *Grief and Genius* (1916), a meditation on how to cultivate 'geniuses' in society, a clear clue that Tsiolkovskii was flirting with then-fashionable eugenicist ideas among certain Western European and American intellectuals and spiritual leaders.[51] Undoubtedly these 'philosophical' brochures that he self-published must have puzzled Tsiolkovskii's more materialistic-minded followers who expectantly followed his mathematics on airplanes, airships, and rockets. But these two strands of his writing—the one seemingly more mathematical and the other more meditative—began to merge into a holistic body of work by the time that the Bolsheviks came to power and intertwined and influenced each other throughout the remainder of his life. And although his 'philosophical' writings were less well-known and certainly less discussed, they form a body of work that *far* exceeds in size his works on aeronautics, rocketry, and space travel.[52] Just as popular Soviet enthusiasm for the science of cosmic travel peaked in the 1920s—helped by Tsiolkovskii's mathematical articles—he simultaneously self-published works with such enigmatic titles as *The Wealth of the Universe* (1920),

---

[50] Maria Carlson (1993) *No Religion Higher Than Truth: A History of the Theosophical Movement in Russia, 1875–1922* (Princeton, NJ: Princeton University Press).

[51] Christine Rosen (2004) *Preaching Eugenics: Religious Leaders and the American Eugenics Movement* (New York: Oxford University Press); Marouf A. Hasian, Jr. (1996) *The Rhetoric of Eugenics in Anglo-American Thought* (Athens, GA: University of Georgia Press).

[52] During his lifetime Tsiolkovskii published 68 philosophical works. A much larger number was never published. See T. N. Zhelnina and V. M. Mapelman (1996) 'K izucheniiu praktiki izdaniia filosofskikh sochinenii K. E. Tsiolkovskogo', *Trudy XXVIII chtenii, posviashchennykh razrabotke nauchnogo naslediia i razvitiiu idei K. E. Tsiolkovskogo* (*Kaluga, 14–17 Sentiabria 1993 g.*): *Sektsiia 'Issledovanie nauchnogo tvorchestve K. E. Tsiolkovskogo i istoriia aviatsii i kosmonavtiki'* (Moscow: IET AN SSSR), 65–87.

The Origins of Life on Earth (1922), *Monism of the Universe* (1925), *Reason for Space* (1925), *The Future of Earth and Humanity* (1928), *The Will of the Universe: Unknown Intelligent Forces* (1928), *Love for One Self or the Source of Egoism* (1928), *Intellect and Passion* (1928), *The Social Organization of Humanity* (1928), and *The Goal of Stellar Voyages* (1929).[53]

Nearly 20 years after having laid out the various mathematical problems of space program, Tsiolkovskii articulated his rationale for this endeavor, bringing a messianic and transformative vision to the *cause* of space exploration, one that echoed Fedorov's ideas about immortality and cosmic unity. And like Fedorov, he elevated above all what he considered the absolute certainty of 'modern science' while at the same time seeking a kind of unity between the human and physical worlds.[54] Tsiolkovskii drew from a whole panoply of influences, only some of which he explicitly acknowledged. For example, his writings show a clear influence of the Bavarian philosopher-mystic Carl Du Prel (1833–1899), famous for drawing a link between cosmic and biological evolution, especially the notion that Darwinian natural selection acted on planetary bodies just as they acted on living organisms. Tsiolkovskii undoubtedly took from Russian theosophists, some of whom already had mentioned 'cosmic' ideas long before he did, and some of whose Russian adherents later lived in Tsiolkovskii's hometown of Kaluga.[55] There were some direct but unacknowledged allusions to the writings of Madame Blavatsky. And as Michael Hagemeister has noted, many of Tsiolkovskii's ideas on matter and universal consciousness were not original; he liberally appropriated concepts from such late nineteenth-century European thinkers as Gustav Fechner (1801–1887) and Ernst Haeckel (1834–1919), whose works were in Tsiolkovskii's library.[56]

---

[53] All of these works—and many others which were unpublished during his lifetime—have been compiled together into one volume: V. S. Avduevskii (ed.) (2001) *K. E. Tsiolkovskii: kosmicheskaia filosofiia* (Moscow: URSS).

[54] For a comparative overview of the philosophies of Fedorov and Tsiolkovskii, see V. V. Kaziutinskii (1997) 'Kosmizm i kosmicheskaia filosofiia', in B. V. Raushenbakh (ed.) *Osvoenie aerokosmicheskogo prostranstva: proshloe, nastoiashchee, budushchee* (Moscow: IIET RAN), pp. 139–44.

[55] For 'cosmic' theosophist ideas in English from the late nineteenth century, see John Fiske's *Outlines of Cosmic Philosophy* (1874) and Richard M. Bucke's *Cosmic Consciousness: A Study in the Evolution of the Mind* (1901).

[56] Hagemeister (2012) 'Konstantin Tsiolkovskii and the Occult Roots of Soviet Space Travel'.

In Tsiolkovskii's worldview, the occult, theories of evolution, and Christianity existed without contradiction, and he expended much energy explaining biblical events with the aid of modern-day science. But his vision principally encompassed two evolutionary projects: first, to enable the migration of humanity off the Earth due to what he saw as irreversible problems on Earth (overpopulation and various geological and cosmic threats including extinction of the Sun); and, second, having left the Earth, to 'perfect' humankind, which would evolve into a kind of universal and immortal energy or radiation representing the best examples of people, lower and defective organisms having been left behind or destroyed. The 'scientific' basis for these two projects depended on an acceptance of two related ideas, *monism* and *panpsychism*, described in detail in *Monizm vselennoi* (Monism of the Universe), a brochure Tsiolkovskii self-published in 1925 that was the most complete published statement of his 'cosmic philosophy'.[57] Taking from Haeckel and Fechner among others, Tsiolkovskii articulated a striking worldview of 'human progress' embedded in universal time and space. According to Tsiolkovskii's monism, all matter in the universe, including organic matter, is made of a single substance, has the same structure, and obeys the same set of laws. Similarly, panpsychism was to him the belief that all matter is made up of 'atoms of ether', smaller than 'regular' atoms, which are in and of themselves *living organisms* or 'happy atoms'. Different living beings with differing abilities for expression result when these atoms combine in different ways. Because these 'ether atoms' are indestructible, there is no such thing as true death since the atoms can be reconstituted in different combinations from the one that gave life to a specific human being. Mortality is thus simply a mirage. In one of his brochures, *Volia vselennoi* (Will of the Universe), published in 1928 in Kaluga, he wrote:

> Death is one of the illusions of a weak human mind. There is no death, for the existence of an atom in inorganic matter is not marked by memory and time—it is as if the latter does not exist at all. The multitude of existences of the atom in the organic form merges into one subjectively uninterrupted and happy life ... The Universe is made in such a way, that it is not only immortal, but that its parts are also immortal in the form of blissfully happy living beings. There is no beginning and no end to the Universe, there is also no beginning and no end to life and its bliss.[58]

---

[57] K. E. Tsiolkovskii (1925) *Monizm vselennoi* (Kaluga: K. E. Tsiolkovskii). Tsiolkovskii republished the brochure with some minor additions and changes in 1931.

[58] K. Tsiolkovskii (1928) *Volia vselennoi: neizvestnye razumnye sily* (Kaluga: K. Tsiolkovskii), p. 7.

This philosophy explained to a large degree Tsiolkovskii's surprising lack of sorrow when several of his children died young, two of them from suicide.

He amplified these ideas in an eight-page meditation titled simply, 'Kosmicheskaia filosofiia' (Cosmic Philosophy) that he wrote months before his death in 1935.[59] Making use of a crude concoction of ideas from post-Enlightenment science, Russian Orthodox Christianity, and Buddhism (such as reincarnation), he ruminated on the goal of 'perfecting' humanity. He believed that once humans—led by 'presidents' who would introduce new ways of living—had gained absolute control over nature and had moved out into space, they would develop ideal relationships with each other. Tsiolkovskii's view on the proper relationship between humanity and nature reflected a belief that 'man' was the most supreme being in the universe. Although humans and the matter around them were essentially made from the same constituents, humans were capable of finer forms of reason that elevated them beyond a rock or a planet. As such, it was man's duty to master nature to introduce order and eliminate chaos. Humans should also harness the power of the universe, such as solar energy, for the improvement of people.[60] His technical works are sprinkled with allusions to space travel representing a 'victory over gravity'. He equated this liberation over nature with liberation over ignorance and imperfection in the human social landscape. Living in open space would allow humans to evolve biologically into a new species, an immortal 'Homo Cosmicus', capable of living naturally in open space, using ambient solar energy.

Tsiolkovskii makes no mention of Fedorov in any of his works although he was more than likely aware of *Philosophy of the Common Task*.[61] While there

---

[59] The essay was first published (in edited form) in 1981. See K. E. Tsiolkovskii (1981) 'Kosmicheskaia filosofiia', *Tekhnika-molodezhi* no. 4, 22–26. The original is in ARAN, f. 555, op. 1, d. 534, ll. 20–27ob.

[60] Tsiolkovsii believed that the short-term goal of humans in space was to use the energy resources of the solar system to improve the lot of the human race. In *Kosmicheskie raketnye poezda* (Cosmic Rocket Trains), he noted that '[c]onquering the solar system will yield not only energy and life that will be two billion times more plentiful than Earth's energy and life, but spaciousness which will be even more abundant'. He also pointed to other sources of energy besides the Sun., particularly, mineral resources on the asteroids (or minor planets) that circle the Sun. beyond the orbit of Mars. He meditated on the technical, managerial, and economical challenges of setting up industry in outer space for extraction and use of various minerals. For the quote, see K. E. Tsiolkovskii (1929) *Kosmicheskie raketnye poezda* (Kaluga: Kollektiv sektsii nauchnykh rabotnikov), p. 8.

[61] The few hints that Tsiolkovskii was aware of Fedorov's 'philosophy of the common task' come from second-hand sources, particularly the journalist Konstantin Altaiskii-Korolev (1902–78) who, in the 1960s, wrote about his private conversations with Tsiolkovskii. Altaiskii (1967) *Tsiolkovskii rasskazyvaet* ... (Moscow: Detskaia literatura). Altaiskii claimed

are obvious similarities—the use of science and technology for the migration of humans off the Earth—there are also key differences. Fedorov, for example, did not seek perfection of humanity in space, nor did he envision human transformation into energy. Both, however, had deeply troubling aspects. Nikolai Fedorov's 'common task', for example, can be interpreted as distinctly totalitarian given its explicit rejection of choice: human freedom was circumscribed by the expectations and limits of cosmic evolution. In other words, each and every last human being would *have* to participate in his project and submit to the 'will of the universe' for the 'common task' to have any meaning. More ominously, Tsiolkovskii's view of the search for human perfection clearly incorporated contemporary racist and eugenicist ideas. In his 'theory of rational egoism', Tsiolkovskii advocated the extermination of imperfect plants and animal life and called for, as he noted in a 1928 booklet, the 'battle against the procreation of defective people and animals'.[62] This view was most strikingly articulated in a piece finished in 1918:

> I do not desire to live the life of the lowest races [such as] the life of a negro or an Indian. Therefore, the benefit of any atom, even the atom of a Papuan, requires the extinction also of the lowest races of humanity, and in an extreme measure the most imperfect individuals in the races.[63]

Was anyone paying attention to these musings of Tsiolkovskii in the 1920s and 1930s when the first enthusiasts of space travel were simultaneously articulating their visions of the future? A few people certainly read them, some of them loosely connected to the *Fedorovtsy*. For example, we know of Aleksandr Chizhevskii, the young intellectual who, influenced by occult writings, wrote extensively on the relationship between cosmic factors (such as sunspots) and social activity on Earth. Chizhevskii wrote a German-language introduction for one of Tsiolkovskii's mathematically inclined monographs on rocketry, but he also had links to Leonid Vasilev,

---

that Tsiolkovskii told him that he heard about Fedorov's ideas about ten years after 'publication' of the *Philosophy of the Common Task*, but it is not clear if he meant publication of the first volume (1905) or the second (1913).

[62] Tsiokovskii quoted in Hagemeister (1997) 'Russian Cosmism in the 1920s and Today', p. 202.

[63] K. Tsiolkovskii (2001) 'Etika ili estestvennye osnovy nravstvennosti' in *K. E. Tsiolkovskii: kosmicheskaia filosofiia*, p. 82. For one of the few Russian commentators to investigate Tsiolkovskii's racist and eugenicists ideas, see N. Gavriushin (1992) 'Kosmicheskii put k vechnomu blazhenstvu" (K. E. Tsiolkovskii i mifologiia tekhnokratii)', *Voprosy filosofii* no. 6. 125–31.

the telepathy researcher who was a member of the short-lived Biocosmists in Moscow.[64] Tsiolkovskii also corresponded with Maksim Gorkii, the erstwhile fan of Fedorov. The writer had apparently heard of Tsiolkovskii during his exile via the latter's 1925 work *Prichina kosmosa* (Reason for Space), a meditation on humanity's spiritual calling to go to space. Although Gorkii intended to visit Tsiolkovskii in Kaluga upon his return to the Soviet Union in 1928, the two never met. Tsiolkovskii, however, sent Gorkii many of his philosophical brochures; unfortunately, there is no evidence to indicate what he thought of them. Their last contact was in 1932 when Gorkii sent a well-publicized congratulatory letter to the 'interplanetary old man' (as he liked to call Tsiolkovskii) on his 75th birthday.[65] Another possible reader of Tsiolkovskii's philosophy was the famous Russian poet Nikolai Zabolotskii although, like Gorkii, he has left very little of note to suggest what exactly he might have gleaned from Tsiolkovskii.[66]

There is no record of published commentary in the 1920s and 1930s on any of Tsiolkovskii's more philosophically minded works in the Soviet media. At a time when ideas beyond the prevailing epistemologies of modern science—such as telepathy, blood transfusionology, the search for immortality, and the noosphere—still remained as entirely appropriate subjects for public discussion, Tsiolkovskii's 'cosmist' ideas (a term he never used) had little or no public resonance, especially among likeminded space enthusiasts. In fact, in the prolific writings of the famous science popularizers Iakov Perelman and Nikolai Rynin, or the brilliant interplanetary theorist Iurii Kondratiuk, or the pioneering rocket engineers Mikhail Tikhonravov, Fridrikh Tsander, Sergei Korolev, and Valentin Glushko,

---

[64] For Chizhevskii's famous introduction, see Alexander Tshijewsky (1924) 'Anstatt eines Vorworts', in K. E. Tsiolkovskii, *Raketa v kosmicheskoe prostranstvo* (Kaluga: K. E. Tsiolkovskii), unnumbered preface page. Chizhevskii's most famous work on 'helio-biology' was published the same year: Chizhevskii (1924) *Fizicheskie faktory istoricheskogo protsessa* (Kaluga: A. L. Chizhevskii). Tsiolkovskii, in turn, published a review of Chizhevskii's book the following year in a local Kaluga newspaper. See *Kommuna*, April 4, 1924. For a recent reading, see V. V. Kaziutinskii (1998) 'Kosmizm A. L. Chizhevskogo', *Iz istorii raketno-kosmicheskoi nauki i tekhniki* no. 2, 98–122.

[65] Gor'kii to Tsiolkovskii (1932) ARAN, fond 555, op. 4, d. 183, l. 1.

[66] For Zabolotskii and Tsiolkovskii, see Sarah Pratt (2000) *Nikolai Zabolotsky: Enigma and Cultural Paradigm* (Evanston, IL: Northwestern University Press), pp. 183–86; Darra Goldstein (1993) *Nikolai Zabolotsky: Play for Mortal Stakes* (Cambridge, UK: Cambridge University Press), pp. 143–48; A. Pavlov (1964) 'Iz perepiski N. A. Zabolotskogo s K. E. Tsiolkovskim', *Russkaia literatura* no. 3, 219–26, which discusses the two letters Zabolotskii wrote to Tsiolkovskii in January 1932.

we see no evidence that they read or were even aware of Tsiolkovskii's millenarian philosophy about space conquest. All of them cite Tsiolkovskii as a genius, a prescient thinker of great truths. All of them quote him in their published works. Perelman and Rynin both even published biographies of the old man in the early 1930s. But none refer to Tsiolkovskii's odd meditations about 'blissful happy atoms'.

When Tsiolkovskii's legacy was fully resurrected in the post-World War II era, commentators skirted around his metaphysical motivations, preferring to publish in copious volumes his more mathematical articles. Museums, statues, conferences dedicated to Tsiolkovskii spoke of him only as a good communist (which by his own admission, he was not), a prophet in the service of modern science and technology, and the patriarch of Soviet space exploration. Statues were adorned with appropriate quotes about the power of modern science. Yet, comprehensive encyclopedias on Russian philosophy published in the 1950s did not mention him.[67] Barring a few sporadic and obscure academic works by graduate students, until the early 1970s, no major figure explored Tsiolkovskii's philosophical works. Then a young graduate of the philology faculty at the Moscow State University, Nikolai Gavriushin, began presenting papers on Tsiolkovskii's philosophy at the annual Tsiolkovskii Readings conferences held in Kaluga every year.[68] This annual conference, held every September, began to draw scores and then hundreds of devotes through the 1970s and 1980s, with Tsiolkovskii as their only cynosure. As a cult grew around Tsiolkovskii, many of the presenters, who included not only historians and archivists but also self-styled mystics and 'non-conformists', began to publish prolifically on Tsiolkovskii's philosophies. This small impulse coalesced and then expanded in the post-socialist era into a massive network of thinkers, writers, and activists who now actively act as the custodians of Tsiolkovskii's legacy. Like the invention of a Russian Cosmist historical tradition around Fedorov, by the 1990s, one could

---

[67] The two most prominent volumes from the 1950s do not mention Tsiolkovskii. See V. V. Zen'kovskii (1948) *Istoriia russkoi filosofii* (Paris; YMCA Press) and N. O. Losskii (1951) *Istoriia russkoi filosofii* (Moscow).

[68] Gavriushin's 1973 dissertation was entitled 'Artistic Creativity and the Development of Science (The Establishment of the Idea of Conquering Space)'. His very first paper on Tsiolkovskii's philosophical work was published in 1972 under the title 'From the History of Russian Cosmism'. He had orally presented the paper 2 years previously at the Tsiolkovskii Readings in Kaluga. For the publication, see *Trudy V i VI chtenii posviashchennykh razrabotke nauchnogo naslediia i razvitiiu tvorchestva K. E. Tsiolkovskogo* (Moscow: 1972), pp. 104–06.

find a similar historical genealogy around Tsiolkovskii's so-called 'Cosmic Philosophy', which, while loosely grouped under Russian Cosmism, has its own particularities and features relevant to the Soviet and later Russian space program. No less an authority than the Russian Academy of Sciences has since published several volumes of Tsiolkovskii's mystical musings with copious annotation, while many Russian intellectuals continue to debate them as an important constituent part of a 'national idea of Russia'. In the post-socialist times, Tsiolkovskii's philosophy was positioned in relation to the writings of Fedorov in an upsurge of jingoistic Russocentric 'cosmist' thinking. One well-known modern day Cosmist wrote in 1998, at a time of great national crisis that:

> the Russian philosophical and cultural tradition, in which a hugely important place is occupied by the creative heritage of K. E. Tsiolkovskii, can become the basis for the rebirth of Russian spiritual identity and its national revival. In Russian ... cosmism, [we see] prevailing the feeling of each [Russian] person's own self involvement as part of a living cosmos.[69]

This seemingly oxymoronic connection between '*Russian* spiritual identity' and *humanity*'s place in the cosmos continues to vivify a new post-socialist generation of Russian intellectuals.[70]

Conclusion

When the first young hero cosmonauts flew into space in the early 1960s, Soviet commentators repeatedly depicted them as emblematic of a modern and technologically sophisticated Russia, overtaking the West. And unlike American astronauts who thanked God for their successes, Soviet cosmonauts were avowedly atheistic; one of the first cosmonauts, the young German Titov, famously declared on a visit to the USA that during his 17 orbits of the Earth, he had seen 'no God or angels', adding that 'no

---

[69] L. V. Leskov (1998) 'K. E. Tsiolkovskii i rossiiskaia natsional'naia ideia', *Zemlia i vselennaia* no. 4, 62–7.

[70] For a lengthy meditation on the role of Russian Cosmism in post-socialist Russia, see V. V. Kaziutinskii (1994) 'Kosmicheskaia filosofiia K. E. Tsiolkovskogo na rubezhe XXI veka' in *Trudy XXVII chteniia, posviashchennykh razrabotke nauchnogo naslediia i razvitiiu idei K. E. Tsiolkovskogo* (*Kaluga, 15–18 sentiabria 1992 g.*): *sektsiia* '*K. E. Tsiolkovskii i filosofskie problemy osvoeniia kosmosa*' (Moscow: IIET RAN, 1994), pp. 4–40.

God helped build our rocket'.[71] As Victoria Smolkin-Rothrock has shown, such exclamations reinforced an alignment between space exploration and campaigns to inculcate atheism among the Soviet population.[72] They also whitewashed 'inconvenient' phenomena dating back to the 1920s when the founding theorist of Soviet space exploration, Tsiolkovskii, was actively writing about space travel in ways that could not conform to orthodoxies about the role of science and technology in post-World War II Soviet society. An article entitled 'Cosmonautics versus Religion' in 1959, for example, could confidently argue that atheism provided the spark for cosmic conquest and that Tsiolkovskii was at the vanguard of this movement.[73] Yet, despite these efforts, through the height of Soviet space exploits during the Cold War, Tsiolkovskii's mystical ideas loomed as an echo of an earlier time.

It is undoubtedly true that when the first space theorists, enthusiasts, and practitioners were articulating a vision for the future of cosmic travel, they were drawing from the modernist and quasi-utopian ideas emerging from particular scientific and technical disciplines rooted in post-Enlightenment rationalism. Expressions of this cult of modern science and technology took the form of fiction (Jules Verne) or popular science (Iakov Perelman) or engineering marvels (e.g. the airplane). All of the principal actors who translated Soviet populist cosmic enthusiasm into the machines that put the first Soviet satellites and cosmonauts into orbit, have paid explicit debt to this tradition. Lest we suspect that they were 'hiding' their true motivations, among the literally hundreds of memoirs about the Soviet space program churned out in the 1990s—many of them highly critical of the Soviet state, Stalin, and Marxism in general—remarkably, not a single one hints that the inspiration to aspire for the cosmos was anything but an outcome of a national and cultural commitment to be the best in science and technology.

---

[71] Anonymous (1962b) 'Titov, Denying God, Puts His Faith in People', *New York Times*, May 7. A variation of this quote is often falsely attributed to first cosmonaut Iurii Gagarin, but there is no evidence to suggest that he said anything about God or his faith in any of his public pronouncements.

[72] Victoria Smolkin-Rothrock (2011) 'Cosmic Enlightenment: Scientific Atheism and the Soviet Conquest of Space' in *Into the Cosmos*, pp. 159–94.

[73] M. V. Mostepanenko (1959) 'Kosmonavtika protiv religii', *Nauka i zhizn'* no. 1, 74–5. See also the invocation of Tsiolkovskii in a book arguing for the inextricable link between technological progress and atheism, G. S. Gudozhnik (1961) *Tekhnika i religiia* (Moscow: Voenizdat), pp. 22–3.

Yet, while the cult of modern science in the twentieth century inspired these actors, it is also undeniable that the central figure at the very root of grand narrative of the Soviet space program, Tsiolkovskii, was deeply influenced by mystical and occult epistemologies. In fact, at the very moment, the 1920s, when the first mass cosmic enthusiasm emerged in Soviet culture, there was a concomitant if less widespread interest in various mystical and metaphysical notions about the cosmic setting of humanity. The existence of this particular historical strand in the early twentieth century is further complicated by a more recent post-socialist discourse that claims 'Russian Cosmism' as a founding imperative of the Soviet space program. But as I show here, this invented tradition, true enough in accounting for the richness of cosmic thought in Russia in the early twentieth century, vastly overreaches in suggesting that it had any influence on the *actual actors* in the Soviet space program. What we are left with instead is a complex story about the ways in which 'scientific' and 'mystical' worldviews in the Soviet context were often difficult to parse out in twentieth century Russia.

The evidence tells us that Fedorov and Tsiolkovskii were enigmatic thinkers who considered with great care the possibility of human migration off the planet Earth and the rationales behind it. They were driven partly by religious imperatives, partly by what they considered 'rational' considerations, and partly by mystical, occult, and Russophilic ideas about the place of humanity in our universe. Their ideas were not heard by many at the time but were implicitly mimicked in a form of cultural echo in the ideas and works of other Russian intellectuals, especially in their evangelical calling to fully conquer and submit nature to the will of humanity. And although they probably influenced no major actor in the Soviet space program of the Cold War, the fact that these ideas existed at all, especially within the worldview of Tsiolkovskii, suggests that while the Soviet march into the cosmos was not inspired by mystical or occultish ideas, it was definitely accompanied by them. And that accompaniment also included deeply unsavory and unsettling ideas about eugenics and racial purity that require a deeper reckoning, especially as 'Russian Cosmism' has taken a more jingoistic turn in the past decade.

Was the Russian cause for the cosmos, as articulated in the twentieth century, an outcome of a modernist imperative? As I have shown here, once articulated in the context of the cult of science and technology, the Russian intellectual framework for cosmic enthusiasm took on an unlikely path, like a Mobius strip of ideas that veered back into a completely separate history to recover its ostensibly religious and mystical roots. The recovery of an imagined movement for the conquest of space became an important project and also a convincing one, given that both the 'scientific' and the 'mystical' existed without contradiction in the life and work of the founding theorist of Soviet space exploration. In one sense then, the intellectual foundations of the space program can be understood as a typical manifestation of the urge of many early twentieth century intellectual movements to respond to the encroachment of modern science into the social order through an active recovery of what they considered 'ancient truths'. As modern science began to explain the world around us, thinkers such as Fedorov and Tsiolkovskii sought to adjust their beliefs into new epistemologies that incorporated what they self-consciously understood as 'ancient' (Christianity) and 'modern' (nineteenth-century scientific disciplines). Both, however, articulated a teleology of social improvement with a central place for the project of 'mastery over nature'. Like a persistent line on each side of the Mobius strip, these aspirations to control of nature—often in troubling ways—remained the one constant in Russian cosmic enthusiasm in the twentieth and now into the twenty-first centuries.

CHAPTER 7

Witchdoctors Drive Sports Cars, Science Takes the Bus: An Anti-Superstition Alliance Across a Divided Germany

Monica Black

Forms of religious and magical healing flourished in many parts of Europe throughout the twentieth century. Not infrequently, they were also bound up with fears about evildoing and witchcraft.[1] In Poland,

---

[1] Generally, see Bengt Ankarloo and Stuart Clark (1999) (eds.) *Witchcraft and Magic in Europe: The Twentieth Century* (Philadelphia: University of Pennsylvania Press). On Denmark: Gustav Henningsen (1989) 'Witch Persecution after the Era of the Witch Trials: A Contribution to Danish Ethnohistory,' *Scandinavian Yearbook of Folklore* xxxxiv: 103–153. England: Owen Davies (1996) 'Healing Charms in Use in England and Wales, 1700–1950,' *Folklore* cvii: 19–32. Finland: Laura Stark-Arola (1998) *Magic, Body and Social Order: The Construction of Gender Through Women's Private Rituals in Traditional Finland* (Helsinki: Finnish Literature Society). France: Jeanne Favret-Saada (1977) *Les Mots, la mort, les sorts: La sorcellerie dans le Bocage* (Paris: Gallimard), translated into English (1980) as *Deadly Words: Witchcraft in the Bocage* (Cambridge: Cambridge University Press), trans. Catherine Cullen. Italy: Thomas Hauschild (2002) *Macht und Magie in Italien: Über Frauenzauber, Kirche und Politik* (Gifkendorf: Merlin Verlag). Poland: Aldona Christina Schiffmann (1987) 'The Witch and Crime: The Persecution of Witches in 20th-Century Poland,' *Scandinavian Yearbook of Folklore* 43: 147–164. West Germany: Joachim Friedrich Baumhauer (1984) *Johann Kruse und der 'neuzeitliche Hexenwahn.' Zur Situation eines norddeutschen Aufklärers und einer Glaubensvorstellung im 20. Jahrhundert untersucht*

M. Black (✉)
Modern European History, University of Tennessee,
Knoxville, TN, USA

folklorist Aldona Christina Schiffmann found in the mid-1980s, 'Sorcery, witchcraft, the evil eye, ghosts, witches, witchdoctors, enchantment by fairies, traditional healing and ... homeopathy ... possession and exorcism [were] not distinguished as different categories.'[2] Something similar may have been true in the German Democratic Republic (GDR), but trying to learn about it is not easy. Unlike agricultural reports, the proceedings of party congresses, or even budgets for puppet theaters, spiritual medicine is hard to locate archivally. When entered into a web portal of the German Federal Archives, even the rather benign (if, obviously, contested) term *Volksmedizin* (folk medicine) returns the message, 'no hits.' Nor does entering 'Superstition', the considerably more pejorative and capacious term under which the practices of spiritual medicine were so often subsumed in East Germany offer many leads. One link sends you to an educational pamphlet produced for youths in 1956, 'Belief in Ghosts, Superstition, Science: How Humanity Learned to Rule Nature and Overcome Superstition'; Another, more mysteriously, directs you to a file concerning the 1983 performance of a state dance ensemble.

Despite a superficial absence of 'superstition' in East Germany, the concept was central to communist self-definition, at least, in a negative sense. The opinion of one Wilhelm Hampel summed the view up concisely. 'Ever since Marx, Engels, and Lenin identified and scientifically demonstrated the fundamental laws of social development,' he wrote in 1961, 'the supposition has obtained of an inherent relationship between the capitalist system and superstition.'[3] That same year, an East German popular science writer used a colorful image to underscore official dogma—that superstition and occultism were products of Western decadence and clericalism run amok and anathema to socialism—The West he declared, was a place where witchdoctors became rich (and drove sports cars), while science was forced to take the bus.[4]

---

anhand von Vorgängen in Dithmarschen (Neumünster: Karl Wachholtz Verlag) and Inge Schöck (1978) *Hexenglaube in der Gegenwart: Empirische Untersuchungen in Südwestdeutschland* (Tübingen: Tübinger Vereinigung f. Volkskunde).

[2] Schiffmann, 'The Witch and Crime,' 148.

[3] W. Hampel (1961) *Schwärmer, Schwindler, Scharlatane* (Berlin: Verlag Neues Leben), p. 298.

[4] Joachim Stahl (1961) '"Renaissance" des Aberglaubens,' *Wissenschaft und Fortschritt* 9: 346–348; here, 346.

In the communist world generally, standing as it did in the tradition of enlightenment and social progress through faith in human reason and science, 'superstition' had long been an object of intense, if not especially coherent, activism.[5] A scientific impulse—one claiming science as the 'ultimate pinnacle of all thought and action,' and asserting its authority even over those existential matters traditionally the purview of religion (life and death, health and sickness)—was particularly strong in East Germany. East German activists and pedagogues saw science as standing absolutely opposed to religion, which was not only 'unscientific' but also the chief instrument of the expropriating classes.[6] Superstition and religion were collapsed into a single, broad, and treacherous category, one that also included 'astrology, medical quackery, fatalism, idealism, etc.' Science was positioned as the chief instrument through which to 'fight superstition, unscientific ideas and mysticism.'[7]

As the image of witchdoctors driving sports cars suggests, the spiritual home of these expropriating classes and their obscurantist attitudes was the capitalist West generally, and West Germany specifically. Yet in the 1950s and 1960s, during what we might call the 'superstition wars,' a dedicated and vocal group of German anti-superstition activists worked hand-in-hand across the East/West German border and across the Iron Curtain.[8] This group of theorizers and combatants—some medical doctors, some criminologists, some armchair ethnographers and sociologists—made unaccustomed alliances. The subject of this chapter is the relationship between two such comrades-in-arms, Otto Prokop and Johann Kruse. Prokop was the GDR's most prominent campaigner against what he called 'medical occultism.' Kruse was a retired schoolteacher from Hamburg who conducted research for decades on witchcraft beliefs in his native Schleswig-Holstein. Kruse gave talks and was widely cited in the press and by anti-superstition critics in both East and West Germany. Prokop's voluminous (if highly repetitive) work on the 'medical occult,' which built on his collaboration with Kruse and other anti-superstition colleagues,

---

[5] S. A. Smith (2008) 'Introduction,' *The Religion of Fools? Superstition Past and Present*, S. A. Smith and Alan Knight (eds.), Past and Present Supplement 3: 30.
[6] Thomas Schmidt-Lux (2008) 'Das helle Licht der Wissenschaft: Die Urania, der organizierte Szientismus, und die ostdeutsche Säkularisierung,' *Geschichte und Gesellschaft* 34:1: 42, 44, 46, 55, 57.
[7] Schmidt-Lux, 'Das helle Licht,' 63.
[8] 'Superstition wars' is a riff on Michael D. Gordin (2012) *The Pseudoscience Wars: Immanuel Velikovsky and the Birth of the Modern Fringe* (Chicago: University of Chicago Press).

East and West, was published—sometimes in identical editions—in both Germanys. Ultimately, the fruit of East/West anti-'superstition' collaboration spread beyond German borders. Theories about witch beliefs with roots in the 'superstition wars' were pursued by a Danish folklorist, who in turn discussed them at an international ethnography congress in Moscow in 1964. They were cited by a noted social scientist in Amsterdam working in the field of mass psychology and eventually made their way into other academic fields like history, if only briefly. In the 1970s, neo-Marxist and social functionalist folklorists in the Federal Republic of Germany (FRG) rediscovered Kruse's work—which had previously been marginalized in the FRG but was well appreciated in the GDR—to learn more about the social consequences of apotropaic healing.[9]

This chapter is not the first work to examine Kruse & Prokop's relationship; folklorist Joachim Friedrich Baumhauer's 1983 book on Kruse was. But this chapter is especially attentive to the Cold War context of the 'superstition wars,' and wishes to highlight a rare instance of intellectual harmony within a historiography that has often focused on East/West conflict. For both Prokop and Kruse, the threat of what they called superstition—particularly folk and magical healing practices and witchcraft beliefs—emanated neither from the communist nor the capitalist world, but from continuities they perceived in German history from the early modern witch hunt to the Holocaust. Obsessed with what they saw as occultism's destructive power, they first found common cause in the 1950s and continued to work thereafter—together and separately—toward eradicating it. Perhaps most surprising of all, and long before the mainstream of academic historiography came to focus on the role of German society in attempting to explain Nazi genocide, both concluded that 'superstition,' and the profound social anxiety to which it sometimes gave rise—had been at the root of the Holocaust.[10]

• • •

Folk medicine in twentieth century Germany has been the subject of a good bit of research by historians, though for various reasons, much of it has focused on the battles that raged early in the century over 'quackery,' and on the National Socialists' brief interest in unifying naturopathy, folk healing, and academic medicine.[11] What work exists on historical folk

[9] Baumhauer, *Johann Kruse*, pp. 168–171.
[10] Baumhauer, *Johann Kruse*, 81–82, discusses this issue specifically with relation to Kruse.
[11] Robert Jütte notes that recent medical historiy has 'preferred to look at structures, agents, trends and developments in biomedicine.' See Jütte (1999) 'The Historiography of

medicine for the era since 1945 seems to have been accomplished largely by folklorists working on West Germany. However, in what follows, and by using a variety of secondary and mostly indirect sources, there is a good bit one can say—at least provisionally—about aspects of religious/ magical healing in East Germany as well.

Michael Simon's 2003 habilitation, which relies on the *Atlas of German Folklore* (*Atlas der deutschen Volkskunde*), is a good starting point. Between 1930 and 1935, German folklorists undertook the enormous *Atlas* project, with the intention of gathering ethnographic information from across the entirety of German-speaking Europe. Based on questionnaires completed by community leaders, including pastors and schoolteachers, the *Atlas* was intended to represent in map form various folk beliefs and practices, language variations, and aspects of religious and material culture, among many other things. One of the practices to which Simon devotes considerable attention is *Besprechen*—the use of verbal charms or formulae to heal illness.[12] Scholars' attempts to date the charms used in these practices have varied widely, but many were derived from popular grimoires like *The Sixth and Seventh Books of Moses*, the first published edition of which appeared in the German-speaking world in the nineteenth century.[13]

On the basis of the *Atlas*, Simon has shown that the practice of *Besprechen*, at least in the mid-1930s, was most heavily distributed in the rough triangle between Hamburg, Dresden, and Gdansk. A significant portion of that triangle lay in what became East Germany. About 80 % of people in the triangle answered 'yes' when asked whether the practices of *Besprechen* were known in their community. Simon's analysis of the *Atlas* further suggests that *Besprechen* was by no means an exclusively rural phenomenon. There was virtually no difference, he found, between Berlin and its hinterlands nor between Dresden or Hamburg and the surrounding countryside when it came to *Besprechen*.[14]

Besprechen was a Christian practice, but not a confessional one. A charm to cure gout, for example, says:

Nonconventional Medicine in Germany: A Concise Overview,' *Medical History* 43: 342–358; here, 342.

[12] Michael Simon (2003) *'Volksmedizin' im frühen 20. Jahrhundert: Zum Quellenwert des Atlas der deutschen Volkskunde* (Mainz: Gesellschaft f. Volkskunde in Rheinland-Pfalz).

[13] Simon, *Volksmedizin*, 147–148, n. 459. Owen Davies (2009) *Grimoires: A History of Magic Books* (Oxford: Oxford University Press), pp. 123. See also Stephan Bachter, *Anleitung Fum Aberglauber: Fauber-bücher und die Verbreitung magischen 'Wissens' seit dem 18. Jahrhundert* (PhD diss.,Universität Hamburg, 2005), 96.

[14] Simon, *Volksmedizin*, pp. 175, 169–171.

In the name of the Father, the Son and the Holy Spirit, amen. I curse you, gout, by the five holy wounds and by the innocent blood of our Lord Jesus Christ, which flowed to earth from those five holy wounds for humanity. I curse you, gout, by the Last Judgment and the bitter sentence that God will pass upon all sinners, that you will not harm me in any of my limbs.[15]

On the basis of this evidence, there would seem to have been little difference between *Besprechen* and prayer. Adolf Spamer, one of the great German folklorists of the early twentieth century, an avid collector of formulae like this one, and a key member of the *Atlas* project, observed in the 1950s, 'There have been constant theological attempts to erect a conceptually and morally sustainable separation between prayer and incantation over the centuries.'[16] None, he suggested, had fully succeeded.

Spamer was a bourgeois-nationalist who began an illustrious career in the interwar period, and endured a fraught relationship with the Nazis. He ended his career in the GDR, where he helped to found the Institute for Folklore at the Technical University in Dresden.[17] For him, as for many folklorists of his day, *Besprechen* was perceived as a largely benign or in any case morally neutral practice. It was a window on to folk religion and the popular medical wisdom of a moment in time. However, within the settings of communities, these practices could sometimes have negative social consequences. Those who undertook *Besprechen* and related healing arts were sometimes also practitioners of evil-averting magic and exorcism.[18] In fact, these things were sometimes of a piece. As we have seen, *Besprechen* entailed soliciting divine power to effect healing, but cures also sometimes involved rituals to banish evil—and evil was sometimes understood to be very directly personified. That is, as Gerhard Wilke has written of the interwar Hessian village of Koerle, people were not always persuaded that sickness 'just happened.' It *was* caused by something—or someone.[19]

---

[15] Simon, *Volksmedizin*, pp. 165–166.

[16] Adolf Spamer (1955) 'Zauberbuch und Zauberspruch,' *Deutsches Jahrbuch f. Volkskunde* 1: 109–126; here, 109. Richard Kieckhefer (2000) *Magic in the Middle Ages* (Cambridge: Cambridge University Press), pp. 69–75, divides charms into prayers, blessings, and adjurations.

[17] Teresa Brinkel (2009) 'Institutionalizing *Volkskunde* in Early East Germany,' *Journal of Folklore Research* 46:2: 141–172.

[18] Robert Jütte (1996) *Geschichte der Alternativen Medizin: Von der Volksmedizin zu den unkonventionellen Therapien von heute* (Munich: C.H. Beck), p. 90.

[19] Gerhard Wilke (1989) 'Die Sünden der Väter: Bedeutung und Wandel von Gesundheit und Krankheit im Dorfalltag,' in Alfons Labisch and Reinhard Spree (eds.), *Medizinische*

So, too, in postwar Germany, were illness and other misfortunes sometimes presumed to be instigated by a particular member of the community, who was further presumed to be in league with the devil—that is, was presumed to be a witch.

In such cases, locals would turn to people—by turns feared and revered—who could identify witches and break their hexes. Such individuals were known by a variety of regional names: *Wiedergutmacherin* (wise woman, propitiator) in some places, *Hexenmeister* or *Hexenbanner* (witch master, witch banisher, witch doctor) in others. In yet other locations, like Franconian Switzerland, the ritual practice of healing by averting evil or dispelling a witch's hex was termed 'the beginning, the commencing' (*das Anfangen*) but the person (usually a woman) who performed the *Anfangen* had no special name.[20]

A group of East German doctors investigating what they called 'superstition and quackery' around Greifswald in the early 1960s found slightly higher numbers of healers—performing various services with the aid of magic books, including *Besprechen* and the laying on of hands—in the coastal areas surrounding the city than in the city itself. On the other hand, they found fewer in the small town of Gützkow than in much larger Greifswald.[21] Again, lest we imagine that magical and spiritual healing practices and the beliefs that gave rise to them were localized in 'isolated' rural places, Dr. Maximilian Meischke, head of the Medical Advisory Council for the East Berlin borough of Friedrichshain—a traditionally working class part of the city—wrote in 1964 that his daily dealings 'with sick patients and their relatives,' had convinced him that 'fears of evil spirits and ominous signs [were] widely dispersed.'[22]

Folklorist Inge Schöck's 1978 study *Witch Beliefs Today* offers us an excellent example—which she called 'Case A'—of the poisonous atmosphere that witchcraft beliefs could generate, and one that also has an inter-German relevance. Case A dealt with a 'refugee from the GDR,' one

---

Deutungsmacht im sozialen Wandel des 19. und frühen 20. Jahrhunderts (Bonn: Psychiatrie-Verlag), pp. 123–140.

[20] Hans Sebald (1978) *Witchcraft: The Heritage of a Heresy* (New York and Oxford: Elsevier North Holland), pp. 81–84.

[21] W. Weuffen et al. (1963) "Aberglauben und Kupfuscherei auf dem Lande," *Das Deutsche Gesundheitswesen* 18: 562–567.

[22] M. Meischke (1962) 'Beobachtungen über Abergläubisches im täglichen Umgang mit Patienten,' in O. Prokop (ed.), *Medizinischer Okkultismus: Paramedizin* (Jena: VEB Gustav Fischer Verlag), p. 111.

Herr C., who landed in a small agricultural community of around 400 persons in the mid-1950s in West German Franconia. Shortly thereafter, he began accusing a local woman, Frau N., of witchcraft. Frau N. belonged to a family that enjoyed a good reputation in matters of housekeeping, child rearing, and as farmers. For generations, they had also been revered for their supernatural competencies. Upon his arrival, Herr C. presented himself to the community as a man with special knowledge, including the ability to know who was to blame for the appearance of illness. He also began spreading rumors about Frau N. He claimed to have seen her through a window using *The Sixth and Seventh Books of Moses*, and told people that while he 'worked with God,' N. 'worked with the devil.' In this community, Schöck noted, 'the devil [was] perceived as a literal and constant threat,' making it 'difficult for many local people to defy [C.'s] accusations.'[23] When two well-regarded local farmers suddenly died in middle age, the climate in the village turned even colder toward N. Local children were told not to accept gifts from her any longer; she became increasingly isolated.

Identifying a local person as a witch—often, but not always, like N., a woman—could and sometimes did produce a dramatic and dangerous climate of paranoia, suspicion, and even violence. Persons so accused sometimes had little alternative but to avail themselves of the courts. Baumhauer briefly mentions such a case from East German Belzig, near Potsdam, from 1956, and in the late 1950s, criminologist Herbert Schäfer collected evidence of dozens of such cases in the FRG. These cases often turned on complaints brought by people who had been accused by neighbors of witchcraft (Frau N. indeed took Herr C. to court) or who had been harmed in some way by neighbors who believed them to be aligned with the devil. While some were victims of slander, others had their houses burned down or were physically attacked.[24]

As noted above, beliefs about witches, the devil, and evil-averting magic were in no way indigenous to Germany, but there does seem to have been an upsurge in witchcraft accusations after the Second World War. Perhaps more important for the present essay, witchcraft beliefs seem to have meant something different in the context of recently divided Germany, and Cold

---

[23] Schöck, *Hexenglaube*, p. 183.

[24] Herbert Schäfer (1959) *Der Okkulttäter. Hexenbanner—magische Heiler—Erdentstrahler* (Hamburg: Verlag für kriminalistische Fachliteratur), gives details about a number of such instances.

War rhetoric flavored all discussions of the 'witchcraft trials,' as the press referred to them in both countries. The Belzig case, Baumhauer writes, was reported on in no fewer than 90 West German newspapers as evidence of East German circumstances 'like the darkest Middle Ages.' Similar cases in West Germany were taken as evidence of the 'clerical-military regime' of the FRG, which was said not only to 'tolerate mysticism and superstition, but actively to promote it.'[25]

• • •

Though the mass media in both East and West Germany used the 'witchcraft trials' as a civilizational cudgel against their Cold War Other, the practices and beliefs that led to those trials in fact became a matter of concern and cooperation across the German–German border, when a former schoolteacher from Hamburg, Johann Kruse, and doctor of forensic medicine Otto Prokop first made common cause in the early 1950s. The direct and indirect cooperation of these activists over the next 20 years would lead them to theorize not only about the meaning of practices like *Besprechen* and apotropaic magic generally but ultimately to trace a line of continuity they perceived from the early modern witch hunt down to the darkest decades of the twentieth century.

An anti-clerical Social Democrat, Kruse had collected materials for an archive on what he called 'the modern witch craze' since the 1920s. Long concerned that folk measures taken to ward off evil were leading to the persecution of women in particular, Kruse published a book in 1951 called *Are there Witches Among Us? Magic and Witchcraft Beliefs in Our Time*. 'Still today, in the middle of the twentieth century,' he wrote, 'every city in Germany has its "witches" and almost every village its "servant of the devil" (*Teufelsdienerin*). Because of their defencelessness, they are ostracized and persecuted, some are mistreated and even killed. Thousands and thousands of women suffer from the effects of this [witchcraft] delusion.'[26]

Kruse was an inveterate opponent not only of witchcraft beliefs and *Hexenbanner* but also of magic books like *The Sixth and Seventh Books of Moses*—which, as we have seen, some people relied upon for healing

---

[25] Baumhauer, *Johann Kruse*, 101.
[26] Johann Kruse (1951) *Hexen unter uns. Magie und Zauberglauben in unserer Zeit* (Hamburg: Verlag Hamburgische Bücherei), p. 7.

charms. Despite the relatively prohibitive price of 10.50 DM, the Moses books, in an edition published by Planet Verlag, sold briskly after World War II. In 1954, Kruse lodged a complaint against the publisher, which led to a spectacular series of trials in the small city of Braunschweig widely publicized throughout East and West Germany. The publishers were tried for dishonest business practices (the book was said to make false claims); inciting the public to commit criminal offenses (including cemetery desecration and the mishandling of animals for use in magical recipes); and for offense against the FRG's 1953 Law for the Control of Sexually-Transmitted Diseases (the books provided information on curing syphilis, as we will soon see).[27]

Called as an expert witness by the court was one Otto Prokop. Born in St. Pölten, Austria in 1921, Prokop belonged to the third generation of a conservative family of doctors. He took up the study of medicine in Vienna in 1940. After serving in the Wehrmacht, he relocated to Bonn, where he finished his studies in 1948 and taught until he was called to an appointment in the GDR in 1957. Prokop's opposite number in the proceedings against Planet Verlag was the famous Göttingen folklore professor, Will-Erich Peuckert, who had been engaged by the defense. On the stand, Prokop denounced magic books, which he saw as instruction manuals for mayhem and criminality. He scoffed at 'magical recipes' against diseases like syphilis, one of which recommended burying a person up to his neck in horse dung, to 'leech the evil fluids out of the body.' The case was 'well known,' Prokop observed, of a man who died trying this cure.[28] In a lengthy essay published first in 1962, Prokop interpreted the book and its effects from a social-psychological perspective. *The Sixth and Seventh Books of Moses*, he wrote, evinced 'a sense of dread, an anxiety-conditioned complex, fed by ... suggestion and ... deeply-rooted hate-tendencies.'[29] The book tended toward the paranoid, schizophrenic, narcissistic, and pathological, Prokop suggested.[30]

For the folklorist Peuckert, by contrast, the Moses books were nothing more or less than *Hausväterliteratur*—a genre of books dating to the early

---

[27] Baumhauer, *Johann Kruse*, pp. 84–86.

[28] Baumhauer, *Johann Kruse*, p. 91.

[29] A. Eigner and Otto Prokop (1962) 'Das sechste und siebente Buch Moses: Zur Frage der Kriminogenität von Büchern und besonders laienmedizinischer Schundliteratur,' in Otto Prokop (ed.), *Medizinischer Okkultismus: Paramedizin* (Jena: VEB Gustav Fischer Verlag), p. 235.

[30] Eigner and Prokop, 'Das sechste und siebente Buch Moses,' pp. 249–250.

modern period, which contained agricultural knowledge and advice about household economics, agricultural production, and medical treatments.[31] On this view, the Moses books represented a stage in the development of early modern German science. After several years of proceedings, the court sided with Peuckert and Planet Verlag: in 1960, all criminal charges against the publisher were dropped.[32]

Legal defeat, however, was by no means the end of the story, nor was it the end of Prokop's and Kruse's alliance. By the time of the court's decision, Prokop had been ensconced for several years in a prestigious appointment as Director of the Institute for Forensic Medicine at the famed Charité Hospital in East Berlin. He was to have an extraordinary career there. Before retiring in the late 1980s, he would train more than a generation of East German doctors, undertake an estimated 40,000 autopsies, and publish dozens of books and articles on aspects of forensic medicine to great acclaim.[33] Prokop was an unexpected figure in GDR history for many reasons. His scientific work (particularly on blood groups) was recognized with the GDR's National Prize for Science and Technology, first class, among numerous other academic honors, and he proudly worked for the elevation of what he called 'materialist science.'[34] Yet even before the Moses books affair, his career was also conspicuous for its preoccupation with the occult, the folkloric, and the fringe. No doubt his chosen medical specialty played some role here. The topic of his dissertation, to give just one example, was 'Murder by Animal Hair.' In it, he recounted various ways in which people had tried (and succeeded, sometimes) in killing family members by lacing their food with bristles cut from livestock or shaving brushes.[35] Equally curious is the relationship Prokop had to the East German state. He won seemingly every accolade the state offered, but was never a member of the ruling Socialist Unity Party and claimed to have no truck with any political ideology.[36] Being Austrian and holding a foreign passport enabled him to travel much more freely than he might otherwise have done. In fact, if his travel was hindered by anyone, Prokop

---

[31] Will-Erich Peuckert (1960) *Verborgenes Niedersachsen: Untersuchungen zur Niedersächsichen Volkssage und zum Volksbuch* (Göttingen: Vlg. Otto Schwartz), p. 124.

[32] Davies, *Grimoires*, p. 259.

[33] Mark Benecke (2013) *Seziert: Das Leben von Otto Prokop* (Berlin: Das Neue Berlin).

[34] Textbooks in English continue to cite Prokop's work on blood groups from the 1960s. Geoff Daniels (2013) *Human Blood Groups*, 3rd ed. (Hoboken, N.J.: Wiley-Blackwell).

[35] Otto Prokop (1952) 'Über Mord mit Tierhaaren,' *Deutsche Zeitschrift f. gerichtliche Medizin* 41: 240–242.

[36] Benecke, *Seziert*, p. 15.

claimed, it was occultist factions in the FRG who wanted to keep him out, rather than GDR officials seeking to keep him in.[37]

A biography of Prokop, published by a former student in 2013, suggests that his 'battle against pseudoscience runs like a red thread' throughout the doctor's life.[38] Yet despite his zeal for that topic, it cannot have been easy for the doctor to write about 'superstition' or 'occult medicine' in the GDR, where such things were supposed not to exist. Prokop held to certain conventions in his work. He would refer to the superstitious nature of 'our *Volk*,' for example, leaving rather ambiguous to which Germans he referred, or he would simply talk about superstition as though it were a timeless and static feature of the human condition rather than a phenomenon with particular historical or local characteristics or meanings.

Kruse, by contrast, faced no such barriers. Because his work and passions had focused much more specifically—even obsessively—on witch beliefs in West Germany, he found in East Germany a ready-made audience for his publications and lectures. In 1959, he was invited to speak at the annual Wossidlo Conference in Rostock by the don of postwar East German folklore, Wolfgang Steinitz.[39] Kruse repaid such honors by publicly commending in West Germany those East German laws that banned the printing of occult works, and he praised various efforts in the GDR aimed at public enlightenment through film, literature, and organizations like the Society for the Dissemination of Scientific Knowledge (Gesellschaft zur Verbreitung wissenschaftlicher Kenntnisse, referred to as Urania).[40] Kruse's work was often cited not only by Prokop but also in such East German popular science journals as *Wissenschaft und Fortschritt* (science and progress) and in various newspapers and books.[41]

• • •

Though the extent of the collaboration between Prokop and Kruse after the Moses books affair is not known, one is very much struck by certain themes common to their thinking over time. When they first met in 1954, Kruse was single-mindedly concerned about witchcraft

---

[37] Baumhauer, *Johann Kruse*, p. 89.
[38] Benecke, *Seziert*, p. 86.
[39] Baumhauer, *Johann Kruse*, p. 102.
[40] Landesarchiv Schleswig-Holstein, Abt. 811, Nr. 3597. Johann Kruse, 'Kurzer Bericht über den Stand des Kampfes geg. den Hexenwahn,' 1966/1967.
[41] Just two examples: Hampel (1961) *Schwärmer, Schwindler, Scharlatane*; Meischke (1961) *Schach dem Aberglauben* (Berlin: VEB Verlag Volk und Gesundheit).

beliefs and the particular toll they took on women. Prokop's interests seemed less precise but centered on pernicious pseudoscience and what he called 'medical occultism.' (In his understanding, this included everything from magical recipes against syphilis to iridology, acupuncture, yoga, homeopathy, parapsychology, and much more besides.) What both men shared, though, was a certain psychological approach to interpreting 'superstition,' which they regarded as a form of *social* pathology. In a 1962 article on the Moses books, which appeared in a volume he edited, Prokop and co-author A. Eigner drew on Freud to understand 'superstition.' It was, they argued, the very 'mechanism of regression'—but one embedded in social structures rather than individual consciousness—and it was nurtured by forms of dread and anxiety and suspicion that pervaded all of society.[42] As a historical example of this, Prokop and Eigner offered the 'witch mania' of the early modern period, which according to their understanding had been 'unleashed by the authoritative representatives of the Catholic Church.' That episode could only be properly grasped, the authors continued, as resulting from an 'interplay between the sadistic drives of the upper and the lower orders.'[43] A glance at 'the ethnological material'—Hutton Webster's 1948 work *Magic: A Sociological Study*, confirmed for the authors that, 'The belief in sorcery ... nourishes the sentiment of fear, hatred, and vengefulness.'[44] As magic fed hatred, so did hatred provide the 'topsoil' in which superstition grew.[45]

It is noteworthy that both Prokop's and Kruse's interest in witchcraft beliefs—past and present—and their focus on social psychology as a tool for understanding that phenomenon was shared by a number of European social scientists in this moment in time. Already in the mid-1950s, Kruse had become an interlocutor of Kurt Baschwitz, who (like Prokop) had a special interest in the history of the early modern witchcraft trials.[46] A Dutch scholar and University of Amsterdam professor of German–Jewish heritage who had fled Germany in the 1930s and narrowly escaped the Holocaust, Baschwitz published in 1963 the well-received *Witches and*

---

[42] A. Eigner and Otto Prokop, 'Das sechste und siebente Buch Moses,' 246.
[43] Eigner and Prokop, 'Das sechste und siebente Buch Moses,' 247.
[44] Eigner and Prokop, 'Das sechste und siebente Buch Moses,' 251.
[45] Eigner and Prokop, 'Das sechste und siebente Buch Moses,' 247.
[46] Landesarchiv Sch.-Holstein, Abt. 352 Itzehoe, Nr. 419, correspondence between Baschwitz and the county court in northern Itzehoe on January 28, 1955, and February 10, 1955. It was in the course of this correspondence that Baschwitz was apparently introduced to Kruse.

Witch Trials: The History of a Mass Hysteria and its Suppression.[47] Focusing in that work on the powerful and educated jurists who prosecuted witches, Baschwitz argued that the witch hunt had been the result of mass hysteria, mass suggestion, and, as one early reviewer put it, 'psychic illness.'[48] Baschwitz made a specific point in the book of rejecting the notion that there were any comparative lessons to be drawn between the early modern witch hunt and 'instances of mass psychosis in our own time.'[49] He also declined to say which such instances he meant, but 'the Hitler movement' had been a key example in his earlier work.[50]

Baschwitz's witchcraft book appeared amidst a flurry of interest, even alarm, among a group of European social scientists concerned, as Kruse was, about *contemporary* witch beliefs in their own countries. Together, they comprised a loose and broad-based 'anti-superstition' coalition that not only spanned the Cold War divide but much of Europe, too. In fall 1964, Gustav Henningsen of the Dansk Folkemindesamling in Copenhagen, gave a lecture at the VIII. International Congress for Anthropology and Ethnological Sciences in Moscow. Citing Kruse, Henningsen's lecture, 'The Witch Craze Today,' described the 'persistence' of witchcraft beliefs and magical practices in many parts of Europe—in Germany, southern Italy, northwest Spain—even, he noted, 'in parts of highly evolved' Denmark. In order to end 'mass hysteria' about witchcraft, Henningsen recommended a general and all-embracing study of the phenomenon in all European countries and proposed that a congress of expert scholars convene in the year 1966 on the topic 'modern European witch superstition.'[51]

As the years went by, Prokop's anti-occult and anti-superstition oeuvre not only became ever more focused, it seems, on witchcraft beliefs, but its tone also became increasingly strident. In 1976, with West German co-author Wolf Wimmer, he published *Modern Occultism: Parapsychology and Paramedicine*. While the book ostensibly dealt with a great variety of practices, it returned to the subject of witchcraft again and again. A chapter

---

[47] Kurt Baschwitz (1963) *Hexen und Hexenprozesse: Die Geschichte eines Massenwahns und seiner Bekämpfung* (Munich: Ruetten & Loehning).

[48] Richard Schmid (1964) 'Review of Kurt Baschwitz, *Hexen und Hexenprozesse: Die Geschichte eines Massenwahns und seiner Bekämpfung*,' Juristen Zeitung 19:10 (15 May): 336.

[49] Baschwitz, *Hexen und Hexenprozesse*, p. 11.

[50] Baschwitz (1951) *Du und die Masse: Studien zu einer exakten Massenpsychologie* (Leiden: Brill).

[51] Landesarchiv Sch.-Holstein, Abt. 811, Nr. 3597. Johann Kruse, 'Kurzer Bericht'.

on psychokinesis, to give an example, dealt in considerable measure with witchcraft. Toward the end of the book the authors turned, as was Prokop's wont, to the example of the early modern witch hunt, offering a striking description of women young and old being 'roasted alive; executioners' lists often registering hundreds of victims on a single day.' Resistance, the authors wrote, 'seemed senseless ... was simple suicide.'[52]

While a complex admixture of stuff is potentially at work here, there is possibly also a bit more to this description of witch burnings than Prokop's earlier and longstanding objection to medical quackery. By the 1970s, his obsession with superstition had not only come to focus ever more intently on witchcraft but also gained a strong undercurrent of what one might call displaced Holocaust memory. To blame for the witch hunt, Prokop and Wimmer now asserted, were the '*Schreibtischtäter*' ('desk-bound murderers') of superstition, '*educated* witch murderers,' whose 'mass murder' was stopped only by a heroic campaign of judicial enlightenment.[53] For Kruse, too, by some point in the late 1960s, talking about witch beliefs and talking about the persecution of Jews had begun to blur— particularly under Baschwitz's influence, Baumhauer suggests. In an unpublished manuscript in the late 1960s, Kruse called the FRG 'a hell for thousands of innocent and defenceless women and mothers, a paradise for war criminals and Nazis,' and in his archive in Hamburg, he continually collected contemporary folkloric 'evidence of vilifying beliefs about Jews, Gypsies, freemasons, and members of various sects.'[54] More obliquely, but at the same time suggestively, Prokop (and Wimmer) observed in the 1976 book, 'We want no more witch burnings. We still have the smell of corpses in our noses ... Who will guarantee that such epidemics don't repeat themselves.'[55]

• • •

As it happened, the social-psychological take on the early modern witch hunt as an instance of mass hysteria arising from a climate of superstition faded almost as soon as it emerged. The questions around which Baschwitz in particular circled concerning the mass psychology of witchcraft beliefs

---

[52] Otto Prokop and Wolf Wimmer (1976) *Der moderne Okkultismus: Parapsychologie und Paramedizin: Magie und Wissenschaft im 20. Jahrhundert* (Stuttgart: Gustav Fischer), p. 188.

[53] Prokop and Wimmer, *Der moderne Okkultismus*, p. 189. Emphasis in original.

[54] Baumhauer, *Johann Kruse*, pp. 81–82.

[55] Prokop and Wimmer, *Der moderne Okkultismus*, p. 190.

were important, but they rested on an inadequate knowledge of the intricate legal proceedings that preceded and accompanied the witchcraft trials. A sophisticated reading of those trial records over the last 40-plus years has revolutionized the study of the witch hunt.[56] Nonetheless, for present purposes, it is significant that such conclusions influenced and were influenced by a diffuse anti-superstition alliance that spanned the Cold War divide just in the years when that chasm was arguably least bridgeable.

Moreover, when we look at the phenomenon of folk healing and magic and the lessons drawn from it across the German/German divide, the Cold War is only one context of which we must take account. The post-Holocaust context is also important. Given the ideological stance of the East German regime toward complicity in the Holocaust—that is, officially, *West* Germans were to blame for it—this is particularly striking. It suggests that if we want further to develop our understanding of popular Holocaust memory in East Germany—a topic on which relatively little research has been done—we need to look at subtexts, not just texts.[57] For activists like the East German Prokop and the West German Kruse, the post-Holocaust context gave the struggle against *Hexenbanner*, witchcraft beliefs, grimoires, and magical healing practices a life or death quality. By the 1960s, both men saw the witch hunt (and, it seems, in some veiled way, Nazi genocide) as instances of *social* pathology, one that Prokop seemed to suggest by the 1970s had specifically German cultural characteristics. 'Not without reason did the excesses of the witch craze reach their most terrible measure in countries where the German tongue is spoken,' Prokop and Wimmer averred.[58] For Kruse, witchcraft beliefs served as evidence of 'mass psychosis,' and a community capable of producing men like Herr C. and of persecuting and isolating women like Frau N. was one that could fall once again under its spell. Over time, in both men's minds, the witch hunt and the Holocaust became more than tangentially linked events—they turned out to be two instantiations of the same phenomenon.

[56] Personal communication with Erik Midelfort. Historians read Baschwitz, though his influence was short-lived. See H. C. Erik Midelfort (1968) 'Recent Witch Hunting Research, or Where Do We Go from Here?' *Papers of the Bibliographical Society of America*, 62: 373–420; Baschwitz cited on 374.

[57] Recent studies include: Thomas C. Fox (1999) *Stated Memory: East Germany and the Holocaust* (Rochester, NY: Camden House); Mark A. Wolfgram (2006) 'The Holocaust through the Prism of East German Television: Collective Memory and Audience Perceptions,' *Holocaust and Genocide Studies* 20:1 (Spring): 57–79.

[58] Prokop and Wimmer, *Der moderne Okkultismus*, p. 190.

More research needs to be done to draw out and understand fully the implications of all of this. But just to add another layer of complexity to the picture, we might also reflect on the state of German historiography on the Third Reich at the moment in which Prokop and Kruse were tentatively making these arguments. In the late 1960s, key structuralist accounts had begun to emerge, in which the mass murder of Jews was seen as arising not from an anti-Semitic society as such, but out of impersonal struggles between bureaucrats for power in a 'dynamic process of cumulative radicalization.'[59] The historiography of that generation often tended to de-emphasize anti-Semitism—sometimes dismissing it altogether as 'merely a propaganda tool.' Even those accounts that were not structuralist, strictly speaking, nevertheless tended to describe an 'impersonal, bureaucratic, "extermination,"' whose motivations were vaguely attributed to 'some sort of relapse into barbarism.'[60]

To be sure, there had always been other ways of explaining Nazism's development and its crimes, and Kruse's and Prokop's views about 'superstition' and social pathology were hardly unique. Ernst Bloch had urged fellow Marxists to try and grasp the significance of fascism's mobilization of ancient myths, fears, and fantasies already in the 1930s.[61] Nor is the point here that Prokop's and Kruse's fears about persecution arising from superstition and social anxiety laid the basis for *good* arguments about the origins of the Holocaust. They did not—they were essentializing, monocausal, ahistorical and laced with toxic metaphors of primitivity and backwardness and civilization (and in that sense were a kind of throwback to some of the earliest, post-1945 analyses of Nazism). Still, if these men's obsessions were neither unique nor especially illuminating, they nonetheless articulated something pretty interesting. When Kruse—who had studied witchcraft and persecution at the local level *since the 1920s*—wrote about the problem of witch beliefs in Germany, he was thinking of something quite close to the ground: the interactions of daily life between

---

[59] Ian Kershaw (2000) *The Nazi Dictatorship: Problems and Perspectives of Interpretation*, fourth edition (London: Arnold), p. 98.

[60] Dan Stone (2003) *Constructing the Holocaust: A Study in Historiography* (London and Portland, Ore.: Vallentine Mitchell), pp. 59–61, 65. The quote about 'impersonal ... extermination' is from Karl Dietrich Bracher (1978) *The German Dictatorship* (Harmondsworth: Penguin), p. 531.

[61] Ernst Bloch (1991) *The Heritage of Our Times*, trans. Neville and Stephen Plaice (Berkeley and Los Angeles: University of California Press; orig. published as *Erbschaft dieser Zeit* in Switzerland in 1935).

individuals in communities—interactions that produced jealousies, fears, enmities, and desires for revenge. Kruse had known actual people who had been cast out of the fold as 'servants of the devil,' and he knew well that a certain set of beliefs could have noxious effects on day-to-day, face-to-face interactions. Similarly, if also naively, Prokop in the early 1960s had focused on social pathology, the literature of mass psychology, the ethnology of dread, and magic as fuel for communal anxiety and unease.

However essentializing and ahistorical their vision, Prokop's and Kruse's fears about persecution and hatred arising from a combination of superstition and social anxiety and unacknowledged patterns of thought at the grassroots were not seen in the mainstream of academic histories of the Third Reich until the turn toward the practices and thoughts and emotions of daily life, which only *began* in the late 1980s and has really gained steam of late. Prokop's and Kruse's analysis of the problem of witchcraft beliefs was much closer, in this sense—and forms an interesting historical parallel to—Arthur Miller's 1950s vision of puritan colonial America in 'The Crucible.' In Miller's play, which indicted the pervasive mistrust of red-baiting 1950s America, those accused of witchcraft stood in for those accused of communism.[62] In a similar way, in Prokop and Kruse's analysis, the practices of evil-averting magic and of identifying witches were frightening analogues of the Nazi menace, where persecuted women stood in for persecuted European Jews.

It should be noted that Kruse's critiques became much more pointed over time, while Prokop's remained rather more vague, more circumspect, and altogether more generalizing (and even more ahistorical). This may have something to do with his position in East German society, which perhaps made him unlikely to find fault with that society on a question of enormous sensitivity. Nonetheless, that an interpretation—even a disguised one—that laid even a modicum of blame for the Holocaust at the feet of 'ordinary Germans' could issue from the pen of a top East German official like Prokop is still quite surprising, and this in a country that, again, admitted no responsibility for the persecution of Jews or for genocide.

Thomas Schmidt-Lux argues that a scientistic impulse conflating religion and superstition was especially strong in the GDR. He further suggests that this impulse and the pedagogical apparatus marshaled to transmit an ideology of scientism to the East German public might serve as a partial explanation for

---

[62] I thank very much an anonymous reader of this chapter for helping me articulate this connection, which I had struggled to do before receiving her/his comments.

a resolute and enduring East German secularism.[63] The evidence presented in this chapter suggests that when we want to understand that complex of related issues—scientism, religion, and 'superstition'—at least two additional factors should be taken into consideration. First, despite East German rhetoric about the 'West German' problem of clericalism and obscurantism, the GDR's anti-'superstition' pedagogy did not unfold in an exclusively communist context. It was a German–German concern that recruited a diverse collection of activists to its banner. Second—and to be sure, this is harder to get at—we might ask how the tacit, spoken-yet-unspoken ways that the Holocaust—and the fear that ideas and practices labeled 'superstitious' might somehow lurk behind it—may have been bound up in that concern. In some ways, for Prokop and Kruse, fear of superstition almost came to appear like superstition 'itself': mostly unseen, but for that, no less dangerous.

[63] Schmidt-Lux, 'Das helle Licht.'

PART III

The Socialist Life-Cycle: Between Science and Religion

CHAPTER 8

Writing Rituals: The Sources of Socialist Rites of Passage in Hungary, 1958–1970

Heléna Tóth

Social anthropologists or historians of religion usually think about rituals as the result of processes that evolve over a long period of time. Although it is difficult to come up with a universal definition for rituals, there is some consensus that they share certain characteristics: they have a specific structure, they are repeatable and they are embedded in particular cultural contexts that they affirm and in turn also (re)generate.[1] Most often, authorship does not belong to these characteristics. There are several reasons for this. Establishing authorship for rituals seems often impossible and also ultimately irrelevant. As a rule, there are no extant sources that could give insights into the genesis of rituals, and thinking about authorship appears to be the wrong question to ask, because authorship seems to imply an element of intentionality that does not do justice to the complex dynamics of functioning ritual practice. In this context, scholars of the social history of socialism are in a unique position. When most socialist states in Central Europe first started introducing alternatives to religious rites of passage in the 1950s, a group of experts emerged in each state whose job it was to come up with specific scripts for socialist baptisms,

---

[1] For a good overview on various definitions: J. A. M Snoek (2006) 'Defining Rituals' in J. Kreinath, J. Snoek, M. Stausberg et al. (eds) *Theorizing Rituals. Issues, Topics, Approaches, Concepts* (Leiden: Brill), pp. 3–15.

H. Tóth (✉)
Otto-Friedrich-Universität, Bamberg, Germany

© The Editor(s) (if applicable) and The Author(s) 2016
P. Betts, S.A. Smith (eds.), *Science, Religion and Communism in Cold War Europe*, St Antony's, DOI 10.1057/978-1-137-54639-5_8

weddings and funerals. It was their task to cast the otherwise abstract concept of 'scientific atheism' in a form that was meant to effectively compete with church rituals. Leaving the question of intentionality briefly aside for now, the study of the sources these ritual experts[2] left behind opens a window into the complex dynamics behind creating national variations of what later came to be called 'the socialist way of life'.[3]

The structures of what we could term emerging 'ritual infrastructures' differed widely from state to state.[4] In Hungary the Institute of Public Education (Népművelési Intézet) became the center of the campaign to create socialist ritual culture, and the main driving force behind the writing and implementation of new rituals was Zoltán Rácz. The fact that we can identify Rácz personally as a key figure behind shaping this particular segment of the socialist everyday in Hungary does not mean that socialist rites of passage emerged ahistorically (without reference to a particular historical tradition) or without precedent. In fact, Rácz and his colleagues drew on a broad variety of sources to create new traditions.[5] Focusing on the early history of the development of socialist rites of passage in Hungary, this chapter examines the sources Rácz and his colleagues relied on as they were writing socialist rites of passage. The chapter argues that in the creative process of translating the atheist state ideology into everyday practice, in other words, as Rácz was coming up with specific forms of applied atheism in Hungary, he was thinking both in national and transnational terms.[6]

---

[2] On the definition of the term 'ritual experts' M. Weber, H. G. Kippenberg (ed.) (2005) *Wirtschaft und Gesellschaft. Die Wirtschaft und die gesellschaftlichen Ordnungen und Mächte* (Tübingen: Mohr Siebech), pp. 92 ff.

[3] Although the term 'socialist way of life' was first explicitly introduced as part of the state ideology in the Soviet Union in 1976, the concept existed already earlier.

[4] On the Soviet Union see V. Smolkin-Rothrock, dissertation. For East Germany, see F. R. Schulz (2013) *Death in East Germany 1945–1990* (Oxford: Berghahn Books).

[5] Although Zoltán Rácz was without doubt a central figure and driving force behind the development of socialist rituals, he was not the only one who published on the theory and practice of rituals. Other prominent authors were Endre Lakos Udvardi, who wrote about name-giving ceremonies, László Szűts who wrote about festivals for the agricultural collective or Károly Varga who wrote about the rhetoric of socialist funerals.

[6] In the growing scholarship on the cultural history of socialism and the history of everyday life in Hungary in the Kádár era secular rites of passage have so far received almost no attention. One exception is the brief mentioning in T. Valuch (2006) *Hétköznapi élet Kádár János korában* (Budapest: Corvina Kiadó), p. 139.

Socialist rites of passage were officially introduced in Hungary in 1958. This measure was in part a reaction to anti-religious campaigns in the Soviet Union. At the end of the 1950s the modus vivendi that had existed between the Russian Orthodox Church and the Soviet state since the Second World War was gradually dissolved, and atheist propaganda was, again, identified as a key task within the broader context of developing socialist culture.[7] Several states within the Soviet sphere of influence introduced socialist rituals precisely around this time. The ritual experts of individual socialist states were far from mere epigones of the Soviet example, however. In fact, as the example of Zoltán Rácz shall show, the actual form and content of the Soviet anti-religious propaganda was not necessarily the most important frame of reference when it came to developing the specific content of socialist rituals in Hungary. The fact, however, that atheist propaganda became a prioritized topic in the late 1950s should be understood within the broader context of the anti-religious campaigns of the Khrushchev era.

Equally importantly, the introduction of socialist rituals in Hungary played an important role in coming to terms with the consequences of the failed revolution of 1956. In the narrative of events that the Communist regime under János Kádár formulated, rituals (especially funerals) and religion also fulfilled strategic roles. In the immediate aftermath of the fighting, as a new regime set up with Soviet assistance resumed Communist power in Hungary, one of the first questions of both logistics and symbolical politics referred to the burial of the dead. While during the fighting the dead were buried as quickly as possible, the newly formed Communist government wanted to capitalize on the political power of the dead by lavishly staging funerals for the defenders of Communist rule as the martyrs of what the regime came to term a counter-revolution.[8] Although these funerals were clearly not part of a socialist everyday, their organization put the issue of a socialist funerary culture more generally also on the agenda.[9] Moreover, in the emerging Communist narrative of the fall of 1956, the Kádár regime cast the complex dynamics behind the uprising that in

---

[7] For an overview on religion in the Soviet Union, T. A. Chumachenko (2002) *Church and State in Soviet Russia. Russian Orthodoxy from World War II to the Khrushchev Years* (New York and London: M. E. Sharpe).

[8] I. Rév (2005) *Retroactive Justice. Prehistory of Post-Communism* (Stanford: Stanford University Press), pp. 126–128.

[9] On the connection between the revolution of 1956 and the emerging cult of socialist 'martyrs' in its aftermath see P. Apor (2014) *Fabricating Authenticity in Soviet Hungary. The*

reality ran the gamut from reform movements within the Communist Party all the way to resistance against Communist rule itself in a simplified, schematic frame and in terms that made sense within the Communist propaganda. In this emerging official narrative the revolution of 1956 was presented as a conspiracy between fascists of the old order, religious reactionary forces and supporters from abroad.[10] Cardinal József Mindszenty, the Archbishop of Esztergom, and the religiosity of workers in particular played an important role in this explanation.[11] Mindszenty had been imprisoned in the late 1940s. In October 1956 he was freed from prison, but as the newly founded Hungarian Socialist Workers' Party asserted political control with the help of Soviet troops, Mindszenty fled to the American embassy in Budapest and stayed there in exile until 1971.[12] While it was a wild exaggeration to cast him and religion as a main motivation behind the uprising, and as little as the emerging official narrative had to do with the actual reality of the fall of 1956, it nonetheless created a framework for a new discussion about the religion as a liability in a socialist society. In this context statistics about church attendance attained an additional level of significance after 1956. Although not a direct reaction to the revolution of 1956, the party directive of 1958 that called for a socialist ritual culture in Hungary opened a new chapter in the so-called conflict between the secular and religious worldviews.[13]

Secular rituals stood firmly in the nexus of science, religion and Communism. Entrusting the development of secular rituals to the Institute for Public Education, thus an institution that was in charge of a variety of popular education projects, underscored the connection between atheistic education in general and the new rituals. Since the end of the Second World War, the Hungarian Workers' Party (the Communist Party) and, after 1949, the newly founded People's Republic, introduced a series of

Afterlife of the First Hungarian Soviet Republic in the Age of State Socialism (London: Anthem Press), pp. 125–164.

[10] I. Deák (1992) 'Hungary', *American Historical Review*, vol. 97, 1058.

[11] M. Pittaway (2012) *The Workers' State. Industrial Labor and the Making of Socialist Hungary, 1944–1958* (Pittsburgh: Pittsburgh University Press), p. 232.

[12] On the complex role of Cardinal Mindszenty in 1956: J. Fuisz (2004) 'Der Beitrag der Religionsgemeinschaften zum Ungarnaufstand 1956'. *Kirchliche Zeitgeschichte* 17:1, 113–132.

[13] 'Az MSZMP Központi Bizottsága Politikai Bizottságának határozata a vallásos világnézet elleni eszmei harcról, a vallásos tömegek közötti felvilágosító és nevelőmunka feladatairól (22 July 1958)' in H. Vass, Á. Ságvári (eds.) (1973) *A Magyar Szocialista Munkáspárt határozatai és dokumentumai, 1956–1962* (Budapest: Párttörténeti Intézet), p. 237.

measures to drive religion out of the public sphere: Church properties were nationalized, religious periodicals were closed, religious schools were appropriated by the state, monasteries were dissolved and religious education was abolished from the school curriculum.[14] Each of these measures had a direct and broad impact on everyday life. The introduction of socialist rituals, however, politicized the private sphere in a new way.[15] In a manual on socialist rituals and celebrations, published in 1960, Zoltán Rácz, summarized their role as follows:

> the main content of the ... rituals consists of the emotional elements, and, equally importantly, of expressing a collective-ideological message, of the striving to establish and root new traditions, and to encourage connections between the family, the intimate inner circle [sic!] and broader communities, and to nurture the cultural demands related to such connections.[16]

Using a language that sounds as convoluted in the original as it does in translation, Rácz nonetheless touched on several key tasks that socialist rituals were meant to fulfill. They were supposed to tie events that marked turning points in individual biographies into the larger political community, thereby harnessing the emotional charge of rites of passage for political purposes. The socialist baptism, wedding and funeral were meant to be occasions, where the abstract concept of scientific atheism (or the 'materialist worldview' as it was more often called in contemporary discourse) would be filled with positive content as it formed an integral part of an emerging socialist culture.[17]

---

[14] Overview of Church–state relations in Hungary in the socialist era in English: J. Wittenberg (2006) *Crucibles of Political Loyalty. Church Institutions and Electoral Continuity in Hungary* (New York: Cambridge University Press), pp. 80–200. More generally on the legacy of the socialist era: M. Tomka (2005) *Church, State and Society in Eastern Europe* (Washington, DC: The Council for Research in Values and Philosophy).

[15] A good overview on the various areas of the politicization of private life in Hungary: P. Apor (2008) 'A mindennapi élet öröme' in S. Horváth (ed.) *Mindennapok Rákosi és Kádár korában. Új utak a szocialista korszak kutatásában* (Budapest: Nyitott Könyvműhely), pp. 13–50.

[16] Z. Rácz (1960) *Családi ünnepek és szertartások* (Budapest: Népművelési Intézet), p. 7.

[17] The term 'atheism' was applied only rarely in scholarly discourse in Hungary in the 1950s and 1960s. Instead, authors wrote about the 'materialist worldview'. In the 1950s, atheism in general was not a central topic in political propaganda. The 1960 foundation of the journal *Világosság* (*Light*) brought a new dynamic into the discourse about the role of religion in a socialist society. However, the authors of *Világosság* were more interested in the

Looking at the sources Zoltán Rácz and his colleagues relied on for working out the content of new rituals invites reflection on the cultural history of socialism in two ways. First, the institutionalization of new rituals was a long process that lasted into the 1970s and beyond. At the end of this long process the socialist name-giving ceremonies, weddings and funerals became inextricably linked to socialist culture and society, and an entire infrastructure emerged around the organization of family events in a socialist fashion complete with courses for orators and a particular aesthetic and material culture. Although the secular rites never completely replaced religious ones, they became established parts of everyday life. At the beginning of this process, however, in the early 1960s, we find a period of experimentation in which local initiatives played an important role. Looking in more detail at the early history of socialist rituals challenges any linear or simplified understanding of the politicization of the private sphere. Notably, although the impetus for the invention of socialist ritual culture clearly came from the state, when it came to working out the form and content of the rituals and to provide infrastructural support for their development, Zoltán Rácz and the Institute for Public Education were increasingly left to their own devices. This, in turn underscores the importance of examining the sources Rácz and his colleagues left behind.

Second, sources from the first decade of socialist rituals in Hungary also reveal that drawing the contours of socialist culture had an international dimension. States in the Soviet sphere of influence in effect created their own, national versions of socialist culture, including specific rites of passage, but socialist experiments elsewhere often served as orientation points. In the first half of the 1960s the Institute for People's Education sent several delegations to study the development of socialist culture within the Eastern Block: the delegations visited the Soviet Union, Bulgaria, Poland, the German Democratic Republic (GDR) and Czechoslovakia. The purpose of these trips was to understand the various infrastructures that emerged around socialist rituals and to look for inspiration and solutions for concrete problems. From the detailed reports stored at the Archives of the Institute for People's Education,[18] a complex picture emerges of the international dimensions of socialist culture: on the

---

history of religion than in applying atheism in everyday life. E. László (1966) *The Communist Ideology in Hungary: A Handbook for Basic Research* (Dordrecht: D. Reidel), pp. 82–83.

[18] Today: Magyar Művelődési Intézet és Képzőművészeti Lektorátus. I am grateful to the head librarian, Mária Hargitai for her help with my research.

one hand an appreciation of the diversity of socialist cultures that emerged based on essentially the same ideology, but, at the same time, also a strong sense of a 'friendly competition'. As in so many other areas, socialist states looked at each other in terms of hierarchies. In the case of socialist rituals, Rácz was thinking in terms of a hierarchy of secularization but also in terms of a hierarchy of qualities of rituals. In other words, Rácz did not merely rely on statistics to evaluate how far specific states were on the road toward a society that was secular in the socialist sense of the word.

Although in many areas of building a socialist society the Soviet Union functioned as a model, this was not the case when it came to secular rituals.[19] Unlike the Soviet space program, this most obviously visible nexus of science, religion and atheism, where Soviet astronauts smiled on behalf of the entire Eastern Block and declared that they did not meet God in space, the Soviet example proved to be only marginally relevant when it came to filling the atheist space with socialist content in everyday life in Hungary. The 'cosmic enlightenment', to use Victoria Smolkin-Rothrock's term, was highly relevant as a broader, literally cosmic, frame of reference for atheistic propaganda, but the specific development of secular rites of passage in the Soviet Union had little impact on Hungary.[20] In fact, if someone had asked Rácz, which of the socialist 'brother states' offered him the most inspiration for his own project, he would have not hesitated to point to Czechoslovakia.

Piecing Together a Socialist Tradition: Experimenting with Applied Atheism

Rituals and celebrations of various types have served as battle ground between Communist ideology (and institutions) and the Churches since the end of the Second World War in Hungary. New holidays were introduced already prior to the establishment of the People's Republic. The three newly established national holidays delivered a clear ideological

---

[19] On secular rites of passage in the Soviet Union see: C. Lane (1981) *The Rites of Rulers: Ritual in Industrial Society. The Soviet Case* (Cambridge: Cambridge University Press). More recently on death and dying in the Soviet Union: C. Merridale (2001) *Night of Stone: Death and Memory in Twentieth-century Russia* (New York: Viking).

[20] V. Smolkin-Rothrock (2011) 'Cosmic Enlightenment: Scientific Atheism and the Soviet Conquest of Space' in J. T. Andrews, A. A. Siddiqui (eds.) *Into the Cosmos: Space Exploration and Soviet Culture in Post-Stalinist Russia* (Pittsburgh: University of Pittsburgh Press), pp. 159–194.

message: May 1 (Labour Day), April 4th (the liberation of Hungary by the Red Army) and November 7th (the commemoration of the October Revolution in Russia) reshaped the commemorative landscape in Hungary. At the same time, the Churches lost not only their lands in the course of extensive land reforms but also had to give up their prominent position in the calendar as church holidays were gradually abolished and turned into workdays.[21] The Communist takeover of holidays and celebrations accelerated after the establishment of the People's Republic in 1949. Rites of passage, however, remained outside the radius of the rapidly expanding circles of the institutionalization of Communist ideology in everyday life.

While the Churches were forced to give up their schools, their charitable institutions and were thus forced out of many areas of public life after the Second World War, rites of passage were one of the very few areas where they were allowed to retain their leading position. Certainly, party members were not supposed to attend mass, have their children christened or have a religious funeral, yet the numerous occasions when such occurrences were discussed within the Party, show that Communists, even the committed ones, still turned to the Church when they wanted to mark important biographical events. When it came to rites of passage, the Communist state did not offer alternatives to religious rituals. In the late 1940s there were no Communist 'ritual experts'. Although there were local initiatives to come up with secular rites of passage, most Communists were able to demonstrate their political loyalty only by abstaining from religious rites. Refusing religious rites, however, functioned ultimately as a weak signifier of one's worldview or political sympathies. Not every family where the children had no baptism was automatically atheist or necessarily a committed Communist one. The introduction of socialist rituals in 1958, thus, fulfilled a specific role. It is no coincidence that, in his first published manual on rites of passage (1960), Rácz explicitly referred to political loyalty: socialist baptisms, weddings and funerals, he wrote, 'provide an opportunity for the participants to express their commitment to socialist society'.[22]

The party directive of 1958 marked a clear turning point in the development of socialist culture in Hungary. With a concentrated effort to develop socialist rituals, the state now claimed an area of the private lives of

---

[21] A. L. Palkó (2013) 'Sírva vigadtunk. Ünnepek és ünneplési szokások Magyarországon az 1956-os forradalom és szabadságharc leverését követően', *Valóság* 2013:3, 90–111.

[22] Rácz (1960) *Családi ünnepek*, pp. 7–8.

its citizens it has hitherto ignored. This was a new front in the Communist party's efforts to break the influence of the Churches and organized religion in general. By the end of the 1950s, most Church properties had been nationalized, religious education had been abolished from the school curricula, most monasteries were dissolved and, as in other socialist states, the Communist Party tried to co-opt a cadre of priests to officially support the building of socialism.[23] The State Bureau for Church Affairs (Állami Egyházügyi Hivatal), established in 1951, managed a complex system of surveillance that reached behind the church walls, keeping tabs on masses and also the private correspondence of priests who were suspected of criticizing the regime.[24] With the establishment of socialist rites of passage, the state introduced the 'competition between the secular and religious world views' into the private sphere of its citizens in a new way.

Rácz and his colleagues were remarkably quick to draw up a plan for possible rituals. Already in December 1958, thus a mere 4 months after the official publication of the party directive, the Institute for People's Education already presented a detailed position paper on socialist rituals, and by 1960 they published both a manual and also visual material on socialist rituals.[25] The manuals supposed to serve as examples for socialist rituals and they also acted as repositories of the experiences of the first experiments. These early manuals presented an ambitious program: urban and rural festivals, public and private celebrations were all to have a possible socialist form. In other words, getting one's first identity card was to be celebrated in a modern, socialist way just as naming babies, getting married or burying one's loved ones were supposed to be.

---

[23] These clerics were referred to as 'peace priests'. On the peace priest movement see S. P. Ramet (1998) *Nihil Obstat: Religion, Political and Social Change in East-Central Europe* (Duke University Press) 110. For contemporary depictions of the peace priests, for example, the correspondence of Cardinal Mindszenty in A. Somorjai, T. Zinner (2013) *Do Not Forget This Small, Honest Nation. Cardinal Mindszenty to 4 US Presidents and State Secretaries 1956–1971. A Documentary Overview* (S. I.: Xlibris), p. 166.

[24] The last wave of clerical arrests took place in the spring of 1960, when a group of priests and students of theology were arrested. The official charge against them was 'agitation against the People's Republic', but in reality their crime seems to have been their popularity among the youth. I. Mészáros (1994) *Kimaradt tananyag. A diktatúra és az egyház, 1957–1975* (Budapest: Márton Áron Kiadó).

[25] First position paper on secular rituals with no author but most likely Zoltán Rácz: 'Javaslat a családi és társadalmi ünnepek, szertartások bevezetésére és elterjesztésére, Budapest, 22 December 1958' in Népművelési Intézet [Institute of Folk Culture] (hereafter NI) 807.

Working out the actual form and content of the socialist alternatives to church rites was a longer process that involved a remarkable amount of experimentation. Already in the first position papers on the subject, the authors of rituals identified at least two possible sources for inspiration. Appropriately for a state that presented itself as a state of workers and peasants, the authors of rituals looked to the urban culture of the workers (including the historical legacy of the workers' movement) and the village for possible content for the new rites. For example, manuals often referred to the 'red funerals' of the nineteenth century, and in particular, to the funeral of Leó Frankel, a key figure in the history of the Hungarian Labour Movement, as a precedent for the introduction of socialist funerals.[26] At the same time, already the first manuals on socialist rites idealized the village as a place, where the 'living traditions' revealed the supposedly true character of rituals. Rácz and his colleagues argued: Rural traditions delivered strong proof that rituals ultimately satisfied a basic human need to mark important events in a biography and were not necessarily religious. Rácz stressed repeatedly that those local traditions that were not superstitious (superstition being a synonym in this discourse for religion) should be preserved and drawn upon for developing socialist culture. Drawing on these traditions had the additional benefit, Rácz pointed out, that it made socialist rites of passage look more familiar.[27]

This line of argumentation about the sources of socialist rituals reflected Communist ideology's understanding of the role of religion and did not do justice to the complexities of actual rural and urban ritual practice. For example, no word was spent on the fact that the urban working class was not necessarily automatically atheistic. Not only did certain professions have an extremely strong religious culture (e.g., the miners) but, especially during the revolution of 1956 it became apparent, how quickly the Churches were able to regain their public appeal, also in the cities.[28] When it came to the ideologically colored depictions of village culture in ritual manuals, the authors downplayed the importance

---

[26] Z. Rácz (1964) *Családi és társadalmi ünnepek. Kézikönyv a tanácsok és a társadalmi szervezetek aktivistái részére* (Budapest: Kossuth Könyvkiadó) p. 9; Z. Rácz, K. Varga (1976) *A polgári gyászszertartásról* (Budapest: Népművelési Propaganda Iroda), p. 15.

[27] 'We have to take care that the new ritual forms should not look too foreign. The participants, especially on the countryside, should recognize as many elements of the rituals as possible as their own, as familiar rites from their own region'. Z. Rácz (1960) *Családi ünnepek*, p. 4.

[28] Pittaway (2012) *Workers' State*, p. 232.

of Churches in the countryside and also did not countenance the possibility that what they presented as a dichotomy (the contrast between remnants of heathen traditions and Christianity) often existed as a sort of synergy in the villages. The idealized depictions of both the working class and the peasantry fitted into broader systems of legitimization on which the Hungarian Socialist Workers' Party relied. First, Rácz and his colleagues tried to establish a historical depth to the socialist rituals by referring to the nineteenth- and early twentieth-century legacy of the Hungarian Labour Movement. This fitted into the regime's overall efforts to use history as a source of legitimacy. Moreover, by keeping the supposedly secular village traditions alive as an alternative to Church rituals, the socialist state cast itself as the protector of the peasantry, in this case in a cultural sense, against the supposed religious impositions of the Church.

The fact that the Party tried to cast itself as the protector of village traditions was particularly significant in the fall of 1958. Just a couple of weeks prior to the party directive on socialist rituals, the Kádár regime initiated a concentrated effort to collectivize agriculture.[29] This was the second wave of collectivization in Hungary. After the establishment of the socialist state in 1949 there were already efforts to establish agricultural collectives. These efforts were in part successful, but a large number of collectives dissolved in the following years. The revolution of 1956 further weakened the position of collectives. Starting in the fall of 1958, the Hungarian government decided to complete the collectivisation of agriculture: the execution of this plan involved the whole spectrum of 'persuasion' ranging from propaganda and 'education' on the countryside to outright violence. 'People's educators' visited the villages, held presentations, published material extolling the supposed virtues of agricultural collectives and when their methods failed the billy club followed. In fact, with quite a bit of black humor, villagers called billy clubs 'condensed people's educators'.[30] In this context, the timing of the introduction of socialist rituals seems particularly significant. The new rituals counted among the ways to undermine the remaining vestiges of the Church everywhere,

---

[29] For general new secondary literature on collectivization in Eastern Europe from a comparative perspective: I. Constantin, A. Bauerkämper (eds.) (2014) *The Collectivization of Agriculture in Communist Eastern Europe. Comparison and Entanglements* (Budapest: Central European University Press).

[30] J. Ö. Kovács (2012) *A paraszti társadalom felszámolása a kommunista diktatúrában. A vidéki Magyarország politikai társadalomtörténete 1945–1965* (Budapest: Korall), p. 366.

including the countryside.[31] With a bit of exaggeration one could argue: Just as the agriculture was being collectivized, also a significant area of private life was being drawn into the socialist collective.

When it came to specific suggestions for the actual content of rituals, however, we find few concrete references to either nineteenth-century labour traditions or to the supposedly pure village customs in the manuals. The proposals for ritual templates contained copious references to contemporary literature (especially poetry) and also contained several references to local initiatives. The depiction of local initiatives served, however, in the vast majority of cases, as a negative example. While Rácz often cited these initiatives as proof for a general demand for a socialist ritual culture, he also considered the local celebrations often tasteless. One returning element of his critique was that there were too many religious references in the local initiatives. In one village the head of a local agricultural collective led a socialist name-giving ceremony and placed his hand on the baby ('as if in blessing'), in another people referred to the socialist orator as 'a socialist priest'.[32]

In fact, in the first couple of years after their introduction, the form of socialist rituals remained quite amorphous. For example, in the propaganda material related to name-givings, a ceremony that was supposed to eventually replace baptism, the manuals and the visual material depicted a whole range of possible scenarios. A short film clip, presented in a news reel in 1959, depicted a name-giving ceremony in an affluent, urban apartment. The circle of guests was small: Only two babies and their immediate families took part in the name-giving ceremony, which was officiated by the local registrar.[33] The film showed expensive gifts and presented the name-giving ceremony as something that would appeal to a cultivated

---

[31] The agricultural collective even received its own celebration in the canon of socialist rituals, the so-called 'closing the financial year celebrations', which combined elements of the harvest festivals with a demonstration of endorsement of the agricultural collective. Based on the suggestions manuals made for this celebration, the event also supposed to include the public criticism of those members of the collective who did not fulfill their quotas or duties. Thus, the occasion was not only a celebration of the achievements of the collective but also fulfilled a normative, regulatory function. Here, too, Rácz laid a heavy emphasis on building on local traditions 'provided they are not antiquated or ridiculous'. L. Szűts, Z. Rácz (1965) *Útmutató ttermelőszövetkezeti zárszámadási ünnepség rendezéséhez* (Budapest: Népmüvelési Intézet), p. 10.

[32] Z. Rácz (1964) *Családi és társadalmi ünnepek*, p. 15.

[33] *Magyar Film Híradó*, 41/1959 October. (The film clip was presented as part of a newsreel).

urban couple. A propaganda slideshow published only a couple of months later, however, conceptualized the name-giving ceremony as truly a mass event, taking place at a so-called house of culture, with the entire village (or factory) present. In this second example, several children had their name-giving ceremony at the same time and besides the families, colleagues and friends, also pioneers took part at the event.[34] One explanation for the diversity of possible ritual scenarios was that Rácz and his colleagues wanted to reach as broad a section of the population as possible. At the same time, however, the authors of the new rituals themselves were not quite sure just how 'collective' such events should or could be, so that they would fulfill their ideological purpose and still be accepted broadly. In these early attempts at creating a socialist ritual culture experimentation played a key role, as did a detailed study of already existing socialist ritual practice elsewhere within the Eastern Bloc.

International Inspirations for a National Socialist Culture: Czechoslovakia as *Primus Inter Pares*

Drawing on the history of the Labour Movement, an idealized version of the village or contemporary Hungarian authors for inspiration for socialist rituals fitted into broader patterns of creating legitimacy for the Kádár regime. Although not apparent on the surface, this process also had an international dimension. In the early 1960s, Zoltán Rácz and his colleagues undertook several trips within the Eastern Bloc to collect information on the development of socialist culture elsewhere. When Rácz and his colleagues visited the socialist 'brother states', they were not merely being cultural diplomats paying courtesy visits. They were looking for solutions to specific problems and for models that they could apply in their own work. Rácz was particularly interested in questions regarding infrastructure. Although in the fall of 1958 socialist culture seemed a priority for the Party, Rácz and his colleagues soon had to realize that they in fact received little support from either the Central Committee or from the local councils when it came to the implementation of concrete measures. Although a government directive called for the establishment of 'committees for socialist rituals' in local councils already in 1960, this remained by and large without much effect. This was not only Rácz's personal experience. The fact

---

[34] Propaganda slide film *Családi és társadalmi ünnepek* (1960) (Budapest: Népművelési Intézet).

that 2 years later a similar directive was passed indicates that the councils were not too eager to accept their new responsibilities.[35] Thus, when Rácz and his colleagues visited Bulgaria, the Soviet Union, East Germany or Czechoslovakia, they were interested in learning about what socialist rituals looked like and also about how to institutionalize them effectively.

The extant sources suggest that ritual specialists from Hungary found the most practical inspiration in Czechoslovakia. The analysis of the extensive study reports of these trips is revealing in some ways but not others. Within the context of this chapter, the purpose of looking at Rácz's notes is not to create a picture of actual ritual practice in Czechoslovakia. Neither do these sources suffice for a systematic comparison of the development of socialist rites of passage in these two states. At the same time, the sources give us insights into the ways in which international comparisons helped Rácz to develop his own view of socialist rites of passage in Hungary and underscore the significance of this broader, international perspective in the process of Rácz becoming a ritual expert.

Czechoslovakia belonged to one of the first states Rácz and his colleagues visited. Already in April 1961 Rácz traveled to Prague to study the work of the so-called citizen committees. Founded in 1953, these committees were local organizations that promoted socialist rites of passage in Czechoslovakia. Starting in 1957 the ministry of interior and the ministry of culture oversaw the work of the committees and also coordinated their work with the local councils. These might sound like dry details on their own but for Rácz and his colleagues this constellation represented a best case scenario: when Rácz looked across the borders, he saw a well-functioning system, in which local initiatives and central regulation seemed to work in harmony, and most importantly this cooperation seemed to create a solid basis of legitimacy for the new rituals. At the time of his visit, Rácz himself was working on a position paper for the ministry of culture, proposing a similar system.

---

[35] Local councils in 1960 had other worries, notably the collectivization of agriculture. While the party directives for collectivization were always formulated in concrete and often militaristic terms, including close deadlines and strict quotas, the language of official directives was significantly softer when it came to formulating goals and action plans for the development of socialist culture. Collectivization was considered a process that should follow a strict timeline and be completed within the foreseeable future, party directives from the very beginning considered the introduction and spread of socialist rituals a long-term process, possibly taking several years. While the language of collectivisation was ridden with military terms, the language of introducing socialist rituals remained that of pedagogy.

The records of Rácz's first visit to Czechoslovakia reveal a complex picture of the international dimensions of socialist rituals.[36] In Prague, Rácz first visited the Faculty of Ethnology at the Charles University, where he met two key figures of Czech ethnology of the time, Antonín Robek and Otokar Nahodil. Robek worked on the role of religious rituals and in particular on their effect on women. For decades, he was a key figure at the Faculty of Ethnology and oversaw several research projects on various aspects of secularization. Nahodil participated in a larger research project that catalogued and systematized the atheistic aspects of Czech and Slovak fairy tales. Nahodil's work on the history of Czechoslovak ethnology was translated into Hungarian already in the mid-1950s and it is possible that Rácz was familiar with it even before his trip.[37] More importantly, both Robek and Nahodil were working on themes in which Rácz himself was interested. In his manuals Rácz often discussed what he considered the controversial role of women in ritual culture: women being liabilities because of their susceptibility to religion but also potential allies because they could be persuaded through a convincing aesthetic of secular rituals.[38] Research to prove the essential secularism of rural culture was also something that came close to the way Rácz portrayed village traditions in his manuals. Although not an academic, Rácz probably would have liked to see himself in that role.[39] Meeting the Czechoslovak doyens of academia who so closely supported socialist rituals must have thus struck a special cord with Rácz, who, by 1961, felt increasingly the lack of practical political support for his project in Hungary.[40]

After Prague, Rácz visited a series of cities in Czechoslovakia, where he observed socialist rituals in practice. Most importantly, his hosts took him to the city of Tábor, where he met Robert Mixa, who, according to socialist

[36] Z. Rácz (1961) 'Jelentés a csehszlovákiai tanulmányutamról, 5–18 April 1961', NI 2110, pp. 1–21.

[37] The translation: Otakar Nahodil (1955) *A csehszlovák néprajz tíz éve 1945–1955* (Budapest: Adakémiai Kiadó).

[38] Z. Rácz (1966) *Családi és társadalmi ünnepek*, p. 66.

[39] In 1974 Rácz even published a history of the Jesuit order, in which he worked through the classical themes of anti-clericalism ranging from the relationship between big business and organized religion to the role of Jesuits at the Second Vatican Council. Z. Rácz (1974) *Jezsuiták tegnap és ma* (Budapest: Kossuth Könyvkiadó).

[40] Rácz notes that he started a regular correspondence with Antonín Robek, but efforts to trace down this correspondence have been unsuccessful so far. In a report on a further study trip to Czechoslovakia that Rácz undertook in the fall of 1967, he wrote about Robek as a yearlong correspondent. NI 4838, p. 27.

lore was the initiator of socialist rites of passage in Czechoslovakia. The city of Tábor in itself was a symbolic place as the center of the Hussite movement and thereby of the reformation in Bohemia in the fifteenth century. Even more importantly for Rácz, it was a group of activists around Mixa, who, starting in the early 1950s, first started to organize socialist rites of passage. Unlike the probably well-meant efforts of Hungarian council members who held name-giving ceremonies that rather embarrassed Rácz, he considered the local initiatives in Czechoslovakia more in line with what he imagined a proper secular aesthetic would look like. For example, Rácz repeatedly remarked on the subdued decoration of the rooms, which was something that he was trying to implement in Hungary as well. Moreover, the engagement of local councils also left a deep impression on Rácz. Visiting the name-giving ceremonies and socialist weddings in Tábor and České Budějovice only strengthened his conviction that the only way to make socialist rituals truly part of everyday life was to engage the local councils. This was a problem with which Rácz grappled through his entire career in Hungary.

If Rácz was excited and also humbled in equal measure by what he saw in Prague and Tábor, the second half of his trip through the Slovakian part of Czechoslovakia, restored his confidence that the Hungarian results after only 3 years of introducing socialist rites of passage were not as weak as he first thought. In the Slovak areas, the influence of the Catholic Church remained comparatively strong, as local party officials explained Rácz.[41] This made the most difference for the socialist funerals: of all rites of passage it was the funeral where the religious rites seemed to remain the strongest. This was a familiar constellation of problems for Rácz: He also considered funerals to be the most challenging ritual for socialists to compete against the religious traditions.[42] Even leaving aside the fact that it was difficult to compete on the emotional level (it was extremely difficult to give comfort to grievers based on the tenets of historical materialism), socialist funerary orators were usually no match for the priests and pastors professionally trained in rhetoric and with extensive ritual experience behind them. In comparison to the Slovak parts of Czechoslovakia,

[41] Z. Rácz (1961) 'Jelentés a csehszlovákiai tanulmányutamról, 5–18 April 1961', NI 2110, pp. 12–16.

[42] It is no coincidence that Rácz wrote an entire separate manual for funerals alone and that he developed and oversaw a training program for socialist funerary orators. R. Zoltán, Z. Rácz, K. Varga (1976) *A polgári gyászszertartásról* (Budapest: Népművelési Propaganda Iroda).

however, Rácz concluded, the Hungarian results of the last 3 years gave reason for optimism for the future.

Moreover, while in Prague and Tábor Rácz felt like a novice, his experience with Hungarian-language secular propaganda made him something of an expert vis-à-vis the representatives of the Hungarian-speaking workers' organizations in Czechoslovakia, most of whom lived in the Slovakian part of the state. Three years after Rácz's visit, in 1964, these organizations invited him to a conference on 'Atheistic Propaganda in Hungarian Language' as an expert lecturer.[43] By that point Rácz had been working on this subject for over half a decade and had established himself in Hungary as the sole most important author on the subject. Unfortunately, Rácz did not keep detailed statistics of his 1961 (or 1964) trip, but the percentage of secular rituals he noted during his third trip in 1967 show that numerically speaking, Hungary kept its middle position in terms of secular rituals between the more secularized Czech territories and the Slovak parts where the influence of the Catholic Church remained strong (Table 8.1).

The mixed experiences of Rácz in Czechoslovakia in 1961 (and in 1964) reveal a characteristic aspect of the international dimensions of the creation of national socialist cultures. Although there was no textbook description what socialist culture should exactly look like, its constitutive elements seemed to be measurable and hence comparable. Rácz constantly thought in terms of hierarchies. At times, these hierarchies were based on statistics (the percentage of religious and secular rites of passage), other times on certain qualities. Rácz's reports on his visits abroad in general suggest that the contrasts he observed between the emerging ritual practice in Hungary and in the neighbor states helped him to focus on specific qualities that he came to consider the strengths of Hungarian socialist rites of passage. With each trip, the contrasts came to play an increasingly important role in his reports.

In 1967 Rácz visited Czechoslovakia for the third and last time.[44] If during his first visit he was essentially conducting basic research and was looking for strategies he could adapt in Hungary, during his third study trip he was observing and studying socialist rituals as an expert himself. By

---

[43] Report on Rácz's participation at the 'Conference of Atheists in Komarno', Z. Rácz (1964) 'A Csehszlovákiában tett tanulmányútról, jelentés', 23–28 March 1964, NI 8067, pp. 1–6.

[44] Two of Rácz's colleagues at the Institute for Public Education, József Fodor and Sándor Heleszta visited Czechoslovakia in May 1966. The main focus of their visit was to discuss 'the cultural interests of rural youth'. NI 4403.

Table 8.1 Rites of passage in Hungary and Czechoslovakia in 1967

1967	Baptisms/Name-giving ceremonies			Weddings			Funerals		
	Church baptism (%)	Name-giving ceremony (%)	No ceremony at all (%)	Church weddings (%)	Secular/socialist weddings (%)	Registration at the civil registry's office (%)	Church funeral (%)	Socialist funerals (%)	Funerals without service (%)
Hungary	80.93	8.34	10.73	51.65	42.26	6.09	82.44	8.28	9.28
Czechoslovakia (Bohemia/Moravia)	41	59	–	26	74	–	68	32	–
Czechoslovakia (Slovakia)	91	9	–	76	34	–	89	11	–

Source: E. Köpeczi Bócz (2004) *Az Állami Egyházügyi Hivatal tevékenysége* (Budapest: Akadémiai Kiadó), pp. 126–129; Z. Rácz (1967) 'Jelentés csehszlovákiai kiküldetésemről, 1967 október 2–11', NI 4838, p. 18

this point, Rácz has not only written a series of position papers on socialist rituals to various ministries in Hungary, he has also published two manuals and various pieces of propaganda material. The statistics showed that socialist rituals were spreading, even if not as quickly as Rácz would have liked to (Table 8.2). Albeit no mass movement emerged around socialist rituals in Hungary comparable to that in Czechoslovakia, slowly, they were becoming part of everyday life.

By this point it has also become clear to Rácz that several aspects of the Czechoslovak example could not be applied in Hungary. For example, Rácz noted with admiration that almost every child seemed to have a socialist name-giving ceremony in Czechoslovakia.[45] At the same time, he now found the ceremonies themselves too impersonal and cold; he criticized that the collective ceremonies took place often in a rather small room at the civil registry's office and that they included too many children, which made it impossible for families to invite the relatives to the name-giving. In Hungary, too, name-giving ceremonies were supposed to be collective celebrations. As we have seen above, both the news footage of the first name-giving ceremony in Budapest and also the propaganda slide show on socialist rituals portrayed collective events. Yet in both cases, both in the sophisticated urban apartment and in the rural 'house of culture', a particular emphasis was laid on making sure that the extended families were also present. And while in reality, unlike the ceremonies presented in the propaganda material, name-givings in Hungary took place neither in the intimate setting of the home nor did it become usual for the entire village to congregate to celebrate name-givings, efforts to invite the broader community (extended family, colleagues, etc.) were nonetheless there. Name-givings in Hungary were thus usually collective ceremonies for several children at the same time but never on the scale and, perhaps consequentially, never with the comprehension as in Czechoslovakia. Even during the 1980s, at the time when this ritual was most widespread in Hungary, only a little more than 30 % of children participated (Table 8.2).

When looking at socialist funerals in Czechoslovakia, Rácz reached a similar conclusion: The Czechoslovak statistics were much stronger than those in Hungary (although the percentage of religious funerals in Slovakia was still high (Table 8.1)), but socialist funerary culture in Czechoslovakia had several characteristics that could not be adopted in Hungary. Most

---

[45] Z. Rácz (1967) 'Jelentés csehszlovákiai kiküldetésemről, 2–11 October 1967', NI 4838, pp. 8–9.

Table 8.2 Rites of passage in Hungary, 1963–1987[a]

	Baptisms/Name-giving ceremonies			Weddings			Funerals		
	Church baptism (%)	Name-giving ceremony (%)	No ceremony at all (%)	Church weddings (%)	Secular/socialist weddings (%)	Registration at the civil registry's office (%)	Church funeral (%)	Socialist funerals (%)	Funerals without service (%)
1963	78.23	6.17		55.65	26.56	11.10	81.33	5.79	11.25
1964	80.88	6.72	12.40	54.85	34.05		82.97	5.78	10.22
1965	82.77	6.70	10.53	55.01	37.18	7.81	83.57	6.21	6.13
1966	82.34	7.60	10.06	54.17	41.93	3.90	86.50	7.37	9.28
1967	80.93	8.34	10.73	51.65	42.26	6.09	82.44	8.28	7.98
1968	80.90	9.82	9.28	51.17	46.06	2.77	82.62	9.40	7.96
1969	78.97	11.25	9.78	47.28	50.60	2.12	82.21	9.83	6.12
1970	79.33	12.55	8.12	46.52	51.02	2.46	83.81	10.07	7.51
1980	63.92	34.78	1.30	28.58	71.42	–	78.09	14.40	5.94
1987	68.24	31.76	–	29.99	70.01	–	77.89	16.17	

[a] 1987 was the last year when statistics were collected about the distribution of rites of passage

The table contains aggregate statistics. The numbers looked, of course, quite different from region to region and there were also differences between cities and the countryside. Nonetheless, even the aggregates show a general trend, especially the fact that by the late 1980s the numbers of secular rituals plateaued.

As comparison: In Czechoslovakia 75 % of funerals were religious in 1955. By 1988, this number was reduced to 39 %. At this point, the percentage of religious funerals in Hungary was still in the high 70s.

Source: E. Köpeczi Bócz (2004) *Az Állami Egyházügyi Hivatal tevékenysége* (Budapest: Akadémiai Kiadó), pp. 126–129; V. Babička (2005) 'Vývoj katolické religiozity v českých zemích v letech 1949–1989' [Development of Roman Catholic religiosity in the Czech lands 1949–1989], *Sborník archivních prací*, 55, 478–480

importantly, socialist funerals in Czechoslovakia were often cremations.[46] Since the nineteenth century, the cremationist movement in Central Europe developed in close association with anti-clericalism. Cremation was considered a modern, enlightened form of funeral which also fitted into a secular world view. Czechoslovak creators of socialist traditions built upon both the cultural legacy of this movement and also on the infrastructure of a network of crematoria built before the communists came to power.[47] Whatever his personal views on cremation may have been, it was clear to Rácz that this was not something that could be forced in Hungary. While in 1961, after his first trip to Czechoslovakia, Rácz suggested to the Committee on Atheistic Propaganda of the Party Committee in Budapest with direct reference to the Czech case that cremations should be popularized in Hungary, 6 years later he seemed less certain in the practicability of his proposal.[48] Not only was there no comparably strong tradition of the cremationist movement in Hungary, the infrastructure was not there either. Until 1962 only a single crematorium existed, in the city of Debrecen, built in the interwar period, which did not start operation until 1951.[49] In fact, cremation in Hungary remained such a contentious issue that the cemetery law of 1970 had a separate clause stating that if the family of the deceased could not unanimously agree on a form of funeral, cremation should not be allowed.[50]

---

[46] On the development of the cremationist movement in the Czech context see the excellent essay of Z. R. Nešpor (2013) 'Cremation Movement in the Czech Lands in the 20th Century' in E. Venbrux, T. Quartier, C. Venhorst, B. Mathijssen (eds.) *Changing European Death Ways* (Berlin: LIT Verlag), pp. 119–140. Interestingly, East German experts on funerary culture looked at Czechoslovakia with admiration precisely for the fact that the percentage of cremations was so high. The members of the Institute for Municipal Economy, the institution that since 1962 was responsible for the management of cemeteries and for managing socialist rituals, studied the distribution of crematoria in Czechoslovakia in detail. Source: Museum für Sepulchralkultur, Kassel.

[47] Even if they found it problematic. The history of the freethinker movement in Czechoslovakia could not be told in simplified terms of class struggle. From the perspective of the Communist Party in the 1960s, it was freighted with the legacy of an essentially bourgeois movement, even if a left-leaning one.

[48] Z. Rácz (1961b) 'A társadalmi szertartások jelenlegi helyzete Budapesten' dated 19 July 1961, NI 2033(z), p. 6.

[49] Z. Xantus (1979) *Hatvan év a kegyelet szolgálatában* (Budapest: Fővárosi Temetkezési Vállalat), pp. 57–58, 89–92.

[50] § 11 (5) on cremation in the cemetery law of 1970. '10/1970. (IV. 17.) ÉVM-EüM számú együttes rendelet a temetőkről és a temetkezési tevékenységről' in *Hatályos jogszabályok gyűjteménye 1945–1987* (1988), vol. 2 (Budapest: Közgazdasági és Jogi Könyvkiadó), p. 209.

Ultimately, Rácz closed his last visit to Czechoslovakia with the conclusion: 'our Czechoslovak comrades were unable to say or show me anything new regarding the ... fight against the religious world view'.[51] This included also the socialist rituals. Between his first and last visit to Czechoslovakia, Rácz developed a significant amount of self-confidence as a 'ritual expert'. In addition, Rácz's trips to other states in the Eastern Block strengthened him in the belief that Hungary occupied a strong position when it came to comparing various forms of socialist secularism. As noted above, Czechoslovakia offered Rácz both inspiration (secular rites of passage as a mass movement) and comfort in the beginning of the 1960s. Other trips generated comparable impressions. For example, reflecting on a trip to East Germany in 1966, Rácz emphasized that the GDR was 'much further along' when it came to ceremonies involving the youth.[52] The East German socialist equivalent of confirmation, the so-called Jugendweihe, was, for example, a ceremony that Rácz found impressive and intriguing. At the same time, while East Germany seemed to be doing really well when it came to mass celebrations, according to Rácz, he observed that they have not found an effective way to develop socialist rites of passage more broadly. For example, although there were state funerals, Rácz was surprised to find that almost no attention was paid to introducing secular funerals for ordinary citizens.[53] After visiting a name-giving ceremony in a small town near the German–Polish border, Brieskow-Finkenheerd, Rácz concluded: This was an 'extremely schematic, bureaucratic formality with a speech lasting almost for an hour, in which everything was discussed starting from the fulfillment of the production plans to the economic-technological basis of Communism, but nothing was said about the family and the individual'.[54] At the end of his trip, Rácz was particularly flattered by a proposal that his East German colleagues

---

[51] Z. Rácz (1967) 'Jelentés csehszlovákiai kiküldetésemről, 2–11 October 1967', NI 4838, p. 31.

[52] Z. Rácz (1966) 'Jelentés tanulmányutamról az NDK-ban, 1–20 March 1966', NI 4403, p. 21.

[53] On the relationship between state funerals and ordinary funerals in the GDR see J. Redlin (2009) *Säkulare Totenrituale: Totenverehrung, Staatsbegräbnis und private Bestattung in der DDR* (Münster: Waxmann).

[54] Z. Rácz (1966) 'Jelentés tanulmányutamról az NDK-ban, 1–20 March 1966', NI 4403, p. 21.

wanted to translate his 1964 manual on secular rituals into German.[55] In comparison to his East German colleagues, Rácz was indeed almost a pioneer. The first East German manuals on secular rites of passage were published only at the beginning of the 1970s.[56]

By the late 1960s, one could argue that a system of socialist rituals was established in Hungary. By this point both the main strengths and weaknesses of the new traditions were in place. The problems that Rácz grappled with in the early 1960s, such as the lack of engagement of the local councils, the poor quality of funeral orations and the lackluster cooperation of cemetery administrators remained endemic until the fall of the Communist regime.

Conclusion

A broad variety of socialist ritual practices emerged in states within Central Europe despite the fact that they shared an essentially secular state ideology. For example, while in Poland the Catholic Church retained a remarkably strong control over rites of passage, in East Germany or in Czechoslovakia the socialist state was able to claim this area of private life as its own to some degree. Long-term historical developments serve at least in part as an explanation. The particular brand of socialist atheism was only one chapter in a much longer history of complex relations between Church and state, and in particular in the cultural history of secular alternatives to religious rituals. For example, the fact that the free thinker movement was popular and strong in Czechoslovakia already in the nineteenth century or that the cremationist movement was widespread created a basis on which socialist writers of rituals could build. This was a complicated legacy, however: After all, the cremationists of the previous century were far from a

---

[55] Efforts to trace down this translation have remained without success so far. If such a translation exists, it was never published.

[56] The Institute for Municipal Economy in Dresden and the Central House of Culture in Leipzig were the two organizations that published manuals on secular rituals. More broadly on the praxis of East German secular funerals see F. R. Schulz, *Death in East Germany*. In addition, on the history of funerals in the Soviet occupation zone immediately after the Second World War, see M. Black (2010) *Death in Berlin: From Weimar to Divided Germany* (New York: Cambridge University Press), pp. 187–229. A detailed analysis of East German manuals on secular funerals: H. Tóth (2013) 'Shades of Grey: Sepulchral Culture in East Germany', in E. Venbrux, T. Quartier, C. Venhorst, B. Mathijssen (eds.) *Changing European Death Ways* (Berlin: Lit Verlag), pp. 141–163.

Communist revolutionary avant garde. Nonetheless, the fact that certain secular cultural practices have already gained acceptance in Czech society prior to the Communist takeover shaped significantly the extent and parameters of socialist rites of passage in Czechoslovakia. Zoltán Rácz and his colleagues in Hungary had no similar basis to build upon.

Rácz and his colleagues counted among a small group of people in Central Europe whose job it was to draw the contours of socialist rituals. The sources they left behind allow a glimpse into the contingencies and complexities of the early history of cultural practices that later became inextricably linked with socialist everyday. They reveal that behind the façade of consistent atheist propaganda, everyday socialist culture rarely received the institutional support the official rhetoric suggested. Although socialist states invested in applied atheism in everyday life in particular historical constellations, these moments never lasted long. As Rácz quickly realized, as the collectivization of the agriculture was finished by 1961 and as the relationship between the socialist state and the Catholic Church became more relaxed in the aftermath of the Second Vatican Council, socialist rites of passage also received much less official attention. Most importantly, Rácz never received the support of local councils he hoped.

Nonetheless, despite lacking effective political support, Rácz grew gradually into an expert of socialist ritual culture both in Hungary and abroad. In the 1970s the State Bureau of Church Affairs, the office that officially acted as the liaison between Churches and states in Hungary (including the surveillance of the Churches), invited him to give a series of workshops on socialist rituals as effective ways to diminish the influence of religion. His colleagues in East Germany asked for his manuals to be translated into German and his colleagues in the Slovak part of Czechoslovakia, where the Hungarian-speaking minority lived, invited him to a conference as an expert in atheistic propaganda in Hungarian.

Rácz's example shows an international perspective played an important role in developing a specifically national version of socialist rituals, especially in the early stage of their introduction. While these rituals were meant to be embedded in local traditions (inasmuch they were not 'superstitious or ridiculous' to use Rácz terms) international comparisons helped Rácz both to draw specific contours and also to gradually gain confidence in his own project. The window of openness to integrate elements of experience from other states within the Eastern Bloc closed by the late 1960s. Roughly a decade after the first experiments with socialist rites of passage in Hungary, Rácz, even if not entirely satisfied with the results, was confident that he knew both the possible extent and the very real limitations of creating and establishing a socialist ritual culture in Hungary. At the same time, the general institutional history of rituals together with the fact that this international openness was there at all destabilizes any simplified top-down explanation about the politicization of private lives in state socialism.

Actual ritual practice rarely corresponded to templates presented in manuals. The history of socialist rites of passage amounts nonetheless to a special chapter in the cultural history of rituals: to a chapter, where the agency of individuals can be studied and analyzed.

CHAPTER 9

In Search of Rationality and Objectivity: Origins and Development of East German Thanatology

Felix Robin Schulz

In 1959 a thoroughly rational man, Herman Feifel, then working as a clinical psychologist for the US Veterans Administration in California, noted that '[e]ven after looking hard in the literature, it is surprising how slim is the systematic knowledge about death'.[1] This insight was both the motivation and the result of putting together a pioneering book, bringing together complex understandings of death from various academic fields. The volume comprised contributions from a large range of disciplines: anthropology, art, literature, medicine, philosophy, physiology, psychoanalysis, psychiatry and religion. The discipline of history and its sister disciplines such as archaeology, those that are concerned with change over time, were curiously absent.[2] Even before the publication Feifel had become a champion for those striving to find meaning(s) in death but

---

[1] H. Feifel (1959) *The Meaning of Death* (New York: McGraw-Hill), p. vii.
[2] There is some mindfulness of change over time to be found, such as the discussion of the gamut of interpretations of death defined by the polar attitudes of stoic acceptance or idealistic glorification of death that mark out Western thought: H. Marcuse, 'The Ideology of Death', in Feifel (ed.), *Meaning of Death*, pp. 64–76.

F.R. Schulz (✉)
School of History, Classics and Archaeology, Newcastle University, Newcastle-upon-Tyne, UK

it was this volume that was pivotal for a new field of study: thanatology. Thanatology—when understood more broadly than the more narrow scientific exploration of death at an electrochemical, neurological or cellular level rendered so complex by the feats of modern technology—is the exploration of individual as well as collective responses to assign meaning to the biological inevitability of mortality.[3] As such the study of dying, death and disposal of the corpse and especially the attempts to understand the human and social response to death 'yield an additional entryway to an analysis of cultures'.[4] This chapter argues this insight by Feifel is particularly true when studying a state such as the German Democratic Republic (GDR) run (at least partially) on the basis of the ideology of Marxism–Leninism, not least because of the added dimension of the antagonism towards organized religion. The way death was construed within the GDR and other Communist states thus offers a further insight into social and cultural practices within these dictatorships. This chapter therefore traces the origins and development of East German thanatology and explores its problems and influences on funeral practices.

In 1983 the *Institut für Marxismus–Leninismus beim ZK der SED* celebrated the 100th anniversary of Marx's death with the publication of the definitive anthology of necrologies to Marx and Engels.[5] Even the title *Their name shall live on through the centuries* epitomizes the East German celebration of achievement as the primary means of dealing with death. Moreover, the volume—as well as its companion volume published early in 1990 to commemorate the death of August Bebel—celebrates rationalism and scientific explanation; for example, the phrase used to describe Engels's death is rather unsentimental: his heart stopped beating and it stressed that he had decided to have his cremains strewn into the North Sea near Eastbourne.[6] While Iring Fetscher is right to stress that for Marx death was not an issue that required much analysis about, his death and the subsequent funeral as well as that of Engels and Bebel were formative

---

[3] The science of death itself is complicated enough, a good starting point despite a lot of further research is still: S. Younger et al. (eds.) (1999) *The Definition of Death: Contemporary Controversies* (Baltimore: John Hopkins University Press).

[4] Feifel, *Meaning of Death*, p. xviii.

[5] A. Malysch et al. (eds.) (1983) *Ihre Namen leben durch die Jahrhunderte fort: Kondolenzen und Nekrologe zum Tode von Karl Marx und Friedrich Engels* (Berlin: Dietz Verlag).

[6] H. Gemkow and A. Miller (1990) *August Bebel—"ein prächtiger alter Adler": Nachrufe, Gedichte, Erinnerungen* (Berlin: Dietz).

for the rhetorics of future socialist burials.[7] Engels chose his words for the graveside address carefully, and one key passage of the brief address clearly sums up Marx's early achievements:

> And he fought with passion, with tenacity, with success, as few others did. *Erste Rheinische Zeitung* 1842, *Pariser Vorwärts* 1844, *Brüssler Deutsche Zeitung* 1847, *Neue Rheinische Zeitung* 1848/49, *New York Tribune* 1852/81—additionally a whole array of pamphlets, work in Paris, Brussels and London, until finally, and as a crowning event, the great *International Workingmen's Association* was founded. That again was possibly a result of which the founder could be proud, even if he had not achieved anything else.[8]

These two sentences contain the key rhetorical elements that would come to dominate socialist funeral oration: a reference to key character traits; an enumeration of publications or achievements; and ultimately an evaluation of achievement. It is, thus, not a surprise that it featured heavily in the handbooks for funeral orators in the GDR as a template.

However, when on that rather cold spring day in 1883, 11 mourners laid Marx to rest they were attending not only the funeral of a journalist and activist but also the funeral of the key member of the school of the Young Hegelians that had emerged in the early 1840s. These young men—among them David Friedrich Strauss, Ludwig Feuerbach, Bruno Bauer and Moses Hess—studied in Berlin under the influence of Georg Friedrich Hegel, the beacon of German idealism. David McLellan suggests that their philosophy is best characterized as speculative rationalism, inasmuch as elements of romantic and idealistic thought were combined with critical and logical elements of enlightened thought.[9] Their main area of enquiry focused on the critique of religion.

A rational critique of religion presented an area that was relatively free of political restrictions at the time and was influenced by the thoughts of

---

[7] I. Fetscher, 'Vorwort', F. Reisinger (1977) *Der Tod im marxistischen Denken heute: Schaff, Kolakowski, Machovec, Prucha* (München: Kaiser), p. 9.

[8] G. Freidank (1975) *Alles hat am Ende sich gelohnt—Material für weltliche Trauerfeiern* (Leipzig: Zentralhaus für Kulturarbeit der DDR), pp. 29–30.

[9] D. McLellan (1969) *The Young Hegelians and Karl Marx* (London: Macmillan), p. 22.

the Young Hegelians and their subsequent rejection of Hegelian ideals.[10] Marx combined Brauer's critique of religion with Feuerbach's radical humanism. He added a material approach and dialectic analysis, thereby arriving at a rejection of religion as a pacifier of the suffering masses, and acknowledging, unlike Feuerbach, the social as well as the individual 'projection'. Marx himself summarizes his critique of religion in a mere two pages in the *A Contribution to the Critique of Hegel's Philosophy of Right* (1844):

> Religious suffering is at one and the same time an expression of real suffering and a protest against real suffering. Religion is the sigh of the oppressed creature, the sentiment of a heartless world, and the soul of soulless conditions. It is the opium of the people.
> The abolition of religion as the illusory happiness of the people, is a demand for their real happiness. The call to abandon their illusions about their condition is a call to abandon a condition which requires illusions. The criticism of religion is, therefore, the embryonic criticism of this vale of tears of which religion is the halo.[11]

Thus religion is not just the visible expression of a more profound human alienation but also the premise of all criticism. What separates Marx from Feuerbach is that he does not see the action of the individual as the solution. In Marx's view, this alienation cannot be overcome by merely understanding this process: the solution is not found in a rational process of thought. What he arrives at is a more comprehensive change of state and society:

> But man is not an abstract being, squatting outside the world. Man is the human world, the state, society. This state, this society, produce religion which is an inverted world consciousness, because they are an inverted world. [...] The criticism of religion concludes with the teaching that, for human beings, man is the highest being.[12]

---

[10] V. Kieran (1991) 'Religion', in T. Bottomore (ed.), *A Dictionary of Marxist Thought*—2nd ed. (Oxford: Blackwell), pp. 465–68; D. McLellan (1986) *Marx* (London: Fontana), pp. 24–38.

[11] T. Bottomore (ed.) (1963) *Karl Marx—Early Writings* (London & New York: McGraw-Hill), pp. 43–4.

[12] T. Bottomore (ed.), *Karl Marx—Early Writings*, p. 44.

Indeed Marx was to criticize Feuerbach further for a too-liberal process of artificial abstraction, as compared with his own emphasis on forced alienation from the material world.[13] God is a projection of ideals, but the need to project one's own ideals is based on the reality of what exists (*Realität des Ist*). It is here that the idea of a historical materialism in Marx's analysis gains major importance; it points to the alienation of the human being from itself by social realities outside of any socialist society. This idea would remain at the heart of all historical materialist thought after Marx. Engels stressed its ground-breaking importance and its scientific nature in his opening paragraph of the graveside eulogy in March 1883:

> Just as Darwin discovered the law governing development in organic nature, so Marx discovered the law of development in human history: the simple fact, hitherto concealed by ideological overgrowth, that humans have to eat, drink, find shelter and clothing before they can engage in politics, science, art, religion, etc.[14]

Marx's critique of religion represents the roots of his analysis and his first field of publication, and was to serve as a model for his later materialist analysis of human nature, state and society. In this context it is vital to understand that, for Marx, the origins of religious alienation lie in the 'matter' surrounding the individual and thus society itself. In the case of religion, the projection of ideals onto an idealized god originated in the craving for social-emancipation. Raymond Williams argues 'there are inevitable difficulties in any serious materialism. In its earliest phases it has a comparative simplicity of definition, since it rests on a rejection of presumptive hypotheses of a non-material or metaphysical prime cause, and defines its own categories in terms of demonstrable physical investigations'.[15] This simple and narrow foundation leads to a 'confused and complex situation, within the interactions and the failures to interact of politics, science, and philosophy', owing to in-built ontological as well as epistemological flaws and in conjunction with a conscious alliance

---

[13] See: K. Marx, Theses on Feuerbach—4th thesis 1845, first published in the year of his death 1886 as an appendix to: F. Engels, *Ludwig Feuerbach and the end of German Classical Philosophy*, http://www.marxists.org/archive/marx/works/1886ecgp/index.htm.

[14] Freidank, *Alles hat am Ende sich gelohnt*, p. 29.

[15] R. Williams (1980) *Problems in Materialism and Culture* (London: Verso), p. 104.

with certain ideologies.[16] Bhaskar's argument that materialism should be treated as 'a series of denials, largely of claims of traditional philosophy—e.g. concerning the existence of Gods, souls [...]; and as an indispensable ground for such denials' is quite helpful when it comes to understanding the foundation of East German (and socialist) thanatology.[17]

Materialist thought replaces nonverifiable assumptions with observable and provable scientific facts. Therefore, in the rite of passage that is to mark the end of life the solution for the materialist is to celebrate life's observable attainments—this is very clear from Engels's oration, since it defines life as an enumeration of achievements, and in Marx's specific case what was to be celebrated were his achievements as a writer, philosopher and thinker. However, as a consequence, materialism is left with a considerable metaphysical vacuum when it comes to the finality of death (moreover this is a vacuum that has often come as a shock to those who have not subscribed to materialism). The new philosophical system needed not only to replace faith and construct epistemological truth and ethics but also to tackle a far more difficult problem. Disallowing for religious beliefs leaves a profound gap, primarily that of the most fundamental ontological problem: What is the meaning of life?

Back in 1883 Engels underlined that one answer was Feuerbach's dictum that life should be embraced enthusiastically and argued that life must be lived fully in spite of death. This set the tone for both the East German socialist funeral and its ideological justifications.[18] When the GDR was founded in 1949 its elite adopted materialism as the state ideology. This had a profound impact on the development of a clearly structured Marxist–Leninist thanatology, because the apparatus of party and state became increasingly effective. Materialism stressed a positive attitude towards life and its possibilities and embraced the ideal of a *'lebendig gelebte gesellschaftliche Dasein'* (lively participation in societal existence) as the very opposite of the proverbial *Karteileiche* (a non-active member of socialist society).[19] Unlike Christian theology, materialism viewed life not as a path—a passage—and death as

---

[16] Williams, *Materialism and Culture*, p. 104.

[17] R. Bhaskar (1991) 'Materialism', in T. Bottomore (ed.), *A Dictionary of Marxist Thought* (Oxford: Blackwell), p. 372, please note this was written before the turn towards critical realism.

[18] See the development of materialism: J. Choron (1963) *Death and Western Thought* (New York & London: Collier), pp. 186–98.

[19] The GDR official liturgy suggested as a possible starting point for any eulogy a line from Otto Grotewohl: 'The highest aim in life is the creative deed'. See: F. Kretzschmar (1982)

redemption—a passageway into a 'better' afterlife—but as a task, a challenge and ultimately as an end in itself. Hence, the ideal of a productive life, in a material as well as a political and social sense, slowly became a central point of reference and was actively promoted by what was to be termed the ideal of socialist humanism.[20]

In East German writings there is initially limited engagement with the problem of death and Marxism. This changed slowly with the proliferation of publications on atheism and critiques of religion from 1955 and especially after 1959.[21] Moreover, this increase of publications also coincided with a 'scientific golden age' when socialism scored one technological success after the other, an age when Sputnik (1957) and Juri Gagarin (1961) were still unsurpassed and scientific success was used to emphasize the factual nature of materialism. The triumphalist spirit of the materialism of the time is nowhere better captured than in a small pamphlet entitled *The Good Lord and the Sputnik* (1958).[22] It was written by Rudolf Rochhausen who, a year later, would gain his first doctorate from the *Institut für Marxismus–Leninismus* at the University of Leipzig and rise through its ranks to become an eminent professor at the institute. Although challenges to religion were common, direct references to death in early writings are rare. Yet, there are two very specific exceptions—both actually views of death that were not East German in origin.

The first is the 1957 translation of Mark Petrovich Baskin's treatise on religion—born in 1899, Baskin was made professor of philosophy in 1930 and taught at the Institute of Philosophy, USSR Academy of Science. It provides a good summary of the Marxist–Leninist understanding of death as a problem in ethical, theoretical and technical terms, combined with the customary characteristic ridicule of 'superstitious' religious beliefs:

Der Tag hat sich geneigt—Zur Gestaltung weltlicher Trauerfeiern (Leipzig: Zentralhaus Publikation), p. 12.

[20] For the instrumental role of the Warsaw-based philosopher Adam Schaff in the postwar engagement of Marxism with the role of the individual see: F. Reisinger, *Der Tod*, pp. 100–29.

[21] The catalogue and the holdings of the *Bibliothek und Druckschriftensammlung des SED-Bezirksparteiarchivs Magdeburg* and material concerning religion and theological question indicates this very clearly, Landeshauptarchiv Sachsen-Anhalt, Magdeburg (LHSA Mageburg), P30.

[22] R. Rochhausen (1958) *Der Sputnik und der liebe Gott Broschiert* (Berlin: Dietz).

> The religious doctrine of the immortality of the soul is not only a wrong doctrine, contradicting reality and long ago disproven by science, it is also a dangerous doctrine. [...] The wisdom of the people recognised long ago the correctness of the materialist point of view when it created the saying: "A healthy mind in a healthy body". No normal human in everyday life takes their bearing from the immortal soul and ends [their] life in the hope that the soul will enter a better world after death. Even those humans that are fanatically religious prefer life to death and in the case of an illness they try to lengthen their life at any price. The advanced scholars of all countries have forever fought a persistent battle to prolong life, and this fight will be successful under Socialism.[23]

The second exception is the 1958 East German reprint of Ernst Haeckel's populist case for atheism *Welträtsel*, first published in 1899. Having initially trained as a medical doctor, Haeckel became increasingly interested in the philosophy of science and the works of Darwin and Lyell. He combined this training with his interests and pursued a more analytical career, being offered a chair in zoology at the University of Jena in 1865. Haeckel combined science—particularly the popularisation of science—with his monistic understanding of nature. The book *Welträtsel* covers 20 main topics in its four main chapters on man, soul, world and god and has one specific section on death. In his preface, the anonymous editor claims—in reference to Lenin's characterization of materialism and Lenin's fondness for this particular work—that Haeckel's scientific *Weltbild* was essentially materialist, but that, due to (reactionary) bourgeois loathing of this term, Haeckel did not dare to ally himself openly with materialism.[24] While Haeckel was not a strict atheist, his strict scientific rejection of god and of creation made this text a prime candidate for re-publication, as it stressed the nexus between the natural sciences and the scientific nature of dialectic and historical materialism.

However, by the late 1950s the challenge to religion has not remain confined to the realm of discourse. These philosophical positions influenced policies and political views; the civil registrar offices (*Personenstandswesen*) throughout the GDR were charged to set up socialist *and* secular alternatives to the Christian rites of passage: baptism, confirmation, marriage and

---

[23] M. Baskin (1957) *Materialismus und Religion* (Berlin: Dietz), p. 110.
[24] E. Haeckel (1958) *Gibt es ein Weiterleben nach dem Tode?—Aus "Die Welträtsel"*(Berlin: Dietz), footnote 1.

funerals.[25] First these ceremonies were based on local recommendations. The tenor of the local advice can be seen in the rather detailed summary drawn up by the registrar of Stalinstadt, the socialist-planned city founded in 1950. It clearly reflects the emerging official thanatology of the late 1950s.

> Birth, the bond between man and woman, and death are originally biological processes that have become socially significant in the development of mankind. The exploitative classes idealized these events and gave them religious meaning and the form of religious ceremonies in order to bind the people to them and to subdue [the people] to their rule. [...] The working class of the GDR and its leading party, derived from the dialectic-materialist world view, and with the help of the power of the state, wants to recognise ceremonially birth, marriage and death as life events of the emerging socialist society. [...] The content of these ceremonies derives from the combination of individual experiences, [with] the intentions of parents, engaged couples and the bereaved as well as the progressive aspirations of mankind. These ceremonies will be rich in content if the specific personal aspirations can be aligned with the general [as in social] aspirations.[26]

Thus, the role of science and the importance of laws of nature and history became tropes regularly emphasized in the official discourse. This was further promoted by some disciplines; Soviet archaeology, for example, on the basis of grave finds purported to prove that Neanderthals, driven by instinct, treated the dead as ill group members. These archaeologists used this to postulate that the idea of clearly differentiating between the dead and the living emerged only gradually.[27] This once more reversed the traditional emphasis so that the event of death is de-emphasized in order to stress the importance of achievement and productivity within socialist society. This was strongly underlined when in December 1961 the Ministry of the Interior sent out general comments on the role of socialist

---

[25] The differentiation of these ceremonies is complex and goes beyond the scope of this chapter. See: F. Schulz (2013) *Death in East Germany, 1945–90* (New York: Berghahn).
[26] Bundesarchiv Berlin (BA Berlin), DO4, No. 326, Grundsätze und Erfahrungen bei der Gestaltung sozialistischer Feierlichkeiten um Geburt, Eheschließung und Tod in Stalinstadt, 24 April 1958, 1.
[27] P. Franz (1980) *Die historische und gesellschaftliche Entwicklung des Bestattungswesens unter der Berücksichtigung der aktuellen Situation in der DDR*, PhD Thesis Martin-Luther-Universität Halle-Wittenberg, p. 1.

rituals. This centrally produced advice succinctly summarizes the foremost functions of socialist funerals as follows:

> [Socialist funerals] contribute to emphasis, on the basis of the Marxist worldview, the dignity of the human being while bearing in mind his attainment and merit. The bereaved should infer from this the obligation to continue the work of the deceased in his spirit and thus fulfil the deceased's legacy.[28]

In this framework, a productive and meaningful existence represented the ultimate sense-making mechanism when it came to reconcile the death (or still better the accomplished life as the true focus) of a deceased family member, friend, colleague or comrade and connect those left behind with the legacy of the deceased. Hence, the ideal of a productive life, in a political as well as a material sense, became a central point of reference and was actively promoted. Socialist rituals were to play their role in this promotion.

In the aftermath of 1961, and the final separation of East and West, a clearly structured East German Marxist–Leninist thanatology began to take shape. More was published on death; indeed, in his 1995 study Thomas Krause rightly argues that Marxist–Leninist philosophers monopolized the problem of dying and death in the 1960s and 1970s as they tried to fill the ethical vacuum left by the negation of Christian ethics and by the increasing optimistic belief in scientific progress (*optimisitischer Fortschrittsglaube*).[29] However, the regime's only answer to the meaning of life and death remained the same as Engels's original answer. Unlike Christian theology, this materialism thus viewed life not as a path—a passage—and death as redemption—a passageway into a 'better' afterlife—but as a task, a challenge and ultimately as an end in itself. Only by exerting positive influence and setting a positive example within society could real significance and meaning be achieved. This view was most suited to those who gave their lives in the name of socialism, those who died or suffered for their ideology and, most importantly, those who were seen as the heroes of socialism (*Held des Sozialismus* or *Kämpfer der Arbeiterklasse*). In 1964 Wolfgang Eichhorn used a formulation that is as emblematic of

---

[28] BA Berlin, DO1, 7500 Letter from Abteilung Innere Angelegenheiten, Sektor II to Deutschen Kulturbund, 22 December 1961, p. 2.

[29] T. Krause (1995) *Der Umgang mit ethischen Problemen des Lebensende in der DDR und die Entstellungen Medizinischen Personals zu Sterbenden und zum eigenen Tod*, MD Dissertation, Medizinische Fakultät der Universität Leipzig, p. 4.

the political antifascism of the GDR as it is for other ideologies that have some characteristics of a political religion.[30]

> Especially the historical-materialist conception imparts, with regard to death, an activity directed towards this life and fearlessness and steadfastness on the part of the proletarian warrior, who gives his life for the cause of peace and socialism, and in the interest of this cause makes even the greatest sacrifice.[31]

Besides the celebration of sacrifice, this demonstrates the necessity of what in 1977 the theologian Ferdinand Reisinger, from a critical Catholic point of view, called the clearly accelerated Marxist search for a meaning in death.[32] However, in Reisinger's view, that search was largely confined to authors from other socialist countries not the GDR. This might have to do with the fact that as with other issues, the GDR stuck more to the orthodox. What emerged as the quintessential thanatological view in the GDR is best exemplified by the early work of Hans Streußloff. The author of the main textbook used to teach materialism, he was one of the most outspoken East German materialist philosophers and, like Rochhausen, taught in Leipzig at the *Institut für Marxismus–Leninismus* until he asked to retire in 1990. In his Ph.D. dissertation he engaged with Christian interpretations of death and employed the clear dialectical approach to contrast Marxist–Leninist notions of death from those of the capitalist 'reality'.

> 'Bad' death is a social phenomenon. It is the product of an exploitative- and class society. This observation even holds true in our time. The realisation of the class character of death remains equally true. Death was, and is, foremost an existential problem for the members of the exploited and suppressed classes and strata of society. [...] We do not entertain the illusion that we can simply abolish death. [...] The object is rather to create and develop such objective and subjective moral values, that, first, death ceases to be a social ill. And, secondly, due to these social changes and the new moral qualities of the socialist personality, to empower the individual to cope with the per-

---

[30] Among others see: J. Brinks (1997) 'Political Antifascism in the German Democratic Republic', *Journal of Contemporary History* xxxii:2, 207–17; S. Behrenbeck (1996) *Der Kult um die toten Helden, Nationalsozialistische Mythen, Riten und Symbole, 1923–1945* (Vierow bei Greifswald: SH Verlag).

[31] W. Eichhorn (1964) *Von der Entwicklung des sozialistischen Menschen* (Berlin: Dietz), p. 76.

[32] Reisinger, *Der Tod*, pp. 268–70.

sonal problem of death, to overcome it, without requiring a religious-opiatic comfort.[33]

In this passage we see the allusions to Marx's writings from 1844 but we also see the strong emphasis on the social component. This is nonetheless achieved without entirely losing sight of the individual in society. Continuing his analysis, Streußloff develops a concept of death in socialism that is succinct. His was a view that represented what was to become the embodiment of the East German thanatology that death is not to be feared in a socialist society and, more importantly, that the fear of death can be overcome by the drive for better socialism. His argument culminates in the following argument that emphasizes and celebrates the strong political antifascism of GDR socialism:

> [B]y overcoming existential uncertainty, characteristic of the class society, the ideology of the fear of death also decisively lost ground for the masses of the population. Therefore, there is within socialism no ideological cult of death, no ideology of fearing death. Not death, but the tasks of life determine the societal awareness of all people.[34]

The unequivocal focus on life and the task of helping to build society in this key formulation of East German thanatology was meant to solve both the all-too-human issue of fearing the end as well as help to resolve the issue of giving meaning to death. This thanatology was quite stark in its ideological purity, and the reality of sepulchral culture in the GDR of the 1960s and 1970s posed some practical problems. For example, when Phillip Daub, the mayor of Magdeburg, died in 1976, both his obituary and the graveside address by Alois Pisnik, the first secretary of the *Bezirk* Magdeburg, made reference to his endurance of great suffering in the name of socialism as the deed of a great individual. In line with the typical biography of higher party functionaries Daub had been interned during the Second World War.[35] Similarly, when the crew of *Sojus 11* died in an accident during the re-entry phase of its flight in 1971, the *Bezirksleitung* Erfurt sent a telegram of commiseration to the local Soviet army com-

---

[33] H. Steußloff (1967) *Zur Kritik der ideologisch-theoretischen Verschleierung der Todesprobleme in der modernen christlichen Theologie*, PhD dissertation, Jena, 90.

[34] Streußloff, *Verschleierung der Todesprobleme*, p. 98.

[35] *Magdeburger Volksstimme* 16 July 1976, 2; *Neues Deutschland*, 15 July 1976, Nr. 167, 2; LHSA Mageburg, Rep. 41, Nr. 1433, Eulogy by A. Pisnik.

mander, praising their heroic and pioneering deeds in the name of mankind stating: 'We bow, full of respect, to the heroism of these great sons of the Soviet people'.[36] They all died as heroes of the working class and merited special treatment; they were granted special guards of honour comprised of friends and dignitaries before the funeral and also military ceremonial such as their medals being carried in procession on a red and golden cushion in front of the coffin or the urn before the latter were buried in *Ehrenfriedhöfe* (cemeteries reserved for the honoured dead).[37] Other forms of death, in stark contrast, were much more difficult to dignify, let alone legitimize.

The funeral eulogy of a young construction worker in the northern city of Güstrow provides a very good example to pinpoint some of the key practical issues faced when translating the thanatological standpoints into words addressing the bereaved. GDR eulogies, like most eulogies, are very schematic; they begin by addressing the bereaved and introducing the deceased. The eulogy was delivered by the mayor of Güstrow who also was the district party secretary. He drafted an initial version, but the typed version shows some deliberated change in pencil that reflect rhetorical and ideological tweaks. In this case, right from the start the issue of age is made the central point of the oration:

> In front of us lies not a man who can look back upon a long life full of joy and suffering, in front of us lies a young man of 23 years who, like a juvenile oak struck by lightning, was wrenched from the midst of life, [wrenched] from the midst of the work of our reconstruction. We bow our heads in awe and respect before this young and blossoming child of this city, we never imagined that in such short a time he would have to be buried in his home town.

Following this acknowledgement of the relative youth of the dead, the eulogy then continues to align the history of the GDR with that of the deceased. In this case this was greatly aided by the fact that the person in question had decided to return to the GDR from the Federal Republic of Germany This was celebrated as an affirmation of both the attraction of home, and the home town, as well as of the political system of the GDR. Following the standard biographical and chronological approach of

[36] THSA Weimar, BPA Erfurt, IV/B/2/5–200, 281: *Fernschreiben*, 30 June 1971.
[37] For more details see: J. Redlin (2009) *Säkulare Totenrituale: Totenehrung, Staatsbegräbnis und private Bestattung in der DDR* (Münster: Waxmann).

most eulogies, the orator traced the promotions and the progress prior to the events that lead to the death. These events were quite characteristically rather bluntly re-narrated:

> Then on Friday—he was possibly thinking about how to spend the weekend in Güstrow—a wall collapsed; he did not succeed in jumping beyond the scaffolding within the matter of seconds at hand; he was struck on the skull and died. It is a terrible blow for his next-of-kin but also for his colleagues: apart from him five more workers died on site. But he had to give his young life.

The last statement is indicative of the in-built fatalism in which death was seen as a natural and unavoidable occurrence. This eulogy is more emphatic than others, and it does not abate; it continues seamlessly:

> Dear relatives! Dear mother and grandmother, dear brother of the deceased!
> It is a tragic, a harrowing event, that disrupted your family and took away your Dieter. Do not expect me to put a mendacious illusion before you. No! The truth demands I tell you: You will never again see your beloved Dieter, neither here nor in another non-existent world.

This strong language and the clear short sentences stress the fervent antireligious and materialist nature of this late 1950s eulogy. It also serves to build up to the modicum of redemption that is to be had when facing the truth.

> However within this bitter truth that burns in your soul arises also the consciousness, the memory of the beloved deceased. We have to accept the death of a human being as his biological end as much as his birth as the beginning of life. Despite this, however, man lives on much more in his family, in his class and his people.[38]

Apart from the use of the term 'soul', what is striking here is that meaning is to be had by living on in the collective—defined in three distinct ways: the personal, the collective and the nation. Hereafter, and not surprisingly, the eulogy turns distinctively political and celebrates the productive life of a construction worker in rebuilding the country and paints

---

[38] Landeshauptstaatsarchiv Schwerin (LHA Schwerin), Rep. 7.11.1. BT/RdB Schwerin, 3995, Eine Ansprache aus Güstrow.

the deceased as an example to follow, thereby trying to give both a personal 'non-opiatic' solace as well as a clear upbeat political message to the mourners. Internally this eulogy was seen as exemplary, and thus it was attached in full length as an appendix to the protocol of the discussion of a regional working party on secular ceremonies in order to demonstrate a successful practical application of a materialist thanatology.[39]

The full brunt of the reasoning behind this approach to fathom the meaning of death, however, only becomes apparent when not addressing the deaths of members of the party cadre and the problems of death and materialism are addressed in official publications. In the handbooks published from the mid-1970s and in the 1980s the problem of stark language and tragic deaths was increasingly accepted and it was acknowledged that one could not, in that context, refer in such a case to a life lived to its full extent. The most widely used handbook was published in 1975. It confirms that the task of the funeral orator was in addressing in tragic circumstances the collision of pure reason with what was illogical and destructive:

> As materialists we know: Nature does not make sense. Death and life "themselves" are senseless, like the sun or the snow. No God, no spirit is the creator of the world. But paradoxically, laws keep stars and atoms in motion.
> Not "God's unfathomable will", but the workings of objective laws govern creation and demise in nature, as well as life and death in humans.
> In tragic situations the knowledge of these related issues cannot—for example the death of a young person in a traffic accident or through a malignant illness—comfort us and help us overcome the painful death; death is and remains without sense. That knowledge, however, can do one thing: it can make even such a deeply nonsensical death become philosophically comprehensible, understandable, and one can draw from this insight, with composed earnestness, strength, in order not to despair in the face of calamity and let oneself drift, but with all the intensity of our tried and tested personality give life what is its due.[40]

Yet, the answer to the question of sense is yet again to affirm the productive life of those left behind. In other words, this is the transformation of the philosophical stance into a practical argument and advice for those who had to face this dilemma when conducting secular and socialist funer-

---

[39] LHA Schwerin, Rep. 7.11.1. BT/RdB Schwerin, 3995, Arbeitsgruppe für weltliche Bestattung im Bezirk Schwerin.
[40] G. Freidank, *Alles hat am Ende sich gelohnt*, pp. 5–6.

als. It traces the development from the seeming embrace of the idea of annihilation via the defence or justification of death as a law of nature, to the attempt to reinterpret, or better redirect, grief and loss into strengthening resolve, thus offering at least a glimmer of a positive insight.

Ultimately, even the chief ideologist of the *Sozialistische Einheitspartei Deutschlands* (SED), Kurt Hager, had to acknowledge this fundamental problem. In a speech, rare for a high-ranking official, as they rarely mentioned mortality outside state funerals in the GDR, Hager addressed a philosophical conference in November 1979 and stressed the challenge that life's borderline situations represent:

> The meaning of life, the multi-layered complex of problems that is health, illness and death and a number of other questions that touch deeply upon the political and moral stance of the socialist personality are a real challenge to Marxist–Leninist philosophy, since the leading principle of its philosophical insight has to be the congruence with life, and thus with practical reality.[41]

Hager is right in pointing out that aligning theory and practical reality [*Praxis*] represented a real challenge. Throughout the 1980s a practical turn in thinking about death can therefore be observed. This practical turn was foremost medical in nature and is linked in particular to one individual, the medical scholar Kay Blumenthal-Barby and his work. In 1982 he founded a working group on perimortal medicine bringing together those who saw a need to think about how to treat those who were dying within the medical system of the GDR. In the same year he also published a book, edited a year earlier, that was meant as a practical aid to those dealing with the daily care of those dying.[42] In the volume he provides a historical, medical, philosophical as well as practical overview of death through the lens of an East German doctor. This is augmented by a review of the key literature as well as an in-depth discussion of and advice on the care of the dying, how to handle the issue of children dying and the difference between dying in hospital, at home or in a care home. The medicalisation of the discussion of dying and death is not a uniquely East German phenomenon: medical personnel obviously has to confront death more regularly then the average individual. What makes the work of Blumenthal-Barby remarkable is that the first work that had been clearly

---

[41] K. Hager (1979) *Philosophie und Politik* (Berlin: Dietz), p. 23.
[42] K. Blumenthal-Barby (1982) *Betreuung Sterbender* (Berlin: VEB Volk und Gesundheit).

aimed at medical personnel was followed 4 years later by a richly illustrated volume clearly aimed at the general public.[43] Similar in scope, the language, the organization of content, the inclusion of a chapter on the humorous side of death and the inclusion of very practical material such as checklists how to handle the aftermath of a death in the family in an appendix clearly suggest the strong desire to reach a wider audience. The book is also remarkably critical of previous practices, for example noting that a 1976 handbook for the training of nurses only devoted a single page out of over 600 to the issue of how to care for those near death. The tenor of lessons that were learned remained familiar:

> It is essential (…) to take a different stance in future, one free of fear and mysticism. The multi-layered complexity of problems of illness and death is a challenge for Marxist–Leninist philosophy. Much remains to be done; the public like the medical personnel have to achieve an attitude to death that is sustained by reason and objectivity. (…) A reasonable attitude to dying and death can make life more beautiful.[44]

Despite the clear practical turn that among other things led to advice on dying and death being more widely published throughout the 1980s, Blumenthal-Barby's plea clearly refers to Hager's acknowledgement of 1979. This shows the persistence of the underlying problem: the lack of an answer to the question about the purpose of death combined with the human desire for greater meaning. There is a strong case for the argument that ultimately East German thanatology characterized by scientific materialism did not embrace the logical conclusion of its materialist understanding of death. It did not celebrate the idea of annihilation, of the unequivocal end, but circumnavigated it at least by laying great emphasis on a meaningful life, the significance of which continued even beyond death and at most by constructing an inspiring ideal thus putting certain individuals on a pedestal as exemplary socialists. It struggled with the problematic nature of tragic or lonely death.

How the population reacted in practice to the slow imposition and practical implementation of the official thanatology still needs to be fully understood. The issues surrounding the social changes in attitude to dying, death and disposal are complex: socialist rule, secularization and growing

---

[43] K. Blumenthal-Barby (1986) *Wenn ein Mensch stirbt* … (Berlin: VEB Volk und Gesundheit).
[44] Blumenthal-Barby, *Wenn ein Mensch stribt* …, pp. 55–6.

scientific understanding all play their role in the changes the GDR saw during its existence. However, what is often overlooked in the light of change is the persistence of tradition. In 1982 Professor Bierschenk, then *Bezirkshygieniker* for the predominantly rural region of Neubrandenburg, wrote a confidential report. It was forwarded with an unusually forceful letter supporting the findings to the Minister of Health. In the report Bierschenk highlights the complexities that arise from the search for rationality and objectivity, as well as the ever present quest for efficiency:

> The fundamental experiences of birth and death, over the last decades and with the coming of age of the new generations under the influence of changing social conditions, have moved from the realm of the family into an area of responsibility of the state, where they have been institutionalized under the auspices of the state. [...] The question of whether the abdication of experiencing the dying of others is an ethical improvement or an impoverishment—in the sense of abetting a lifestyle that is increasingly free of conflicts and in the sense of a further relinquishment of elementary familiar duties of care into the care of the state—has, to the best of my knowledge, been posited neither from a socio-theoretical point of view in the context of our condition nor has it been more than partially answered to any extent other than on a most ostensibly pragmatic level. The vast majority of those whom I questioned [...] actually repress [this question] until they have to face the situation more or less prepared. [...] The remaining denominational pastoral care—understood as a humanitarian and active attention to the individual who conscious of dying, the guidance provided to the family member and the keeping of the dignity of the dead—according to the reports received from the districts within our region where denominational institutions still existing and thus allow for a comparison, are strong evidence that nothing of comparable or better quality exists within the realm of state provision. Successful attempts to create dignified funerals are only one part of a whole set of problems.[45]

This highlights a key issue for those who think about the complex social problem of death and the individual, social and cultural meaning of death: Death is simply situated more in the world of experience and plays out in the physical environment. Death's after-effects, besides direct personal grief, are explored more through architecture, art and poetry than in dry advice in books. GDR thanatology permeated the pamphlets

---

[45] BA Berlin, DQ1, No. 11613, Letter Möwius to Mecklinger, 17 March 1982, attached report, 5–6.

by and handbooks for secular funeral orators; it even influenced the work of local and regional officials; it changed how the medical profession saw death and it might even have caused the intermittent attention paid by the ZK and some of the ministries to the question of sepulchral culture—but ultimately death remained a marginalized topic. The East German state simply did not understand the complexity of the sepulchral culture, and merely focused on some aspects. Yet, the widely discussed secularization of the GDR did not stop at the door of the church (if one allows for a very wide definition of the term belief). There was another intriguing disengagement. By the 1980s a radical form of funeral was gaining ground. This new ceremony eschewed speeches and embraced silence, nothing was said and the remains or cremains were buried in silence. Detailed statistics are not available, but according to one source silent ceremonies had reached a level of 20.8 % throughout the entire GDR.[46] The motivations behind choosing silence are difficult to pin down, but they offer the intriguing range from trying to keep the state out of private grief to simply wanting to be left alone. Maybe Professor Bierschenk was right and the population, despite the Churches and the official thanatology, was left bewildered with the reality of mortality until they had to face it directly or indirectly through illness or loss.

In conclusion, on the one hand it is absolutely right to note that for Marx and those who built upon his thought from the mid-nineteenth century, death did not represent a central problem. There is little mention of death in Marx's writings. This is remarkable on three separate fronts: (1) the importance of religious critique as the very basis of his reasoning; (2) the tragic deaths in his family and his personal inability to find solace; and (3) the quite remarkable Victorian celebration of the dead around him. On the other hand, it is more than telling that despite this absence, especially theologians have stressed the centrality of Marx's critique of religion (and Freud's thesis of denial) in framing both the modern religious understandings of death as well as the development of modern secular understandings of it. For theologians, such as Ferdinand Reisinger or John

---

[46] BA Berlin, DY 30/IV B 2/14/46, Arbeitsgruppe Kirchenfragen am ZK der SED: O. Klohr/W.Kaul/K.Kurth, Über Wirkungsfelder und Wirksamkeit kirchlicher Institutionen in der DDR. Kirchenstudie 1981 (Rostock Warnemünde). See also: I. Lange (2004) 'Von der Wiege bis zur Bahre: Zur Geschichte Sozialistischer Feiern zu Geburt, Ehe und Tod in der DDR', *Kulturnation: Online Journal für Kultur, Wissenschaft und Politik*, xxvii:1. The area of funeral rites of the GDR warrants further study, ideally utilizing oral history projects to unlock the many regional differences.

Bowker, Marx's critique of religion and by extension, Freud's psychoanalytical critique, is a useful foil since in their view, death to Marx and Freud represented the very source of religion, while for Bowker and Reisinger the relationship is ultimately inverted. In their view religion is the very origin of death and even more so religion is the meaningfulness of death.[47] This means, for example for Bowker, that death is invariably linked with the act of sacrifice, for he argues that '[i]t is a human privilege, just as surely as it is a human suffering, to acquire consciously the necessary condition of death, and to confirm it as sacrifice, as the means through which life is enabled and secured'.[48] This re-introduction of sacrifice as a trope is significant since in this theological challenge to the critique of religion it is stressed that ultimately the works of Marx or Freud and those inspired by them offer scant solace and do not offer much of a satisfactory answer to a fundamental question: what is death for? Indeed, this is what East German thanatology tried very hard to answer but ultimately in practical terms struggled with greatly. The cold materialist and scientific truth of annihilation is an answer but not one that is easy to relate to for those who grieve—and the prevailing materialist thanatological answer in the GDR did not provide a satisfactory answer outside the celebration of the lives and deaths of great men (and a few great women) in advanced adulthood.

---

[47] J. Bowker (1993) *The Meaning of Death* (Cambridge: Cambridge University Press), pp. 3–42; Reisinger, *Der Tod*, pp. 18–91.
[48] Bowker, *Meaning of Death*, p. 227.

PART IV

Socialism and the Problem of Religious Heritage

CHAPTER 10

Religion and *Nauka*: Churches as Architectural Heritage in Soviet Leningrad

Catriona Kelly

Since the 1960s, models of the secularisation process going back to Bentham, Comte, and Marx, which assumed that the rise of science would foster the decline of religion, have been challenged both from a conceptual and a practical point of view. Alongside dissatisfaction with the traditional assumption that modernity was certain to bring rationality and progress comes evidence that secularism is by no means an inevitable result of scientific and technological development. The case of the USA, where religious belief and practice remains widespread, is one key example.[1]

I draw in this article on material collected for my history of churches as architectural monuments in the Soviet period, *Socialist Churches: Radical Secularization and the Preservation of the Past in Petrograd and Leningrad, 1918–1991* (Catriona Kelly 2016), Socialist Churches: Radical Secularization and the Preservation of the Past in Petrograd and Leningrad (DeKalb, IL: Northern Illinois University Press). My thanks to the AHRC, the University of Oxford, and the Ludwig Fund, New College for supporting the research involved.

[1] A study that attempts to place such interpretations in perspective, while continuing to subscribe to these, is B. Wilson (1982, 2002) *Religion in Sociological Perspective* (Oxford University Press), which contends, for example, that religious authority was able to persist only where development was particularly rapid (p. 39), and that 'the social system of advanced societies functions on rational premises' (p. 47). Among influential presentations of a market-driven model of US religious sociology, where the multiplicity of creeds is

C.H.M. Kelly (✉)
New College, Oxford University, Oxford, UK

© The Editor(s) (if applicable) and The Author(s) 2016
P. Betts, S.A. Smith (eds.), *Science, Religion and Communism in Cold War Europe*, St Antony's, DOI 10.1057/978-1-137-54639-5_10

Former socialist countries, which underwent a significant revival of interest in religion after the collapse of the Soviet bloc in 1991, are also regularly adduced as challenges to established secularisation theory.[2] These counterarguments—along with the fact that successful scientists are by no means always atheists, and that nonprofessionals have proved remarkably good at selecting which religious and scientific beliefs and values they wish to combine—are so regularly rehearsed that there is no need to repeat them in detail.[3] That religious belief and practice, and ideologies of secularisation, have interacted and transformed each other over the last 250 and more years is now the informed consensus among sociologists, anthropologists, philosophers, and historians. As Charles Taylor has put it, the rise of science is not an adequate explanation for the decline of religious belief, even if some historical subjects have been minded to interpret historical change in exactly this way.[4]

There is, though, a danger of losing from view exactly the perspective of those historical subjects who firmly believed that science *was* a path to the disappearance of religion. As *The Atheist's Handbook* put it in 1975, 'By sanctifying the primitive views held by human beings in the past, and presenting them as divine revelation, religion sets its face against science, which is in a state of perpetual development, and which clarifies and deepens our understanding of the objectively existing world'.[5] The defining assumption of Soviet culture was that as science and knowledge spread, religion would disappear: 'There is nothing surprising in the fact that, as a result of socialist

---

seen as a sign of strength, rather than of fragmentation and decline, is R.S. Walker (1993) 'Work in Progress: Towards a New Paradigm for the Sociological Study of Religion in the United States', *American Journal of Sociology*, 98:5, 1044–55.

[2] For a clear, if one-sided, account of how Eastern European history calls the secularisation model into question, see P. Froese (2008) *The Plot to Kill God: Findings from the Soviet Experiment Secularization* (Berkeley: University of California Press). On the other hand, Steve Bruce (2002) *God is Dead: Secularization in the West* (Oxford: Blackwell) attempts to account for the specificity of the Eastern European situation by claiming that the social changes there were forced on the population, rather than representing organic social transformation, which begs all kinds of questions.

[3] See for example Charles Taylor (2007) *The Secular Age* (Cambridge, MA: Harvard University Press), which sets its face against what Taylor calls 'subtraction stories' in the discussion of cultural history (p. 22).

[4] Taylor, *Secular Age*, p. 4.

[5] *Nastol'naya kniga ateista* (1975) (4th edn.; Moscow: Izdanie politicheskoi literatury), p. 399.

[economic and social] transformation, religion is dying out faster in the USSR than in other countries'.[6] In fact, religion was not 'dying out' (or not as fast as Soviet administrators hoped). But the sense that the successes of 'scientific atheism' might be limited did not lead—publically at least—to a re-examination of tried and tested goals and strategies.[7]

The correlation of enlightenment and atheism on the one hand, and ignorance and religion on the other, was a foundational principle of Soviet ideology and education.[8] Official publications such as *Nauka i religiya*, a leading journal from the Khrushchev era, pressed home the link. But a point sometimes missed in Anglophone discussions of the subject is that *nauka* signified not just 'science' in the narrow sense (the natural sciences) but the entire range of intellectual and academic activity that is conveyed by the word *Wissenschaft* in German, but by the clumsy phrase 'science and scholarship' in English. The natural sciences—or at any rate, physics and mathematics—were in some respects less subject to close ideological control than the humanities or social sciences. Research that appeared to be orthodox in its materialist implications could be assigned an invisible significance by the person doing it. It was considerably trickier to research medieval history without engaging with the issue of whether religion was 'progressive'. Precisely history stood at the centre of the academic conflicts and purges of the late 1920s and early 1930s.[9]

This did not mean specialists in history were always at loggerheads with Soviet power. At stake was not just the 'conformism' of which the St. Petersburg historian Sergei Yarov wrote.[10] Rather, many practitioners of *nauka* believed in the values of scholarship and science as themselves

---

[6] V.A. Kuroedov (1982) *Religiya i tserkov' v sovetskom gosudarstve* (Moscow: Izdanie politicheskoi literatury), p. 128.

[7] For the embarrassment, see for example *Nauchnyi ateizm: Uchebnik dlya vuzov* (1973) (Moscow: Izdanie politicheskoi literatury), pp. 256–78 (this section on 'Scientific Atheist Education', with the exception of some minor alterations to statistical material, also appeared in later editions, such as those of 1974 and 1978, etc.).

[8] See for example Michael Froggatt (2006) 'Science in Propaganda and Popular Culture in the USSR under Khrushchëv (1953–1964)', D.Phil Thesis, University of Oxford.

[9] For a first-hand account of the purge against medievalists in Leningrad, see D.S. Likhachev (1995) *Vospominaniya* (St Petersburg: Logos). A good account of the strained relations between literary historians and Soviet government organisations in the early years is Mikhail Robinson (2004) *Sud'by akademicheskoi elity: otechestvennoe slavyanovedenie (1917-nachalo 1930-kh godov)* (Moscow: Indrik).

[10] Sergei Yarov (2006) *Konformizm v sovetskoi Rossii: Petrograd 1917–1920-kh godov* (St Petersburg: Evropeiskii dom).

transcendent—independent of ideology, whether of a religious or a political kind.[11] They aimed at preserving overall political neutrality, but this did not preclude allegiances on a strategic basis. P.P. Veiner, a local historian who was passionately committed to honouring and conserving Petrograd's past, was a case in point. He recollected in the early 1920s:

> When, immediately after the October Revolution, we began working for the preservation of monuments of art and history, we did not hesitate to engage in cooperation with a party that was alien to us, since we considered that the demands of the cause lay beyond all politics, and on the other hand, we are well aware that mistakes will never be forgiven; the devastation of the country's economic life can still be put right, but no action of any kind, no efforts, however great, can bring back what has been lost in this area [i.e. demolished and destroyed monuments].[12]

In Veiner's case, the attempt to cooperate was to prove a tragic miscalculation: arrested repeatedly between 1918 and 1925, he received a 3-year prison sentence in 1925. The pattern of repression began again when he returned; he was again arrested in June 1930, and the sentence of execution by shooting was carried out the following January.[13] But as a gifted amateur, Veiner was particularly vulnerable. For others in the heritage lobby, their status as practitioners of *nauka* who were recognised professionals could offer political shelter. Among these were the practitioners of scholarly architectural history who attempted to direct the preservation of heritage.[14] Arbitrating which ecclesiastical buildings might

---

[11] For a critical discussion of this, see two round-tables organised by the journal *Antropologicheskii forum/Forum for Anthropology and Culture* (2005) 'The Research Object and the Subjectivity of the Researcher: Forum 2', *Forum for Anthropology and Culture* no. 2, and 'Fieldwork Ethics: Forum 5' (2007) *Forum for Anthropology and Culture* no. 4, http://anthropologie.kunstkamera.ru/en/index/8_1/.

[12] G.A. Kuzina (1991) 'Gosudarstvennaya politika v oblasti muzeinogo dela v 1917–1924 gg'., *Muzei i vlast': Sbornik nauchnykh trudov NII kul'tury* (Moscow), vol. 1, p. 155. Cited from E. V. Minkina, 'P.P. Veiner: muzeinyi rabotnik', in *Ot Muzeya Starogo Peterburga k Gosudarstvennomu muzeyu istorii Sankt-Peterburga. Trudy GMI-SPb. Issledovaniya i materialy* (1997) (St Petersburg: GMI-SPb.), pp. 28–9.

[13] E.V. Minkina (2011) *P.P. Veiner: Zhizn' i tvorchestvo* (St Petersburg).

[14] Certainly, professional architect-conservators were not invulnerable. Aleksandr Anisimov, a leading restorer who worked particularly in Novgorod and Moscow, was arrested in 1930 as part of an alleged 'counter-revolutionary' group of 'idealists and anti-Marxists' at the Central Restoration Workshops; in 1931, he was sentenced to 10 years in labour camps, and in 1937, was executed (I.L. Kyzlasova (2000) *Aleksandr Ivanovich Anisimov (1877–1937)*,

be worthy of elevation to the status of national monument (*pamyatnik*) left them open to charges of hostility to faith, on the one hand, and covert support for 'obscurantism' on the other. But some conservators managed to negotiate this process successfully, precisely because they had recourse to the values and vocabulary of *nauka* in order to make their case.

The crucial problem behind the arbitration of the architectural status of religious buildings lay in mediating between three positions that were all in their own ways coherent but also irreconcilable. For religious believers, any church was a 'monument': to the celebration of the Eucharist on the one hand, and to the dead who were honoured within its walls—the saints celebrated in icons, and those actually entombed, including the famous dead, such as General M.I. Kutuzov, buried in the Kazan Cathedral, St. Petersburg, in 1813. More broadly, any church was a sacred place, to be kept free of desecratory [*poganye*] beings and phenomena, from secular paintings and music to dogs.[15] For militant atheists, churches were 'havens of obscurantism' [*ochagi mrakobesiya*], ripe for demolition precisely because of their function.

From the point of view of architect-conservators when speaking as professionals, on the third hand, the value of churches was entirely related to their architectural merit or lack thereof. The demolition of churches by famous architects was to be resisted at all costs—but so was the building of hideous churches and (a far more topical concern during the Soviet period) the retention of hideous churches. Professional architect-conservators considered themselves entitled to step in and direct the ways in which believers made use of churches for worship—for example, to impede reconstruction or rearrangement of a kind that they saw as

---

Moscow: Izdanie Moskovskogo Gosudarstvennogo Gornogo universiteta, pp. 57–80). His arrest came during a nationwide crackdown on 'bourgeois specialists', and when city planning was undergoing Sovietisation (see Heather DeHaan (2013) *Stalinist City Planning: Professionals, Performance and Power in 1930s Nizhnii Novgorod*, Toronto: University of Toronto Press). At this point, Konstantin Romanov and Aleksandr Udalenkov, two leading Leningrad restorers, underwent a milder form of purging (see their personal files in the Institute of History of Material Culture, St. Petersburg), after which they played only a marginal role in preservation within the city. However, this period of history (as with museums) was exceptional: there is no evidence for singling out of conservationists even during the purges of 1937–1938, let alone in the post-war era.

[15] Father Petr Nechaev (1893) *Prakticheskoe rukovodstvo dlya svyashchennosluzhitelei* (see e.g. 5th edn.; St. Petersburg: no publisher given), a guidebook for parish priests, gives a good sense of the day-to-day life of the Orthodox Church in the period just before the Revolution. It is still used for seminary courses today.

inimical to the architectural integrity of a given building. But they also felt so entitled when the users of the building came from some other social group—including the administrations of museums.[16] The relations between architect-restorers as practitioners of *nauka* and users of religious buildings were not at all straightforward and did not really conform to any of the patterns (enthusiastic atheist activism, going through the motions, or espousing certain views because conformity or prudence required this), which were broadly typical of Soviet practice in a nonprofessional context.

The evidence in this paper comes from Leningrad, and it is possible that people working in other locations had different attitudes. But from 1918 right up to 1991, it was Moscow and Petrograd/Leningrad professionals who set the tone for others across the country (as was true of museum work also), so that their behaviour patterns can be taken as having national significance, in terms of ideals, if not necessarily of reality.[17]

The so-called Cold War era does not overlap with a neat division either in the history of Soviet anti-religious rhetoric and practice or in the history of the relations between *nauka* in the sense of 'scientific restoration' and religion. Conflicts between the 'scientific' (professional-academic) view of church architecture and the religious view started to surface well before 1917. Churchmen by no means always took the conservation and protection of historical items as part of their mission. Keeping a church in good repair was deemed essential by both ecclesiastical and secular legislation and practice—but this could work

---

[16] A case in point was when the State Inspectorate of Monuments in Leningrad nagged the 'post-box institute' doing secret defence research about repainting the façade of the Saints Simeon and Anne Church, or the director of the Museum of the Arctic and Antarctic about the hooks for Soviet banners that had been drilled in the walls of the Church of the Co-Religionists. See the material in files no. 173 (Saints Simeon and Anne) and no. 481 (Co-Religionists), NA UGIOP [Scholarly Archive of the Board of the State Inspectorate for the Preservation of Monuments], St. Petersburg.

[17] It is an open question how much restoration work actually got done at all in Soviet Russia, outside the famous historic cities and individual sites, at least before the War. Judging by the documents in the Institute of History of Material Culture in St. Petersburg, in north-western Russia, activity was, in the 1920s, mainly going on in Novgorod, and even here, little got done in the late 1920s and 1930s. Files in the State Archive of the Russian Federation (GARF) indicate that work was also done in Central Asia, particularly Samarkand. From the late 1940s, the picture altered, but even then, the vast majority of normative documents (brochures about notable restorations etc.) concentrated on Moscow and Leningrad. A usefully broad survey of Soviet restoration practice that confirms this picture of late development in the so-called periphery (the Russian provinces and non-Russian republics) is the massive collection of articles ed. A. S. Shchenkov (ed.) (2004) *Pamyatniki arkhitektury v Sovetskom Soyuze: ocherki istorii arkhitekturnoi restavratsii* (Moscow: Pamyatniki istoricheskoi mysli).

to the detriment of archaeology and history. It had been recognised back to at least the 1840s that efforts to ensure decorous appearance [*blagolepie*] often involved the overpainting of historic icons and rebuilding of ancient churches. The Archaeological Commission of the Academy of Sciences controlled work on pre-1725 structures, but anything later was in the hands of the ecclesiastical authorities. Levels of frustration among professionals were high. As a contributor to the Odessa newspaper *Russian Speech*, speaking on behalf of the Society, put it in 1913, 'One of the negative features of the Russian nation is its total indifference to its own historic past. While other peoples, moving along the path of progress, never cease closely studying the fascinating life of eras long vanished, and carefully protect from destruction everything that testifies visually to the gradual development of their culture and art, we Russians, blindly caught up in everything foreign [*inozemshchina*] do not love or value our many-faceted artistic antiquity'.[18]

As a result, architect-conservators, along with many other professionals, including physicians, educationalists, and some academics, welcomed the Russian revolution of 1917, assuming that at last their hopes stood some chance of realisation. Already after the February Revolution, a statement on 'the great heritage of our ancestors' from the Union of Art Activists, a broad front for artists and writers, put forward a strong case that, when the separation of church and state took place, ecclesiastical property should be transferred to state ownership:

> The best decision is the recognition of all buildings and movable property belonging to churches, monasteries, chapels, eparchies (dioceses), and the vestries and treasuries attached to these as the property of the state in general [*obshchegosudarstvennaya sobstvennost'*], and what is more, this rule should be extended to all objects, whether they are included in inventories or not, and whether they are currently listed or have been removed from the lists. Any restriction of this rule will mean that a whole series of monuments may perish, often of first-class importance, since not all monuments are currently known about, and some of them are to be found on the rubbish dumps of property that has been removed from church inventories.[19]

---

[18] Anatoly Vol'nokhodov (1913) 'Okhrana Russkoi Stariny', cutting from *Russkaya Rech'* no. 2179, held in the files of the Defence and Preservation in Russia of Artistic and Historic Monuments, RA IIMK, f. 68, ed. khr. 1, ll. 42–42 ob. As a note in the file indicates (ibid.), the article was written with the direct cooperation of the Society.

[19] A draft of this statement is held in the personal archive of Konstantin Romanov, RA IIMK, f. 29, ed. khr. 12, l. 7–9. It was likely written by Romanov himself, who was also the author of various draft statutes on monument preservation. Typewritten in the old

This stand meant that conservators did not oppose the nationalisation of church property that took place in January 1918, but on the contrary, welcomed this. Nor did they in principle oppose the removal of property from churches that took place in the wake of the Decree on the Removal of Church Property in 16 February 1922. Rather, their efforts were directed at trying to make sure that works of art—including both churches and their movable contents—did not suffer. In this process, the authority of *nauka* was invoked as vital to the preservation of the country's heritage. In the words of a Special Commission of the Museums Department which met on 4–5 March 1922 in Petrograd:

> Churches both in Petrograd and in the provinces preserve within them a large quantity of monuments that are of exceptional scholarly and artistic value. Given the special degree to which church architecture in Rus[s]ia was developed, ecclesiastical monuments of Art and History have a special interest, and if these perish, irreparable damage will be done to world knowledge of history and art. Therefore, the confiscation of such monuments must be carried out with special care. At the same time, in many churches that are overall monuments of an elevated artistic kind are to be found objects of later date, which are often of high material value, but distort the fundamental character of the monument. The removal of such objects would only increase the scientific and scholarly and artistic significance of the churches containing them. For this reason, all confiscations of valuable items from churches must take place only after a careful scientific and scholarly and artistic examination carried out by the relevant organs of preservation.[20]

Conservators at this period were gifted with historical expertise and respect for the canons of good taste but not necessarily political acumen. The removal of church property was regarded by a significant constituency within the Bolshevik party, including, notoriously, Lenin, as an excellent opportunity to act against the power of the church itself. At the local level, many officials were not only hostile to religious 'cults' and those who espoused them but also desperate to meet targets for quantities of precious metal confiscated. They

---

orthography, the text probably dates from the early autumn of 1917. The Union of Art Activists [Soyuz deyatelei iskusstv], which held its inaugural conference in March 1917, and ceased its existence in September of the following year, is best-known as a force for the defence of artistic autonomy, but as the present case shows, state-public relations were more complicated than this interpretation might suggest.

[20] 'Zapis'' obsuzhdeniya Osoboi Komissii, sozdannoi po porucheniyu Zav. Petrogradskim Otdeleleniem Aktsentra [...]', 4–5 March 1922, RA IMMK, f. 67, ed. khr. 73, l. 79–79 ob.

accused conservators of acting as a 'fifth column' for religious believers. On 8 December 1922, officials from the Management Department of the Moscow District Soviet in Petrograd wrote to accuse Department of Preservation staff of obstruction. Every time they visited a church in order to sequestrate property, the parishioners would claim that the contents were under state protection, and monuments preservation staff would back them up. 'A Soviet organisation, [the Department of Preservation nevertheless obstructs another Soviet organisation [...] in the execution of the duties conferred on it by the state'.[21] Yet church believers were unlikely to see the 'scientific' process of filtration as less sacrilegious than the removal of church property by officials of the 'Aid to the Starving' bureaucracy, particularly as the two sets of officials were supposed to carry out confiscations in collaboration.

The emphasis on the need to perceive church architecture from a point of view informed by historical and artistic knowledge drove a wedge between the supporters of architectural conservation and religious groups in later years also. In 1924, preservationists at the Old St. Petersburg Society refused to support worshippers at the Trinity-Izmailovsky Regiment Cathedral who had protested about the conversion of a chapel outside the Cathedral into a tobacco kiosk.

> [The Society] has no intention of taking the Trinity Cathedral under its supervision and protection, and still less could that apply to the chapel, which was built in 1894–1896, has no architectural connection with the Cathedral and no artistic significance.[22]

As attitudes to religious faith hardened, so concepts of 'historical and religious significance' began to narrow. By the late 1920s, some preservationists were starting to hold the view that asserting the importance of church architecture was counterproductive. In 1928, Aleksandr Udalenkov, then head of the Leningrad Restoration Workshops, wrote candidly to Sergei Ol'denburg, Permanent Secretary of the Academy of Sciences, urging caution. It would not be possible to preserve everything, he contended, and in the circumstances, selectivity was requisite. In Moscow, efforts to preserve some churches had simply led to the demolition of the (by implication much more interesting) baroque monument, the Red Gates:

---

[21] See the report of 15 December 1922, TsGA-SPb., f. 1001, op. 7, d. 19, l. 20.
[22] Ibid., l. 36.

And so I would see it as essential to avoid any protests this time round and to, as it were, firm up the ground that we will have to stand on when it comes to another, more serious, case. One has to resign oneself to the fact that, in order to save the most valuable monuments of history and art, we will have, and probably on more than one occasion, to sacrifice a whole series of interesting and noteworthy, but still in sum less valuable, ones.[23]

If churches were saved, the problem of what to do with them was ideologically vexing. One possible response was the siting of museums of 'scientific atheism' precisely in churches that were considered to be of the first architectural importance. The agitational displays were organised as explicit, aggressive interventions into the devotional landscape.[24] But even so, there was always the risk that the church architecture might have trumped its setting. At a blunt discussion in the Leningrad City Committee of the Communist Party on 3 January 1939, Aleksei Kuznetsov, then Second Secretary of the Committee, acknowledged the problems of combining, in the St. Isaac's Cathedral, a respect for the 'colossal value of the building as a historic work of architecture' with agitational ends (the building then housed the Anti-Religious Museum, run by the Union of Militant Atheists):

> Whatever you might say about anti-religious propaganda in St Isaac's Cathedral, the architecture itself and the construction of the building are bound to dominate. Why should we deceive ourselves?

In the event, Kuznetsov's conclusion was neither that the museum should be shut nor that a purely architectural display should be mounted. Instead, the anti-religious side of the museum's work should be strengthened and the excursion leaders given a more thorough grounding so that their comments about Foucault's Pendulum expressed the due political orthodoxy (Fig. 10.1).[25]

A month later, the Union of the Militant Godless reported that—perhaps in response to the criticism?—'in the Museum, the exhibition 'The Origins

---

[23] NA IMMK, f. 67, ed. khr. 86, l. 13.
[24] For an excellent discussion of the Anti-Religious Museum in St. Isaac's Cathedral, see Adam Jolles (2005) 'Stalin's Talking Museums', *Oxford Art Journal* 28.
[25] TsGAIPD-SPb., f. 25, op. 2, d. 2000, ll. 10–11.

RELIGION AND *NAUKA*... 237

Fig. 10.1 Foucault's Pendulum in the Museum of Atheism, St. Isaac's Cathedral, early 1930s. The slogan to left reads, 'THE STRUGGLE FOR A SCIENTIFIC WORLD-VIEW' (Frontispiece to *Iz ochaga mrakobesiya v ochag kul'tury* [Leningrad: OGIZ-Priboi, 1931].)

of Religion and Its Harmfulness in Class Terms' has just been opened'.[26] While the siting of anti-religious exhibitions in 'top category' architectural monuments represented a concession by the Party and city authorities that these were historically significant, the primary function of anti-religious museums definitely remained the struggle with 'obscurantism'.[27] In the mid-1920s, there had been plans to create a historical museum of church interiors, declaratively known, from 1925, as the Museum of Dying Cult (its first title was the Museum of Church Life). By 1930, this plan had been shelved and the collections of the planned museum dispersed; at this period, *nauka* was expected to play an explicitly engaged role.

Architect-conservators were not the only group who had to patrol this tricky front line. Anthropologists, or *etnografy*, to give them their usual Soviet title, were also expected to perform complicated acts of juggling: preserving historical material, representing it in politically expedient ways, and using it to the ends of cultural transformation (and sometimes all at the same time).[28] This was not a straight binary conflict between *nauka* and *religiya* (or more broadly, material associated with alternative, non-socialist, belief systems), or indeed, *nauka* and expediency. Rather, ideological rectitude, the Soviet civilising mission, the effort to educate in an intellectual sense, and a commitment to the custodianship of cultural material for its own sake were all constituents of the mandate. Museum curators who performed only part of the brief were liable to castigation—but this applied just as much to those who neglected the curatorial role as to those who failed in propaganda terms. The criticism aimed at the Anti-Religious Museum in St. Isaac's Cathedral during 1938 was aimed not just at the neglect of serf labour in the displays about the construction of the building and not just at the fact that portraits of Lenin and Stalin were placed far too near the icons. The curators were also lambasted because the reserve collections

---

[26] TsGAIPD-SPb., f. 24, op. 10, d. 421, l. 20.

[27] This mission is set out, for instance, in the planning materials for the Museum (see TsGALI-SPb., f. 276, op. 1, d. 37, passim; ibid., d. 45, passim). An aggressively anti-religious policy was also supported by rank-and-file 'militant atheists': see for example the letter from Yuly Blok, 2 June 1940, which contrasts the light and crowds in the Leningrad working churches for Easter 1940 and the power-cut state of St. Isaac's: TsGAIPD-SPb., f. 24, op. 2в, d. 4410, ll. 43–44.

[28] The St. Petersburg anthropologist Dmitry Baranov has dealt with the ethnographical museum in an excellent series of articles; see for example his contribution to Bassin and Kelly (eds.) (2012) *Soviet and Post-Soviet Identity*, and '"Imparting" Traditions: How the National Past Was Written into the Narrative of the Soviet People', *Forum for Anthropology and Culture* no. 9, 184–99.

were housed in damp and chaotic basement areas and because no proper inventories had been carried out. The many and confusing lists had totally unprofessional descriptions of the items: 'Oil painting 158 × 120 cm' and so on. Among the jumble of files were folders titled 'Sum ole robbish' [*vsyakaya irunda*] and 'Look inside for contents'.[29]

This sense of fragmented and contradictory objectives was in part resolved by the shift in attitudes to the national heritage, including both historic buildings and the Orthodox Church itself, that emerged during and after the Great Patriotic War. Under the First and Second Five-Year Plans of 1928–1938, churches that had undergone meticulous restoration during the first years of Soviet power were summarily torn down.[30] In 1941, buildings that had previously been threatened with demolition were carefully protected from air-raids.[31] As heritage came under threat, its value increased. If the compatibility of 'Soviet' identity with an attachment to the past had always been controversial, even after the rise of 'national Bolshevism' in the late 1930s, once the patrimony began to be destroyed in enemy action, its ideological significance rose.[32] This included historic buildings constructed for the purposes of 'cults'.[33]

But if it had now become much easier to argue for the national importance of church architecture, not all its possible meanings were equally valid. One of the leading younger figures in the campaign to preserve heritage was Nikolai Voronin (1904–1976), a graduate of the Academy of the History of Material Culture in Leningrad. Forced to mask his devotion to Old Russian buildings in the mid-1930s, Voronin returned to his subject at the end of the decade, in more clement times. In 1945, he published two studies of medieval architecture, and in the spring of the same year, championed heritage at the All-Soviet Architectural Congress. A patriot who in the late 1940s sometimes adopted 'anti-cosmopolitan' rhetoric

---

[29] TsGALI-SPb., f. 276, op. 1, d. 14, l. 2.

[30] An example was the wooden Trinity Cathedral on Revolution Square in Leningrad, demolished in 1933, though it had been extensively restored in the 1910s and 1920s after a fire in 1922.

[31] For example, the Trinity-Izmailovsky Regiment Cathedral.

[32] During 1942–1943, the illustrated weekly *Ogonek* regularly carried pictures of heritage sites, such as Tolstoy's estate at Yasnaya Polyana, that had been destroyed by the 'fascist invader'.

[33] See for example Academician Igor Grabar, 'Vosstanovlenie pamyatnikov stariny', *Sovetskoe iskusstvo* 28 November 1944, p. 2. I discuss this wartime change more extensively in 'The Shock of the Old: Architectural Preservation in Soviet Russia', forthcoming in *Nations and Nationalism* (special issue on European heritage, ed. Mark Thatcher), 2016.

(as when denying Belorussian influence on seventeenth-century Russian culture), Voronin still remained loyal to Communist ideals, including the belief that religion and rational thought were incompatible. The idea that 'an intellectual' might also be a believer made him incredulous.[34]

All the same, the *products* of religious thought (if not the thought they had stemmed from) now had an enhanced status—provided their original purpose was tactfully downplayed. Policy at the local level was also affected by the nationwide emphasis on the sacrosanct character of the buildings damaged by the invaders. In the words of a report sent by local party officials in Novgorod to the Leningrad Provincial Committee of the Communist Party:

> The German barbarians have destroyed all the institutions of political education and objects of material value. For instance, in the city of Novgorod, the monument to the Millennium of Russia has been destroyed, and also the eighteenth-century Palace Tower, the twelfth-century Church of St Theodore the Stratilate, etc., also the library of the city of Novgorod, which had more than 80,000 books in it. The golden domes of the St Sophia and St George Cathedrals have been carried off to Germany.[35]

Medieval church architecture could now be mentioned in one breath with books, and even 'institutions of political education'. Precisely such architecture, too, was at the centre of a survey of war damage organised by the Board of Architecture of the Leningrad Provincial Soviet in June 1944.[36]

As well as national resonance, the changes had international resonance. One of Stalin's motives in reaching rapprochement with the Orthodox Church in 1943 was the difficulty of justifying to his American

---

[34] See the critical, but balanced account of his life in A. A. Formozov (2004) 'Rol'' N.N. Voronina v zashchite pamiatnikov kul'tury Rossii', *Rossiiskaia arkheologiia* no. 2, 173–80.

[35] 'Spravka o sostoyanii narodnogo obrazovaniya v raionakh oblasti, osvobozhdennykh ot nemetsko-fashistskikh zakhvatchikov' January 1944 (exact date not given), TsGAIPD-SPb., f. 24, op. 11, d. 198, ll. 1–2.

[36] 'Tablitsa stoimosti ushcherbov, prichinennykh otdel'nym pamyatnikam arkhitektury, nakhodyashchikhsya v gorodakh Novgorode, Gdove, Staroi Russe, Tikhvine, Gatchine, Ropshe, Usad'be, Gruzino i sostoyashchim pod gosudarstvennoi okhranoi', TsGANTD-SPb., f. 388, op. 1–1, d. 2, l. 1–3.

and British allies what they saw (with good reason) as the persecution of religious believers.[37] In the period after the War, transnational contacts in the Russian Orthodox Church were to become significantly more important.[38] Amid the cultural contests of the Cold War, church–state relations became the subject of propaganda conflicts.[39] In 1948, for example, TASS, the official news agency, put together a photoreportage about the Catholic Church of Our Lady of France, the only working church belonging to that faith in the city. As the TASS staff put it, 'These photographs will be used to disprove the slanderous rumours about the Soviet Union being put about abroad'.[40]

It was not just believers who had propaganda value. What foreigners might think of decayed churches mattered too. In 1955, Metropolitan Grigory used that as an argument when pushing for the reopening of the Trinity Cathedral in the Alexander Nevsky Monastery (Fig. 10.2):

> Large numbers of delegations from foreign churches have noted, when they are visiting the Metropolitan of Leningrad, the wretched condition of the Monastery's Cathedral (which is located right next door to the Metropolitan's quarters) and have openly expressed their astonishment with regard to this, and particularly given the information that is invariably provided to guests by the Metropolitan of Leningrad about the inner freedom of the Church and the favourable inclination towards it of the Soviet Government.[41]

---

[37] See for example Nathaniel Davis (1995) *A Long Walk to Church: A Contemporary History of Russian Orthodoxy* (Boulder: Westview Press).

[38] A case in point was Metropolitan Grigory (Chukov) of Leningrad, whose diary of his 1947 visit to the USA is available in L.K. Aleksandrova (ed.) (2012) *Dnevnik; Fragmenty. Arkhiv. Istoriko-bogoslovskoe nasledie mitropolita Grigoriya (Chukova)* (St Petersburg, http://www.bogoslov.ru/text/2908919.html.)

[39] This was also true in the late 1920s, but at that period, there was less effort on the part of the Soviet authorities to prove that believers were being humanely treated. Wheeling out clerics of various hues (rabbis and imams as well as priests) to make claims that there was religious freedom in the USSR became a much-repeated routine in the Cold War era, as the files of the Plenipotentiaries on Religious Affairs make clear.

[40] TsGA-SPb., f. 9620, op. 1, d. 9, l. 25. For an indication of the possible prompting of this—a story in the French press about the arrest on charges of spying of Madame Bovard, a parishioner of the Church of Our Lady of France, and a resident of the city since 1911, as well as a holder of two medals for wartime bravery and 'noble labour' (*doblestnyi trud*), see ibid., ll. 1–2.

[41] ASPbE, f. 1, op. 7, d. 40, l. 12.

Fig. 10.2 Metropolitan Grigory of Leningrad, late 1940s (http://www.bogoslov.ru/text/2908919.html)

In the case of the Trinity Cathedral, this argument seems to have carried weight; permission was given for the reopening shortly after this missive was sent. And as the numbers of foreign travellers rose in the following decades, concern about the condition of churches also rose incrementally.[42]

---

[42] By the start of the 1980s, the Council on Religious Affairs was closely monitoring the state of all religious buildings that stood along so-called tourist routes (i.e. those which foreigners were officially allowed to use on visits to the USSR). Grigory Zharinov, in response to a telephone enquiry, filed a list of such buildings in April 1982, in which he claimed that

The perception that conservation, previously seen as an obstacle to 'socialist construction', was vital to post-war reconstruction brought improved resources for key restoration projects, including the palaces outside Leningrad but also, for instance, the medieval churches of Novgorod.[43] Believers quickly grasped the new attention being given by the secular authorities to church architecture and began to invoke the historic significance of churches in their petitions for reopening. They might also emphasise that they intended to contribute to the preservation of heritage by taking a given church into their care. The clergy also oversaw this process. In 1948, when Father Ioann D'yakonov, a parish priest in Tikhvin, wrote to enquire whether the Cathedral of the Dormition might be opened for worship, the Metropolitan's secretary, Father Evgeny Lukin, advised him in considerable detail about best approaches:

> In the application, you should particularly emphasise that your Religious Community requests the Uspensky Church should be transferred to it for use simultaneously and sequentially with the Tikhvin Church, and that *this request is prompted mainly by the need to support and protect from destruction a monument of history.*[44]

Father Lukin could not have made such a suggestion had he not conferred first with Metropolitan Grigory of Leningrad and sought his blessing for it.

At the same time, interest in church architecture remained selective. All over Soviet Russia, churches that were deemed to be of secondary importance stood unrestored into the 1980s.[45] The principles of 'scientific restoration' and the traditions of ecclesiastical architecture were still distant, if not antagonistic. The rise in the Orthodox Church's influence was offset by a considerable growth in the institutional status of the monuments

---

all such churches in Leningrad province were in good condition, including those seldom or never visited by tourists anyway: TsGA-SPb., f. 2017, op. 1, d. 77, ll. 54–55.

[43] The Novgorod reconstruction works are extensively discussed in Shchenkov (ed.), *Pamyatniki arkhitektury*. See also Victoria Donovan (2011) '*Nestolichnaya kul'tura*: Regional and National Identity in Post-1961 Russian Culture', D.Phil Thesis, University of Oxford.

[44] ASPbE [Archive of the St. Petersburg Eparchy], f. 1, op. 11a, d. 10, l. 3. Emphasis added.

[45] For example, in Voronezh, where I lived as a student from 1980–1981, several churches in the centre were still burnt-out shells, and one of these, dating from the sixteenth century, was being used as a rock-climbing range.

protection authorities, granted new authority by nationwide legislation in 1948. In the tussle between religious communities and heritage preservers, the heavy weaponry was on the side of the latter, since users who failed to carry out repairs could be evicted for breach of contract—a threat that carried weight in the case of 'cultic groups', which were socially marginal, indeed, formally 'non-Soviet'. In Leningrad, Nikolai Belekhov, head of the State Inspectorate of Monuments, was an especially formidable figure. When he directed the Orthodox parishes who were using the Cathedral of the Transfiguration and the St. Vladimir Cathedral to redecorate the churches' domes with gold leaf, the parishes had to do this, despite the high costs of the operation and the fact that the diocesan authorities and the Plenipotentiary of the Council on Church Affairs took their side.[46]

The monuments preservation bodies continued to have the edge in later decades of Soviet power as well. Their institutional power declined in the Khrushchev era, a period of radical modernisation in the fields of architecture and planning, but so did the leverage of believers. The institutional power of the church rose under Brezhnev, but so did the capacity of the heritage lobby to influence policy. While there were some signs of a greater rapprochement towards religious groups on the part of the monuments inspectorate, no effort was made to expand the number of religious buildings that were placed at the service of 'cults'. Rather, the assumption was that allocating churches for 'cultural and educational purposes' (which had also been the strict letter of the regulations in the early Soviet period) would continue to be the appropriate policy.

What in practice this often meant was using churches to house museum displays, or sometimes, though less often, using them as museums in their own right. Usually, the premier architectural monument in a given city or town would be a church–museum, in other words, somewhere to which visitors were taken on excursions that primarily focused on the architecture of the building as an expression of the creative genius of 'the people' (*narod*).[47] Sometimes—as in the case of the St. Sophia Cathedral in Novgorod, the St. Sophia Cathedral in Vologda, or the churches of the Pskov, Pereslavl'-Zalessky, and Old Ladoga Kremlins—the church–museum would form part

---

[46] For the case of the Prince Vladimir Cathedral, see ASPbE, f. 1, op. 7, d. 18, ll. 29–34. On l. 34 is a note: 'Gold leaf issued. 30 July 1949'.

[47] I remember this obligatory element in excursions when visiting Russia as a student in 1979–1981: we were told, for example, that the architecture of Novgorod was a tribute to the glorious achievements of the *narod*, while at the same time being told that the architecture on the far bank of the river was an expression of merchant taste.

of a 'museum reserve', in which an entire old city quarter was preserved as a testament to its time.[48] Certainly, the attitudes to church architecture did change at this period, both in the intelligentsia generally and among campaigners for heritage. In the 1970s, the All-Russian Society for the Preservation of Monuments of History and Culture, an association that linked whose members included professional conservators and museum-workers (in Leningrad, its chairman was Boris Piotrovsky, distinguished archaeologist and the director of the Hermitage) devoted most of its public campaigning to secular buildings (Fig. 10.3).

But below the surface, representations were made about the fate of important churches, including, for example, Vasily Kosyakov's Mother of God of Kazan' Church in the New Maiden Convent, in a neo-Byzantine style that had previously been considered out of keeping with Leningrad's best architectural traditions. Some heritage campaigners even pressed for the rebuilding of the Church of the Saviour on the Haymarket, demolished in 1961. But it was as cynosures of the urban landscape that the churches were championed not as buildings that were used for worship. A VOOPIiK drive to return buildings to the uses for which they had been constructed left churches off the list.[49]

The 'museum reserve' drive too was primarily concerned with preserving the exteriors of historic buildings. The interiors, on the other hand, were meant to be used for exhibition space (in provincial towns, the preferred recourse was to hand them over to the local history [*kraevedcheskii*] museum). But what to do with interiors that were regarded as too architecturally sensitive to house glass cases and other alien objects? An article by M.E. Kaulen published as late as 1992 pointed to the fact that attitudes to churches had changed in the post-war era: 'Placing the

---

[48]'Museum reserves' began to be set up in the 1940s, but the pace of creating them increased from the 1960s, till, by the early 1980s, historic areas in most major towns and cities were so denominated, both in the RSFSR and beyond (e.g. the old centres of Samarkand, Bukhara, Baku), alongside individual sites such as Etchmiadzin Cathedral in Armenia, and others. For a discussion of one particularly important museum-reserve, the wooden architecture museum on the island of Kizhi, see Sanami Takahashi (2009) 'Church or Museum? The Role of State Museums in Conserving Church Buildings, 1965–1985', *Journal of Church and State*, 51:3, 502–17. For a memoir from a local historian directly involved in the creation of the museum reserve at Yaroslavl', see M.G. Meierovich (2004) *U menya poyavilas' mechta* (Yaroslavl': Aleksandr Rutman).

[49] For further details of this period, see Catriona Kelly (2013) 'From "Counter-Revolutionary Monuments" to "National Heritage": The Preservation of Leningrad Churches, 1964–1982', *Cahiers du monde russe* 1, 1–30.

Fig. 10.3 Boris Piotrovsky (1908–1990) photographed in 1965 (www.hermitage.ru)

church–museum on display as a work of art of an autonomous kind and an exhibit in its own right was a strategy that was 'rehabilitated' in the 1950s, 1960s, and 1970s, as restoration and reconstruction practices developed, and museum-reserves were organised, and as new attitudes to museum work emerged'. But Kaulen also questioned whether the use of historic church architecture for its original purpose was necessarily the best outcome, noting that there was still controversy about 'whether returning a church to its original functions would have positive effects'. On the one hand stood the fact that visitor surveys had indicated people interpreted empty buildings as 'cold' and simply could not relate to them. On the other there was the problem that placing alien (i.e. modern) objects in a historic interior was quite out of order and that ecclesiastical use risked damage to frescoes, wall-paintings, and icons. Above all, there was the fact that the architecture partly vanished from view. 'In churches that are actually being used for worship, factors such as the artistic merits of the architecture and its characteristic features, and indeed its antiquity, cease to strike the viewer; the decorations in the church and the characteristic atmosphere of the Orthodox place of worship tone down the impact of the architecture itself, impede appreciation of the church as a work of art'.[50] Kaulen's solution was that the church should be reanimated in a 'virtual' way. She commended cases where audio–visual installations had been set up or information provided in the form of a short film; even using lighting creatively could be very helpful.[51] Well into the twenty-first century, in church buildings that were considered to be of the first architectural importance, services would be held at most occasionally, but an impression of more regular use would be generated by, for example, laying on live choral singing when parties of tourists came round.[52]

In this perspective, it is not surprising that the new legislation in 1992 allowing believers the right to request the return to them for worship of 'cultic buildings' generated especial friction when these buildings were in

---

[50] M.E. Kaulen (1992) 'Problemy sovremennogo ekspozitsionnogo pokaza kul'tovykh inter'erov i perspektivy ispol'zovaniya kul'tovykh sredstv', in E.A. Shulepova (ed.) *Voprosy okhrany i ispol'zovaniya pamyatnikov istorii i kul'tury* (Moscow: Nauchno-issledovatel'skii institut kul'tury), pp. 117–21.

[51] Kaulen, 'Problemy sovremennogo ekspozitsionnogo pokaza', pp. 123–6.

[52] This was the case, for example, in the summer of 2013, when my sister and I visited a large number of churches in Moscow, Uglich, Yaroslavl', and Kirillovo-Belozersk; most were non-active, but in the majority of those, there was live choral singing (and, in a specifically post-Soviet touch, sales of the choral group's CDs at a stall alongside).

use for the purposes of *nauka*. For example, one major eighteenth-century church in St. Petersburg had been ceded, in the 1970s, to a scientific institute. A parishioner who had been involved with the parish since soon after it was founded in the late 1980s conceded that the institute staff had ('well done them!') succeeded in saving the iconostasis from destruction. But this did not mean they were at all keen to share the building with worshippers:

> We almost didn't have services back then. Why? Because the institute didn't want it. [...] They tried anything they could, fire regulations and stuff. They said we had candles burning, and all that—lots of complications. And actually the fire precautions wasn't there in any case, they'd never bothered about any of that stuff, not a bit. And at first it was tough going, it really was. Every time—every time!—we had to ask permission from the folks at the institute to have Mass.

In fact, it was only when the local film studio used the church to shoot scenes from a bio-pic about eighteenth-century court life that icons were re-hung on the walls—a situation that incidentally benefited believers. While they were able to keep the icons even after the film crew had left, it took several more years before finally, in 1993, the institute moved out. 'And then, for the first time, we took the awful partition down that was separating us, and we saw the whole cathedral, and the state it was in'.[53]

A scholar working at the institute when all this happened put a different spin on everything. The institute had remained there only because alternative premises were at first not available. But he too emphasised the incongruity of the situation. 'Well, to be honest, we felt, so to speak, a bit ... awkward, really, occupying a building like that ... it probably shouldn't have been storage space to begin with [...] We'd be running round with our bones—the scientific ones, I mean—and there'd be a service going on'. His underlying attitude was defensive; sure, the place was in terrible condition, but that was the fault of another institute, which had been using it before. Insisting that relations with the parishioners had been perfectly good, he expressed real human warmth only when it came to an inspector from the monuments department: 'He became a really good friend'.[54]

---

[53] From an interview with a long-standing parishioner of the St. Petersburg church concerned by Alexandra Piir, Oxf/AHRC-SPb-09 PF16 AP. For a similar account, see her interview with the church elder, Oxf/AHRC SPb-08 PF12 AP.

[54] Interviewed by Alexandra Piir, Oxf/AHRC SPb-08 PF8 AP. The informant was at all points extremely cautious about giving the impression of conflict, and was only 'off the record' prepared to cite a completely different case—a famous monastery—where he had heard from a

The church under contention here was, while of acknowledged architectural importance, not one of the buildings that had been considered the city's premier 'monuments' in the Soviet era. In the case of these, both in Leningrad-St. Petersburg and all over the country, the claims of *nauka* endured into the post-Soviet period. In St. Petersburg, three of the city's most important historical sites, the Smol'nyi, St. Isaac's, and St. Peter and St. Paul Cathedrals, were all used mainly (in the first case, exclusively) for 'cultural purposes'.[55] One deciding factor was the state of preservation of the interior. So, the Kazan' Cathedral, also a building of enormous architectural importance, but stripped of most of its decoration in the late 1920s, was returned for worship, while the Cathedral of the Saviour on the Blood, which had retained its mosaics, was not, though the former was regarded as the more significant architectural achievement.

Alongside the widespread questioning of the value of *nauka* both inside academia and beyond, the tenacity of the museums lobby was striking. It came in the face of sharp criticism of scholarly approaches to church architecture from rank-and-file believers. In Moscow, these were particularly fierce. As a contributor to a St. Petersburg believers' forum lamented:

> How many more so-called mon[u]ments of architecture all over Russia have still not been handed over to the Church! When everyone's known for ages—culture never saved anyone, or the people generally, from sin and degeneration. You can go further: a concert hall, a gallery or art museum, still more a 'multi-functional whatnot' set up in a church, are simply sacrilegious.[56]

---

friend that relations were much more vexed than in the church his institute had used. Another informant (Oxf/AHRC SPb-08 PF9 AP) was considerably more explicit about conflicts with the Church. 'There was one of the priests in charge back then. He behaved... well, not very courteously and politely'. As well as presenting his memories of this case, he mounted a wide-ranging attack on the Church's enhanced social status (its access to tax concessions, its obstruction of museification projects, e.g. in Tobol'sk, and its privileges in terms of government subsidies for restoration). While not always factually accurate (the informant had the impression that government subsidies for restoration were commonplace in the Soviet period, whereas they were reserved for buildings considered to be of extraordinary merit), the account given here was highly characteristic of the corporate values held by architect-restorers, museum-workers, archaeologists, and others professionally involved in the national heritage industry.

[55] Smol'nyi and St. Isaac's formed part of the 'Museum of Four Cathedrals', along with the Church of the Saviour on the Blood and the Sampson Cathedral; the St. Peter and St. Paul Cathedral was run by the Museum of the History of St. Petersburg.

[56] See the post signed Lyudmila Il'yunina on the site, 'Russkie tserkvi—O khrame Presvyatoi Bogoroditsy na Rybatskom prospekte', http://russian-church.ru/viewpage.php?cat=petersburg&page=48.

But such views met equally vehement criticism from other sections of Russian society. The journalist Dmitry Verkhoturov, writing for the anti-religious site scepsis.ru in 2007, accused the Church of concentrating on historic buildings for purely pragmatic reasons:

> Over decades, museums have carried out restoration work and kept buildings in good condition. It is not surprising that it is these buildings that the Russian Orthodox Church has been eying up in the first instance. Unlike big companies, museums don't have large amounts of funding to lobby their interests and to defend themselves from attacks, which is why they have been the first places where conflicts with the church have come about.[57]

Members of the church hierarchy, who were no doubt aware of the sensitivity of restitution politics, tended not to get involved in disputes. Indeed, they were not necessarily eager to sign up to the exacting and costly requirements of 'scientific and scholarly restoration' that were still required of those using historic buildings.[58] Despite legislation in 2010 that returned ownership of all religious buildings in use before 1917 to the faiths that had once been their owners, there was no wholesale transfer to worshippers of the major historic churches that remained in use by museums—a situation that seemed to be tacitly accepted by those at the top of both lobby groups, if not necessarily at the grassroots.

As this paper has shown, the social effects of the secularisation movement in Soviet Russia went well beyond the crudities of 'scientific atheism' and their failure to make much impact on significant sectors of the population. The supporters of heritage preservation made up an influential and articulate group of professionals who, while rejecting the early Soviet propaganda image of the church in an architectural sense as a 'haven

---

[57] Dmitry Verkhoturov, 'Bor'ba tserkvi za imushchestvo, nevziraya ni na chto' (20 September 2007). First published on 'Regiony Rossii' site; see http://scepsis.ru/library/id_1486.html. Cf. L. Vorontsova, 'Razrushat' li muzei radi tserkovnogo vozrozhdeniya?' (extract from her book *Religiya i demokratiya*, Moscow, 1993), http://scepsis.ru/library/id_904.html (accessed 17 March 2015): 'Only a few cultic monuments were handed over to museums, and it goes without saying that exactly these were the ones that have survived to the present in the best shape'. This assessment leaves out of consideration the fact that churches used for worship were also kept in very good condition.

[58] In March 1990, Metropolitan Alexii declined to accept a further 20 churches on behalf of the Leningrad Eparchy, including the Royal Stables Church and St. Isaac's Cathedral, on the grounds that this was simply not practical (ASPbE., f. 1, op. 11, d. 99, l. 50; one reason was a shortage of priests, the other that the interior of the Royal Stables Church was presumably in terrible condition).

of obscurantism', at the same time took a detached attitude to the Church in a theological sense.[59] Defending the claims of *nauka*, they aimed to protect heritage, including historic churches, in a manner worthy of its transcendental ideals, opposing themselves alike to demands for wholesale destruction and to indiscriminate retention. The widespread claims that the project of secularisation in Russia and the former Soviet Union was a 'failure' should be offset against the striking success of Russian architect-conservators in pushing through their own vision of how historic churches should be used—with respect above all for their aesthetic and architectural features, rather than their religious history. And this vision, unlike much else in the church–state relationship, survived with remarkable tenacity into the twenty-first century. Many Russian intellectuals who were not 'militant atheists' were deeply concerned about the likely fate of religious buildings in the hands of believers and convinced that the rebuilding of demolished churches was more likely to detract from the appeal of the urban landscape than to add to this.[60] Though city planning was an ideologically sensitive and politically central area in Soviet institutional relations, many of the values that it propounded and disseminated were accepted as appropriate by academics, and the reflective public, both under Soviet power and afterwards. Thus, the conflict between *nauka* and *religiya* did not lead to an all-out one-sided victory in 1991; in significant ways, the Soviet understanding of such 'sciences' as architectural history continues to shape the life of the Russian church today.

[59] This was undoubtedly related to the long history of anti-clerical attitudes to the Orthodox Church in the Russian intelligentsia: as Ariadna Tyrkova-Williams put it in her memoirs, first published in 1954, 'the only thing that stopped the intelligentsia from storming the Church was police regulations' (*Vospominaniya* [Moscow: Slovo, 1998], p. 242). This recollection has all the more authority in that Tyrkova-Williams was a liberal who wrote her memoirs in emigration, in other words, was not contributing to a Soviet legitimation narrative.

[60] See, for instance, the debate on reconstructing churches in St. Petersburg that grew up round the Live Journal entries by a widely-read local historian in 2011 (http://babs71.livejournal.com/486024.html?page=1#comments), accessed 17 March 2015. One participant, for example, observed: 'it's all subjective, of course, but: it's not a good idea to pull churches down, or anything else for that matter—but if the architecture's bad, then you have to, only replace it by something better'. In 2013, when I took part in a debate about conservation broadcast on local TV, all the participants (who included an architectural historian and a theatre historian as well as a member of an architectural preservation lobby group) agreed off-air that the rebuilding of churches was not a good idea, with the possible exception of the Church of the Saviour on the Haymarket, whose demolition has, since the 1960s, been generally acknowledged as a disaster (for more on this topic, see Kelly, 'From "Counter-Revolutionary Monuments"').

CHAPTER 11

The Antireligious Museum: Soviet Heterotopia between Transcending and Remembering Religious Heritage

Igor J. Polianski

Throughout the Soviet period, science and atheist propaganda and the practical struggle against organised religion were an integral part of the Bolshevik project aimed at the 'disenchantment of the world'.[1] In the years following the Bolshevik Revolution innumerable forms of persuasive media and techniques were used to elevate Marxism over religion: booklets and calendars, radio and films, public lectures and city tours, planetaria and observatories, antireligious exhibitions and 'museums of atheism'. Yet in regard to the third decade of the post-Communist era, it seems to be a common perception among the public and academics these days that all of these antireligious campaigns were for naught and failed to achieve their aims. In everyday life and in the political sphere, battles between religion and its opponents still flare up in the societies of the former Soviet bloc. In Russia's case, this assessment has bred political scandal, which culminated

---

[1] M. Weber (1994) 'Wissenschaft als Beruf: 1917/1919' in: *Studienausgabe der Max-Weber-Gesamtausgabe* (Tübingen: Mohr), vol. 17, pp. 1–25, p. 9.

I.J. Polianski (✉)
Universität Ulm, Ulm, Germany

© The Editor(s) (if applicable) and The Author(s) 2016
P. Betts, S.A. Smith (eds.), *Science, Religion and Communism in Cold War Europe*, St Antony's, DOI 10.1057/978-1-137-54639-5_11

in 2003 when Orthodox protestors defaced the art exhibition 'Caution, Religion!' at the Andrei Sakharov Museum in Moscow. Ironically, the USSR was constantly criticised for its restrictive attitude towards the Church; today the Russian Federation is criticised for the growing influence of Orthodoxy.

In recent years, there has been growing academic interest in the means and strategies that allowed religion to survive during the Soviet period. The present chapter deals with a very particular niche in which religion persisted: the so-called antireligious museums, later known as museums of atheism—institutions that preserved religion even as they were devoted to destroying it. The central concern of this chapter is to question the still predominant view that reduces the communist struggle against religious belief to the polar oppositions of the Cold War, ignoring the central debates in the history and sociology of religion in Western societies.[2] Using the case of the museums of atheism, it assesses the specific transformation of belief imposed by state-sponsored atheism in the broader transnational context of religious reinforcement and dislocation of belief during the last decades of the twentieth century.

BETWEEN CONSERVATION AND EXTERMINATION: THE DILEMMA OF THE MUSEUM OF ATHEISM

Until the late 1980s, museums of atheism were characteristic landmarks in every major city of the USSR. By 1990, Ukraine alone had 11 major institutions of this kind.[3] The museums were usually housed in the most important local cathedrals. Leningrad's museum resided proudly on the Nevsky Prospekt, the city's main boulevard, in the magnificent Kazan Cathedral.

In the Crimean city of Yevpatoria, the central mosque Juma Jami (*Cuma Cami*) served as a museum; in the Buryatia region, the museum was housed in a large Buddhist Datsan. Even the legendary Ukrainian town Dikan'ka, forever associated with devilish machinations thanks to Nikolai Gogol's fantastic stories, had its own museum of atheism. The magazine *Nauka i zhizn'*

---

[2] See, for instance, J. Anderson (1994) *Religion, State, and Politics in the Soviet Union and the Successor States 1953–1993* (Cambridge: Cambridge University Press); S.P. Ramet (1993) *Religious Policy in the Soviet Union* (Cambridge: Cambridge University Press).

[3] See N.N. Zlatsen and E.N. Demina (1990) *Muzei SSSR, Spravochchnik* (Moscow: Ekonomika).

(*Science and Religion*) reported that its foundation was in keeping with the spirit of the great writer: the brave Cossacks from the works of Gogol would always poke fun at the devil. Accordingly, officials chose to house the museum in the church which in Gogol's *The Night Before Christmas* annoys even the devil.[4] By 1929 there were 30 antireligious museums; in 1932 there were 45; by the 1970s there were hundreds.[5] Yet all these museums disappeared in the political upheavals of the 1990s.

How can we situate these intensive efforts of collection, preservation and exposure of the religious material culture in the history of museums in general? The starting point of this paper is a consideration of the French sociologist Maurice Halbwachs (1877–1945) and his concept of collective memory, which he expresses as follows:

> If society preserves elements of ancient rites and beliefs in its religious organization, this is not just to satisfy its most undeveloped groups. But to appreciate a religious movement or religious progress exactly, people must recall, at least in rough outline, the point from which they took their departure long ago ... This is why the light shed by Olympian cults on the universe and within the innermost aspects of the human soul appeared more splendid to the extent that nature offered certain places of shadow and mystery still haunted by monstrous animals or evil spirits born of the earth and that there existed in the soul terrors through which the civilized men of that time were still allied to primitive tribes ... In order to depict the superiority of the Olympian powers it becomes necessary to evoke however vaguely the ancient assault of the giants and the crushing or enslaving of the old gods.[6]

Similarly, during the years following the establishment of the Soviet Union the Bolsheviks efforts to eradicate religion highlighted the old cults—against which the atheist heroes and the supremacy of the secular worldview shone all the more brightly.

---

[4] N.V. Gogol' (1831) 'Noch' pered Rozhdestvom', in: *Vechera na khutore bliz Dikan'ki* (Evenings on a Farm near Dikan'ka). V. Evseev (1976) 'Dikan'ka dalekaya i blizkaya', *Nauka i religiia* 6, 13–27, 13. See also K. Stepin (1976) 'Taina komnaty No. 101', *Nauka i religiia* 6, 17–20.

[5] M.S. Butinova and N.P. Krasnikov (1965) *Muzei istorii religii i ateizma* (Leningrad: Lenizdat), p. 8; M.E. Kaulen (2001) *Muzei-khramy i muzei-monastyri v pervoe desyatiletie Sovetskoi vlasti* (Moscow: Luch), p. 135.

[6] M. Halbwachs (1992) *On Collective Memory* (Chicago: The University of Chicago Press), p. 85.

Phases of Museological Atheism

Three phases can be roughly distinguished in the history of Soviet 'musealization' of religion. The first period followed the October Revolution lasting until 1927. Immediately after the Bolshevik seizure of power innumerable churches, monasteries, mosques and synagogues were closed and disbanded. Though from 1920 some outstanding monasteries or churches were turned into museums. The detection, assessment and musealization of the Church was performed by the Museums Department of the People's Commissariat for Enlightenment (Narkompros), under the direction of Natalia I. Sedova (1882–1962), the wife of the powerful People's Commissar Leon D. Trotsky (1879–1940). Cathedral and monastery museums (so-called Museum-Monuments) were established to retain the Church's inventory of objects classified as historically and artistically valuable.[7] Among the very first monastery museums was the gorgeous Trinity Lavra of St. Sergius in Sergiev Posad near Moscow, which was musealized on 20 February 1920.[8] Interest in it had been since 21 October 1918, after which the Commission for preservation of monuments of arts and antiquities in Trinity Lavra was founded at the Museums Department of Narkompros.[9] In November 1918, the Member of the Commission, renowned theologian and art historian Pavel A. Florensky (1882–1937) protested these plans vehemently. He claimed that the sanctuaries had fallen into the hands of 'artistic-archaeological predators'; an 'honest death' at the hands of demolition crews would have been better than this 'museological lethargy'.[10] Instead, Florensky suggested his own concept of 'synthetic arts'. According to this proposal the monastery should be placed as a whole including the monks, candlelight and scent

---

[7] See 'Instruktsiia Kollegii po delam muzeev i okhrane pamyatnikov iskusstva i stariny Narodnogo Komissariata po prosveshchcheniiu ot 3 yanvarya 1919'. *Revoljutsiia i tserkov*' 1, (1919), p. 30. See B. Kandidov (1929) *Monastyri-Muzei i antireligioznaya propaganda* (Moscow: Bezbozhznik).

[8] Dekret SNK RSFSR (20 April 1920) 'Ob obraščenii v muzej isoriko-chidožestvennych zennostej Trioce-Sergievoj Lavry', *Sobranie uzakonenij i rasporjaženij rabočego i krest'janskogo praviteľstva*, 27, 133.

[9] O.N. Kopylova (2000) 'Istorija Svjato-Troickoj Sergievoj Lavry i Moskovskoj Duchovnoj Akademii v dokumentach Gosudarstvennogo Archiva Rossijskoj Federacii', in: *Tezisy dokladov 2 meždunarodnoj konferencii 'Troice-Sergieva Lavra v istorii, kuľture i duchovnoj žizni Rossii'* (Sergiev Posad), pp. 6–7.

[10] P.A. Florenskii (1993) 'Khramovoe dejstvo kak sintez iskusstv' in: *Florenskii, Izbrannye trudy po iskusstvu* (St. Petersburg: Mifril, Russkaya Kniga), p. 289.

of incense under monument protection. His favourite prototype was to be called Białowieża National Park, where the last wild European bisons could survive.¹¹ Apparently his proposal was taken seriously.

Priests and monks were left as living exhibits and curious relics, or renamed as research assistants, untouched for the moment.¹² In the Museum of the Trinity Lavra of St. Sergius, for example, Konstantin A. Bondarenko alias hieromonk Ksenofont (1886–1937) served by the mid-1920s as chairman of the Historical Commission. His former boss Archimandrite Kronid (1859–1937) headed the security guards.¹³ Although some churches and cathedrals had been renamed as 'museums', they were still used for regular worship. In addition the priests used this label as a new income source by selling tickets.¹⁴ Thus in the words of Michel Foucault these museums were heterotopias of deviation: 'those in which individuals whose behavior is deviant in relation to the required mean or norm are placed'.¹⁵ The specialty of these heterotopias, however, was the public visibility of deviation. The main purpose of opening up these religious buildings to the public was apparently to demonstrate the deviants as defeated and disempowered 'ex-persons' (*byvshie ljudi*). Karl Josef Pazzini called this kind of deviation-exhibitions (prototypically implemented by criminal museums) 'institutions of the risk-free presentation' of scary objects.¹⁶

The early Soviet period of heterotopic 'national parks of religion' was short lived, however. Around 1927, the Soviet League of the Militant Godless subjected cathedral and monastery museums to increasing ideological pressure.¹⁷ They were accused of being shelters for counterrevolutionary priests and monks.¹⁸ It was only then that the history proper of the antireligious museum began. The collection, preservation and exhibition of religion were still allowed, but only from a communist point of view. 'In the

---

¹¹ Ibid.
¹² See I.V. Tarasova and G. A. Chenskaya (2002) 'Iz istorii muzeinogo dela v Rossii: Muzei tserkovno-arkheologicheskii i antireligioznyi', *Trudy Gosudarstvennogo muzeya istorii religii* 2, 17–30.
¹³ See B. Kandidov (1929) *Monastyri-Muzei i antireligioznaya propaganda* (Moscow: Bezbozhnik), p. 121.
¹⁴ See B. Kandidov (1929), p. 180.
¹⁵ M. Foucault (1986) 'Of Other Spaces' (trans. J. Miskowiec), in: *Diacritics* (16), pp. 22–27, 24.
¹⁶ K.J. Pazzini (1989) *Bilder und Bildung* (Hamburg: LIT), p. 124.
¹⁷ See 'Protokoly plenarnykh zasedanii 1–5 dekabrya 1930 g'. (1931) in: *Trudy Pervogo, Vserossiiskogo muzeinogo sezda*. Vol. 1, ed. Luppola I.K., p. 9.
¹⁸ Kandidov (1929), pp. 126–130.

musealized church we are immediately in the citadel of our class enemies', wrote Boris P. Kandidov (1902–1953) in his propaganda pamphlet of 1929 for the League of the Militant Godless.[19] On Kandidov's telling, the antireligious museum was a fighting arena of two rival ideologies.

The central concern of antireligious museums during that period was to desacralise sacred ideas, primarily by unmasking apparent miracles. This included exposing the working of phenomena such as icons that weep blood, examining relics and explaining natural mummification. In this way the former anticlerical campaign was continued. Between October 1918 and the spring of 1921 in countless churches and monasteries, public exhumations were staged of the supposedly 'incorruptible' relics of Russian Orthodox saints.[20] By 1921 at least 63 bodies in total were exhumed in this way.[21] The sacrilege had as its goal to demonstrate that these holy relics were no different than the remains of ordinary mortals; they had not been spared the ravages of time and had gone the way of all flesh. The campaign peaked in the spring of 1921 at the Hygienic Museum of the People's Commissariat of Public Health in the centre of Moscow with the opening of a major 'Exhibition for Social Medicine and Hygiene'. Alongside organic tissues in various stages of decay and saponification, the exhibition smugly presented the range of what the nationwide exhumation campaign had accomplished to date.[22] In the late 1920s, most of the relics became the property of new established antireligious museums.

The secondary approach was to contrast religion with science. Antireligious museums sought not only to attack the clergy but also to show the 'real story' behind Genesis in accordance with Marxist dialectic stages of evolutionary progress of nature and society. They collected and displayed panels, diagrams, dioramas and models explaining comparative anatomy and embryology, fossils from dinosaurs and hominids, portraits of socialist 'miracle doctors' and meteoric stones, 'Paradise apples' bred by Ivan Michurin and statues of Ivan Pavlov and Pavlovian dogs. Crucial to the propagation of scientific atheism was the re-signification of natural objects or artefacts through a process whereby they were taken out of a

---

[19] Ibid., p. 175.
[20] Cf. the list of exhumations of the People's Commissariat for Justice in: Kandidov (1929), pp. 216–227.
[21] Robert H. Greene (2010) *Bodies Like Bright Stars: Saints and Relics in Orthodox Russia* (Illinois: DeKalb), p. 146.
[22] N.A. Semaško (1922) 'Nauka I sharlanstvo. O vystavke "moshchei"' in: *Revoliutsiia i tserkov*' (1/3) 30–32, 30.

sacred context and relocated in a narrative of evolutionary development. This 'disenchantment' of the objects had to be applied even to seemingly non-ideological objects, such as for example a house mouse. As a starting point for the exposition, mice were featured in a painting, which illustrated the legend of the plague of mice, described in the Books of Samuel. An atheistic counterpoint served as a diorama depicting a group of living and dead mice in the colourful company of stuffed owls, hawks and foxes and explaining the natural factors influencing the populations of mice. This should convey a clear message: the feared phenomenon of regular recurring 'mice years' is not the result of divine punishment, but rather can be traced back to natural causes.[23]

Among the main attractions of antireligious museums which served this purpose was Foucault's pendulum, typically accompanied by a painting or diorama showing Galileo Galilei before the Roman Inquisition, or Giordano Bruno being burned at the stake. Like a wrecking ball toppling religious belief, the pendulum hung on a long string under the dome, its movement proving the Earth's rotation (see the illustration in Chap. 10 by Kelly). The antireligious museums marked the first time that Haeckel's idea of a 'scientific altar' was realised on a mass scale. This new category of museum could be described as a socialist *Kunstkammer* or a secular temple of Marxism–Leninism. As 'institutions of the risk-free presentation' of scary objects, they became places of blasphemous conquest in which science overwhelmed religion. Incidentally the antireligious museum recapitulated the proto-museal tripartite structure and pretence of the *Kunstkammer* or 'theatrum sapientiae',[24] described by Francis Bacon: It had to provide a comprehensive worldview that included 'history of Creatures, history of marvels, and history of Arts'.[25] Following this prototype the antireligious museum combined in a unique way natural objects, artificial things and boundary phenomena of 'marvels' (e.g. Apples of Michurin) in order to highlight the artistic competition between nature and man.

---

[23] A.V. Kondratov (1968) 'Nauchno-ateisticheskaia i antireligioznaia propaganda v kraevedcheskikh muzeiakh', in: *Trudy nauchno-issledovatel'skogo instituta muzeevedeniia i okhrany pamiatnikov istorii i kul'tury.* Vypusk 21. Moskva, pp. 3–72, 33.

[24] For the tripartite structure of the *Kunstkammer* see H. Bredekamp (1993) *Antikensehnsucht und Maschinenglauben. Die Geschichte der Kunstkammer und die Zukunft der Kunstgeschichte* (Berlin: Wagenbach), p. 88.

[25] F. Bacon (1857) 'Of the Proficience and Advancement of Learning Divine and Human' in: *Works* (ed.) *James Spedding*, vol 3. (London: Longman), p. 330.

The most important foundations of this new museum policy were the Moscow Antireligious Museum in Donskoi Monastery (1927), the Moscow Antireligious Museum in Strastnoi Monastery (1928) and the State Antireligious Museum in St. Isaac's Cathedral of Leningrad (1931). Other museums opened in Kostroma (1927), Tyumen (1929), Odessa (1930), Baku (1935) and Yaroslavl (1938).[26] By the end of 1930s with the consolidation of the Stalin cult, however, the militant attitude towards the religious cult declined. All antireligious museums were liquidated in the immediate pre-war period by Stalin or during the war in territories occupied by the German Wehrmacht.[27]

A third phase in the history of museological atheism began during the tenure of Nikita S. Khrushchev. Ironically, the ending of Stalin's cult of personality coincided with renewed struggle against religious cults. In the summer of 1954 'old Bolshevik' and historian of religion and atheism Vladimir D. Bonch-Bruevich (1873–1955) claimed in a personal letter to Khrushchev that Lavrentii P. Beria had colluded with pastors to inhibit atheist propaganda.[28] Only a few months later ensued an offensive on all fronts of the anti-Church policy, which lasted until the end of Khrushchev's tenure. Alone in 1960, 1392 Russian Orthodox churches were shut down.[29] The wave of closures resulted in numerous new antireligious museums. In this era, the flagship of Soviet museological atheism was the museum in the great Kazan Cathedral of Leningrad. This institution, established in 1932, was the only museum of atheism to survive the wave of closures under Stalin, as it was under the auspices of the Academy of Sciences and not the People Commissariat for Education.[30]

Typically, the new museums were referred to as 'Museums of the History of Religion and Atheism'. The name change signalled a new historical perspective and a reassessment of religion as a closed chapter of history. For instance, the exhibition in the Kazan Cathedral presented an

---

[26] Tarasova and Chenskaya (2002), p. 24. See also V.N. Mordvinova (1930) *Muzei Moskvy i Moskovskoi oblasti* (Moscow: Gosizdat RSFSR), pp. 38, 41.

[27] See *Rossiiskaya muzeinaya entsiklopediia* (2001). Vol. 1 (Moscow: Progress), p. 229.

[28] V. Bonch-Bruevich Vladimir to N. Khrushchev, letter dated 12.6.1954, in: Archive of Academy of Sciences of Russian Federation, Saint Petersburg branch (Arkhiv RAN): fond 221, Op. 2, d. 255, l. 14.

[29] See D. Pospelovskii (1995) *Russkaia Pravoslavnaia tserkov' v XX veke* (Moscow: Respublika), p. 283.

[30] See M. Shakhnovich (2007) 'Kak zakryvali Leningradskii Muzei Istorii Religii AN SSSR (1946–1947)', *Istoriia Peterburga* 5, 13–17.

overview of the development of different denominations, beginning with the origins and primitive forms of religion like animistic and totemic cults. Innumerable fetishes, voodoo figures, objects of witchcraft and amulets were included to provoke disconcertment, disgust or even fear and to illustrate the well-known aphorism attributed to Publius Papinius Statius: 'Fear created the gods'.

Highlighting the cruelty of religion culminated in the cellar of the former cathedral, where the exhibition housed the section devoted to Roman Catholicism during the Inquisition (Fig. 11.1). The visitors were presented with a diorama showing a torture chamber where the sufferer was bound on the rack while waiting for punishment. The executioner, shown wearing a red hood, presented to him a wide choice of terrible instruments of torture: branding iron, thumbscrew, Spanish boot, heretic fork, Judas chair and the Scold's bridle, to name just a few. The high number of visitors to Leningrad's Museum of the History of Religion and Atheism, where annual totals reached 400,000 and more during the 1970s, was not least because of its torture chamber exhibition (a feature later adopted by other museums).[31] Naturally, a large space was also devoted to Russian Orthodoxy. The exhibits were designed to serve a deterrent function. In Leningrad, for instance, a wall painting from the village church of Tazovo portrayed Leo Tolstoy as Judas before the Final Judgment (Fig. 11.2).

The deterrent function was coupled with temporal distancing. An important innovation was to transform heterotopia into heterochronia by restaging the Church and its believers as relics of the past (Fig. 11.3). Although the monks were already in the 1920s staged as the 'last bisons', they were perceived as being still artificially kept alive in a parallel reality of heterotopia. By the 1970s the 'bisons' were presented as long extinct and stuffed as a degenerate species.

This temporal distancing was the central narrative in the section of Leningrad's museum known rhapsodically as 'The Overcoming of Religious Survivals in the Period of the Large-Scale Construction of Communism in the USSR'. In Leningrad it was located—where else?—in the former sanctuary of the Kazan Cathedral. This exhibit presented the creation of the New Soviet Man by contrasting him with images of the 'Old Man', *homo religiosus*. The museum visitors encountered a veritable bestiary of different sects with curious names like Churikovtsy, Pustynniki,

---

[31] See Report for 1970, in: Central State Archive of Literature and Art St. Petersburg (TsGALI), fond 195, op. 1, d. 443, l. 8.

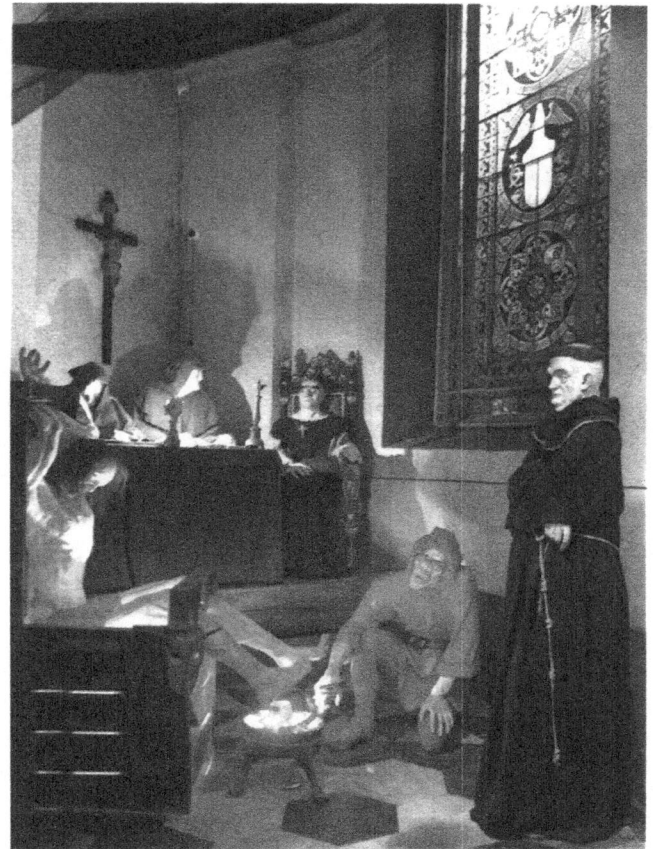

Fig. 11.1 Torture chamber, Diorama (*Source*: Dnepropetrovskii Muzei istorii religii i ateizma. Kratkii putevoditel'. Dnepropetrovsk 1980)

Innokentevtsy, Molokany or Tryasuny. Religious faith was a sign of morbidity, a result of physical deterioration or mental illness. In order to show that the disease of religion is curable, the exhibit presented examples of repentant clerics who had converted to atheism.

In this context the juxtaposition of religion and science was crucial. Emblematic here was a painting by Petr Mikhailov titled *Thinking* (1964),

Fig. 11.2 Leo Tolstoy as Judas before the Final Judgment. Wall painting, church of Tazovo (*Source*: Ya.I. Shurygin, *Kazanskii sobor*, Leningrad 1987)

displayed in the Kazan Cathedral, which showed a pastor reflecting sadly the newspaper with a portrait of Gagarin (Fig. 11.4). The triumph of science over religion was also shown on a far deeper anthropological level as man's victory over nature and over his primal fears. The Museum Guide declared in 1965:

> Under communism, the productive forces will reach such a high level that the man will make himself free from the domination of the elemental forces of nature. People will lose their fear of the unknown. And with it, the breeding ground of religion will disappear. Even in our present man creates such conditions under which he is no longer a servant of God more, but the ruler and creator of nature.[32]

This triumphalism tied into the final section of the exhibition, devoted to so-called positive atheism. The aim of positive atheism was not merely

---

[32] M.S. Butinova and N.P. Krasnikov (1965) *Muzei istorii religii i ateizma. Spravochnik-putevoditel'* (Moscow, Leningrad: Nauka), p. 187.

Fig. 11.3 R. Rakauskas, 'The shadow of the past'. Third prize in the atheistic photo competition 1962 (*Source*: Sovetskoe Foto 1964; 2)

to liquidate old religious traditions and holidays but also to resume a trend that had been there in the 1920s in order to infuse them with new secular meaning and to propagate optimism about the bright future of atheist society. To achieve this aim after 1960, 'State Commissions for New Civil Rituals' were created all over the Soviet Union. The museums of atheism received the order to draw on scholarly publications in order to popularise and implement these programmes. Under their aegis many 'palaces of

Fig. 11.4 P. Mikhailov, 'Thinking', Oil painting 1964 (*Source*: N. A. El'shina, N. I. Lattik, Russkaya i sovetskaya zhivopis' v sobranii Muzeya istorii religii i ateizma. Katalog. Moscow, Leningrad 1965)

happiness' devoted to the celebration of marriage as well as 'palaces of the newly born' to administer a kind of socialist baptism were founded, and civilian funeral liturgies were developed.[33] Within the museum setting, atheism was portrayed not only as a negation of religion but as a positive creative force bringing happiness. A typical example presenting health and salvation as a secular promise of communism is the sculpture composition by G. Glikman. Following its famous biblical legend, it showed the well-known ophthalmologist V.P. Filatov giving sight to a blind girl.

Aesthetic Rivalry and the Paradigm Shift in the 1970s and 1980s

Positive atheism confronted museum experts with perhaps their greatest challenge. On the cognitive level the struggle against religion had already tipped the scales in favour of atheism by the 1970s. Science had clearly disproved religion. Yet in atheist circles the disturbing idea emerged that

---

[33] Reports on extra-museum activity, in TsGALI, fond 195, op., 1, d. 274, l. 2–30.

the cognitive dimension of religiosity was only one of and perhaps not even the most important dimension of religious belief. The worry was that the current approach of the masses and especially the intelligentsia to historical memory would open the country\s national cultural heritage and aesthetic traditions to clerical influence.[34]

The question was whether atheism was able to overcome religion on an aesthetic level as well. Did it possess enough emotive force? In the 1970s and 1980s, what was once a cognitive fight between religious and atheistic objects eventually became an aesthetic contest. In 1971 one member of the advisory board of Leningrad's museum declared, 'We want to represent atheism bigger, better and more beautifully than religion. But how?'[35] The solution was to learn from the Church about how to address the feelings of the people and to adopt its aesthetic techniques.[36] This aestheticisation and historicisation trend in the museums of atheism can be traced back to developments in Leningrad. As I noted above, the museum there served as a coordination and consultation centre for all museums of atheism in the Soviet Union.[37] In the spring of 1981 the museum in Kazan Cathedral re-opened after a fundamental redesign of its permanent exhibition. Marcus Bach, an American author, expressed his surprise towards the 'refined' new image: 'The contrast between the 1974 exhibit and today's is dramatic. No whips, thumbscrews, etc. The approach now is much less obnoxious, and less blatant. (...) The section on Father Gapon and his part in the 1905 revolution could even be construed as a partially positive statement on one religious figure'.[38] The historian Mark R. Elliot asked in 1983: 'What, if anything, can these impressions of a museum before and after renovation tell us about possible changes in the Soviet approach to religion?'[39] In retrospect, it is plain that the redesign signalled nothing less than a paradigm shift.

What was once a temporal distancing strategy—religion portrayed as a relic of the past—became a *modo nostalgico*, in which religion was

---

[34] See P.F. Filippova (1974) 'Propaganda ateizma na materialakh kul'tovykh pamiatnikov', *Nauchno-ateisticheskaya propaganda v muzeiakh* 1, 17–20.

[35] Minutes of the meeting of the advisory board of the museum from 17 November 1971, in TsGALI, fond 195, d. 469, l. 9.

[36] See G. Proshin (1987) *Muzei i religiia* (Moscow: Sovetskaya Rossiia), p. 208.

[37] See Ya.I. Shurygin (1987) *Kazanskii sobor* (Leningrad: Lenizdat), p. 111.

[38] M. Bach, quoted in M.R. Elliott (1983) 'The Leningrad Museum of the History of Religion and Atheism', *Religion in Communist Lands*, 11:2, 127.

[39] M.R. Elliott (1983), 128.

represented as part of cultural heritage. This shift put the focus on the history of Russian Orthodoxy in the nineteenth century. Church artefacts were displayed respectfully in the symbolic order of collective memory and national identity, imbued with the sacred aura of the past. As the temporal axes were rearranged in late Soviet culture in general, it was precisely this change that undermined atheistic intent: the 'future' and 'progress' faded, giving an extraordinary dignity to 'history', 'tradition' and 'memory'. At the same time, the museums of atheism tried to exploit the new memory boom to revive the future-oriented Marxist idea of progress. This approach received its highest expression in the former sanctuary of the Kazan Cathedral. In Mstyora, a town famous for its lacquer miniatures and icon paintings, symbols of everyday life and communist eschatology were represented on large-format iconostasis-like panels (Fig. 11.5). The obvious didactic purpose of this combination of icon painting and 'socialist realism' was to co-opt Russian Orthodox aesthetics.[40]

This 'change through rapprochement' between two rival powers was accompanied by the fear that atheism could never win the aesthetic arms race. A senior staff member of the Leningrad Museum warned:

> Musealized objects of the cult are always double-edged. [...] Religion's system of sensual-emotional methods of influence, which has been perfected for centuries, leads in the museum to resistance by using the architecture, painting and sculpture of the interior.[41]

Six years later the staff member claimed that under the guise of objective historiography and interest in the history of art, the museums of atheism mutated into platforms of religious propaganda.[42] These prophecies came true a few years later in a series of spectacular 'de-musealizations' during Perestroika. In 1991, television stations reported live from the attic of the Leningrad Museum of Atheism about miraculous 'recoveries' of lost relics from Saint Seraphim Sarovsky and Saint Ioasaf Belgorodsky. And Vladislav Sherdakov, director of the Leningrad Museum of Atheism from 1967 to

---

[40] Conversation with the contemporary witness Marianna Shakhnovich, on 17 April 2008.
[41] G. Proshin (1981) 'Novaya ekspozitsiia Muzeia istorii religii i ateizma (Ot metodologii k praktike)', in: *Muzei v ateisticheskoj propagande. Sbornik nauchnykh trudov* (Leningrad: GMIRiA), p. 43.
[42] See Proshin (1987), p. 212.

Fig. 11.5 'Working in Soviet Society', Mstyora panel, fragment (*Source*: Ya.I. Shurygin, Kazanskii sobor, Leningrad 1987)

1978, confessed to having become a pious Christian many years previously. The many hours he spent at his workplace in the former cathedral—in close proximity to the spiritual influence of sacred objects—allegedly caused the transformation.[43] The only museum of atheism that survived the political changes of the late 1980s and early 1990s was Leningrad's—but only after being located from the Kazan Cathedral into a secular building and undergoing a name change. As its official new name indicates, it is devoted solely to the history of religion; today it contains not a single exhibit on atheism and makes no mention of its previous function.

'RELIGION AS A CHAIN OF MEMORY': CONCLUSION

During the 1980s, the USSR's Marxist–Leninist worldview and its paradigm of atheism imploded. This was the apparent onset of the religious re-enchantment of the (Soviet) world. 'Karl Marx was driven into flight by Holy Vladimir' is the American sociologist Andrew Greely's somewhat maudlin description of the religious reawakening in the upheaval of perestroika and glasnost.[44] But when we look at the competition between the 'communist future' and 'national heritage' in the late decades of the Soviet period in general and at the ligature of spirituality and antiquity in museological atheism in particular, then a more nuanced view of this religious revival comes to light. Following the functional view on religiosity—which sees organised religious life as a multidimensional behaviour-shaping social phenomenon defined not in terms of what it is but of what it does—a wide range of phenomena can be considered forms or functional equivalents of religion. Religion fulfils innumerable sociocultural functions as ritual and knowledge, devotion and belief; moreover, it operates on a communal and particular level[45] and emerges in modern societies as a so-called invisible religion even at football stadia and in spa salons.[46]

---

[43] A. Skvortsov (2002) 'Delo Bozhie na zemle. Pamyati Vladislava Nikolaevicha Sherdakova', *Podem* 1.

[44] See A. Greeley (1998) 'Religiöses Wiedererwachen in Russland?' in: D. Pollack, I. Borowik and W. Jagodzinski (eds.) *Religiöser Wandel in den postkommunistischen Ländern Ost- und Mitteleuropas* (Würzburg: Ergon), p. 539.

[45] See C.Y. Glock and R. Stark (1965) *Religion and Society in Tension* (Chicago: Rand McNally).

[46] See T. Luckmann (1974) *The Invisible Religion* (New York: Macmillan).

From the perspective of this functional definition, Marxism–Leninism is a phenomenon that contains manifest civil religious elements.[47] Accordingly, the USSR's museums of atheism can be seen as crystallisation points or even temples of the civic or political religion of the Soviet State. But as we have seen, state efforts to promote atheism were focused on science's intellectual triumph over religion. Although several attempts have been made by Soviet officials during the 1970s and 1980s to address the deep-seated emotive sphere of religion, they were obviously unsuccessful. Ironically, by disproving religion with science, Soviet atheists triumphed in an arena in which the Church had already lost, both in the West and in the East.

The vast majority of sociologists agree that social changes brought about by modernity eliminated the dominance of faith and the experience of God in everyday religious life.[48] Yet in certain areas, organised religious traditions have shown remarkable tenacity in the face of secularisation by reinventing themselves and shifting their focus. According to Danièle Hervieu-Léger, modernity caused not a disappearance but a 'complex redistribution ... in the sphere of believing'.[49] One field in which conventional denominations have being flourishing since the 1970s is the field of national heritage. This area is becoming increasingly important because of the challenges associated with rising social acceleration in late modernity.

The French historian Pierre Nora goes so far as to argue that archives and museums represent a new 'religion of preservation', a 'cult of trace' that our society of fluidity, mobility and desecration creates as compensation.[50] This idea seems plausible, especially with regard to museums. As an institution, museums have their historical roots in religious contexts.[51] And if museums have become a kind of temple, then churches have increasingly taken on a museum-like dignity in reaction to accelerated change, the erosion of tradition and the disintegration of collective memory. Hervieu-Leger, who has analysed these relationships in astonishing detail, argues in the monograph 'Religion as a Chain of Memory' that

---

[47] See J. Thrower (1992) *Marxism–Leninism as the Civil Religion of Soviet Society* (Lewiston: Mellen Press), p. 164.
[48] See Luckmann (1974).
[49] D. Hervieu-Léger (2000) *Religion as a Chain of Memory* (Cambridge: Polity Press), p. 132.
[50] P. Nora (1990) *Zwischen Geschichte und Gedächtnis* (Berlin: Wagenbach Klaus), p. 21.
[51] E. Sturm (1991) *Konservierte Welt: Museum und Musealisierung* (Berlin: Reimer), p. 55.

the religion of modern societies emphases the consciousness 'of belonging to a particular chain of belief', which at the same time is a chain of collective memory, an envisioned past.[52] Later she continues:

At the source of all religious belief ... there is belief in the continuity of the lineage of believers. This continuity transcends history. It is affirmed and manifested in the essentially religious act of recalling a past which gives meaning to the present and contains the future.[53] While intact 'societies of memory' in the past had no need to recall the past, the rapid pace of change in post-traditional society 'paradoxically gives rise to appeals to memory'.[54]

With Hervieu-Léger's thesis in mind, let us reconsider the cultural shifts in the late Soviet Union. Throughout the Soviet period the socialist society defined itself as society of change par excellence. The loss of historical memory was not only the product of the ideology of progress; it was also a result of the painful social disruptions that occurred during collectivisation, industrialisation and the Stalinist purges. In the mid-1960s, a fundamental shift in the relationship between progress and memory took place in the socialist world.[55] The socialist 'expectation societies' changed into 'societies of deadened future'.[56] In the USSR memory became hotly contested across the political spectrum and was invoked simultaneously by groups as divergent as 'Memorial', a democratic-liberal association, and 'Pamiat', a national-patriotic Russian Orthodox group. Moreover, the reversal of the temporal axiology put nondenominational organisations, particularly museums, in an ambivalent position.

During the 1920s the museums of atheism served as heterotopic 'institutions of the risk-free presentation' of religious artefacts; in the 1930s and partly in the 1950s they became places where religious thought was defiled; during the 1960s they were reconceived as heterochronies, in which a temporal distancing of 'religious survivals' took place. Finally, in

---

[52] Hervieu-Léger (2000), p. 82.
[53] Ibid., p. 125.
[54] Ibid. p. 141.
[55] See N. Mitrokhin (2003) *Russkaya partiia: Dvizheniie russkikh natsionalistov v SSSR 1953–1985 gody* (Moscow: NLO).
[56] R. Gries (1997) ""...und der Zukunft zugewandt" oder: Wie der DDR das Jahr 2000 abhanden kam' in: E. Bünz, R. Gries and F. Möller (eds.) *Der Tag X in der Geschichte: Erwartungen und Enttäuschungen seit tausend Jahren* (Stuttgart: Deutsche Verlags-Anstalt), pp. 309–378.

the 1970s and 1980s religious objects once dismissed as prehistoric garbage were promoted to dignified relics.

This folkloristic reinvented 'religion of preservation', however, seems to have little to do with its conventional origin. Religion conceived as a multidimensional offer of service leaves the monopoly of conventional churches and becomes dispersed among countless other providers on the religious market. The late and post-Soviet society is no exception. The Russian Orthodox Church is only one religious offering among many, one primarily focused on national commemoration and the spirit of the past. In 1992, the metropolitan of St. Petersburg, Ioann, argued that the task of reviving Russians' 'historical memory' can be fulfilled only by the Russian Orthodox Church, as the Orthodox Church was the embodiment of 'the collective memory of the people'.[57] As we know from survey results for 1990–1991, the majority of respondents, including most self-proclaimed atheists, see religious belief as an indispensable part of the national historical consciousness.[58] The recent media scandal surrounding the so-called punk prayer of Pussy Riot in the Cathedral of Christ the Saviour shows less the opposition of 'religious' and 'secular' than the political dissent between 'patriotic' and 'pro-Western'. The historian Martin Schulze Wessel observed a structural analogy between nation and religion for the entire post-communist space. The 'nationalization of religion and the sacralization of the nation' represents a 'modern justification' of the Church.[59] It is this process that took place in the museums of atheism. Indeed, such a justification can achieve its full suggestive power only within the horizon of the national Church's history.

Nevertheless, during the 1990s and perhaps still today, the traditional Church has barely been able to satisfy the manifest functions of religion, such as providing a comprehensive worldview or helping individuals cope 'with these ultimate problems of human life' in the face of death and illness, suffering and risk, disappointment and acceleration, evil and injustice.[60] When it comes to such matters, psychologists, physi-

---

[57] Ioann Mitropolit Sankt-Peterburgskii i Ladozhskii (1992) 'Boites' lozhnykh prorokov. *Sovetskaya Rossiia*', 29 August.

[58] See D.J. Furman (1998) 'Religion und Politik im postkommunistischen Russland' in Pollack, (Borowik and Jagodzinski 1998), p. 503.

[59] M. Schulze Wessel (2006) 'Die Nationalisierung der Religion und die Sakralisierung der Nation im östlichen Europa' in: M. Schulze Wessel (ed.) *Nationalisierung der Religion und die Sakralisierung der Nation im östlichen Europa* (Stuttgart: Steiner Verlag), p. 7.

[60] Y.J. Milton (1957) *Religion, Society and the Individual* (New York: Macmillan), p. 9.

cians, insurances, digital games, meditation groups and travel agencies are much more in demand than priests. So while state atheism is no longer, its efforts of temporal distancing have left a permanent trace: its anticlerical work is being nonviolently completed by anonymous forces of modernisation. Looking at the dynamics of the (post-) Soviet religious revival, the Russian–American cultural theorist Mikhail Epstein correctly wrote: 'Religion was left behind in the past—today it zooms into the future, as a phantom of the phantom ... No, God is not dead. He's become a metaphor'.[61]

---

[61] M. Epstein (2001) 'Ateizm—eto dukhovnoe prizvanie. O Raise Gibaidulinoj', *Zvezda* 4, 159–174, 174.

CHAPTER 12

Religion, Science and Cold War Anti-Communism: The 1949 Cardinal Mindszenty Show Trial

Paul Betts

While the face-off of religion and science has been a central element of modern history for centuries, there has been surprisingly little scholarly attention to how this rivalry played out during the Cold War, especially with regard to Eastern Europe. Historiography on religion in the East Bloc largely addresses the uneasy and often tense relationship between organized religion and communist states, usually with an eye toward documenting political repression, the development of an alternative civil society or the roots of East Bloc resistance, reform and revolt.[1] Scholarly literature on science under socialism is more limited, generally concerned with chronicling technical achievements, maximizing industrial production or its ability to raise the standard of living in resource-strapped

---

This chapter is dedicated to the memory of Charles R.W. Betts, 1934–2014.

[1] For example, G. Weigel (2003) *The Final Revolution: The Resistance Church and the Collapse of Communism* (New York: Oxford University Press); B. von der Heydt (1993) *Candles Behind the Wall: Heroes of the Peaceful Revolution that Shattered Communism* (London: Mowbray) and P. Mojzes (1981) *Christian-Marxist Dialogue in Eastern Europe* (Minneapolis: Augsburg).

P. Betts (✉)
St Antony's College, Oxford University, Oxford, UK
e-mail: paul.betts@sant.ox.ac.uk

Eastern European economies.² However, the specific interplay between these two forces remains relatively unaddressed, even if the confluence of religion, science and communism was an integral element of Eastern European Cold War culture in its own right.³ This chapter explores this theme, albeit from an oblique angle. Unlike most of the contributions in this volume, I discuss the dramatic relationship between religion and science in Eastern Europe from a decidedly different vantage point, namely from the perspective of the West.

A revealing episode of this historical encounter was the arrest, show trial and imprisonment of the conservative anti-communist Hungarian prelate, Cardinal József Mindszenty. Even if his name may remain largely forgotten today, his case provoked remarkable international controversy at the time. No issue exercised the international Catholic world more in the early Cold War, as Roman Catholic leaders and publicists around the world raised their voice in protest, as numerous demonstrations were organized on both sides of the Atlantic. Yet Mindszenty's fame as an international media event went far beyond the Roman Catholic faithful, to the point that his plight became a Cold War cause célèbre of Soviet aggression, religious persecution and human rights. Books covering the cardinal's trial were produced by the dozens across the Western world and were translated into Japanese, Chinese, Spanish and Arabic. Mindszenty was even the subject of several full-length feature films produced in the USA and Britain in the 1950s, including *Guilty of Treason* (1950) and *The Prisoner* (1955), the latter starring Alec Guinness.⁴ While the trial mobilized the transnational Catholic community in unique ways, it had equally important effects in the world of social science. Arguably the trial's most ardent

---

² E.P. Hoffmann and R.F. Laird (1985) *Technocratic Socialism: The Soviet Union in the Advanced Industrial Era* (Durham, NC: Duke University Press); D. Hoffmann and K. Macrakis (1999) *Science under Socialism: East Germany in Comparative Perspective* (Cambridge, Mass: Harvard University Press); and I. Vladimirovich Bromleĭ (1986) *Main Features of Science Organization in Socialist Countries* (Moscow: Institut istorii estestvoznaniĭa i tekhniki).

³ *Christianity and Modernity in Eastern Europe* (2010), B. Berglund and B. Porter-Szucs (ed.) (Budapest: CEU Press).

⁴ So pervasive were these media images of the martyred archbishop that Mindszenty himself opened his 1974 memoirs by correcting *The Prisoner* film's inaccuracies and misleading scenes, ranging from the wrongly featured 'luxurious' furniture in his prison cell to what he called the film's wrongful emphasis on the courtesy of the sentries, the prison's good food and supposedly 'pleasant and congenial' conversations with guards. Cardinal József Mindszenty (1974) *Memoirs*, trans. R. and C. Winston (London: Weidenfeld and Nicolson), pp. xxvii–xxviii.

student was no other than the US government, as the Central Intelligence Agency (CIA) became particularly fascinating by how the trial seemed to showcase the USSR's new scientific techniques to extract confessions. The widespread Cold War anxiety about 'brainwashing' had its origins in the Mindszenty trial, spurring subsequent US government funding into psychological research in the name of state security. How and why the 1949 Budapest trial emerged as an international Cold War flashpoint for politicizing both religion and science in the Cold War West is the subject of this chapter.

The 'Crucifixion of Mankind'

One of the most remarkable things about the mountain of Cold War historiography is the relatively meager attention accorded to the role and power of religion, at least until very recently.[5] The ideological differences between East and West have long been—and often still are—portrayed as a colossal confrontation between rival secular utopias, conventionally characterized as a half-century face-off between an American Empire of Liberty and its counterpart Soviet Empire of Justice. In this rendition, the spheres of economics, politics and culture emerged as the key superpower battlefields, as each bloc worked to present itself as uniquely capable of providing the good life to the masses and delivering on promises made to weary home fronts during World War II.[6] Religion hardly figures in these dominant narratives, or only does so uncomfortably. Yet religion was fundamental to the Cold War from the very beginning across the Cold War divide.

To be sure, the nexus of religious fundamentalism and American Cold War politics is hardly news. It is well known that Truman, Eisenhower and John Foster Dulles's muscular Christianity shaped a good deal of US foreign and domestic policy in the early Cold War and helped build a strong case for a new interventionist American globalism after 1945. Truman's high-profile series of epistolary exchanges with Pope Pius XII in 1947 about the need to defend what was perceived as an imperiled Christian civilization is one ready example; Eisenhower, too, made no bones about his belief that, as he put it, 'spiritual weapons [are] our country's most powerful resource'. In 1955 the US Congress passed a law adding the phrase 'under God' to the American 'Pledge of Allegiance' to the flag

---

[5] Key in this respect is D. Kirby (ed.) (2002) *Religion and the Cold War* (London: Palgrave).

[6] O.A. Westad (2007) *The Global Cold War* (Cambridge: Cambridge University Press).

for all American schoolchildren,[7] and 2 years later the US government changed its official motto from 'E Pluribus Unum' to 'In God We Trust', which was inscribed on all US paper currency after 1957.[8]

Things were not so dissimilar in Western Europe, though in this case Catholicism—and less so Protestantism—was the preferred language of Christian transnational solidarity and moral mission. The Italian electoral contest of 1948 highlighted the new West European marriage of religion and politics after 1945, culminating in the Vatican's own stark electioneering slogan: 'either for or against Christ, this is the principal question'.[9] West German Chancellor Konrad Adenauer was also not shy in speaking about 'spiritual weapons' and the urgent defense of a Christian *Abendland*.[10] The postwar period gave rise to a new 'political Catholicism' across Western Europe, as Catholics successfully buried the interwar legacy of anti-liberalism and confessional fighting with Protestants in favor of a new ecumenical Christian front dedicated to upholding a new liberal version of Christian democracy after 1945. Other Catholic pro-Europeanist political leaders, such as Alcide De Gasperi, Adenauer and Robert Schuman, saw the budding West European Economic Community as in part a Christian bulwark against Soviet atheism.[11] Even Britain's Labor Foreign Secretary Ernest Bevin called for a 'spiritual union' of nations in 1948 based on shared Western European values.[12]

How Christianity was politicized in early Cold War Eastern Europe is much less known, however. To be sure, religious persecution was nothing novel to Eastern Europe in the aftermath of World War II. The 1946 arrest and imprisonment of Archbishop Stepinac of Zagreb on trumped up charges of fascist collusion during the war was a kind of dress rehearsal of Mindszenty's fate, as the Catholic community worldwide reacted angrily to this communist assault on the church. Other cardinals held in

---

[7] S.J. Whitfield (1991) *The Culture of the Cold War* (Baltimore: Johns Hopkins Press), pp. 89–90.
[8] W. Inboden (2008) *Religion and American Foreign Policy, 1945–1960* (Cambridge: Cambridge University Press), p.257.
[9] R.A. Ventresca (2003) 'The Virgin and the Bear', *Journal of Social History* xxxvii, 439.
[10] P.T. Jackson (2007) *Civilizing the Enemy: German Reconstruction and the Reinvention of the West* (Ann Arbor: University of Michigan).
[11] P. Chenaux (1990) *Une Europe Vaticane?* (Brussels: Editions Ciaco).
[12] Quoted in M. Duranti (2012) 'Curbing Labour's Totalitarian Temptation: European Human Rights Law and British Postwar Politics', *Humanity: An International Journal of Human Rights, Humanitarianism, and Development*, iii:3, 365.

captivity in the East Bloc, such as Cardinal Stefan Wyszyński of Poland, who was placed under house arrest for 4 years between 1953 and 1956, were key rallying points for Catholic mobilization. Yet it was Mindszenty's defiance and subsequent show trial that captured the West's imagination and played a key role in forging a new identity for the Cold War West.

At this point it is worthwhile to provide a little background on the case. Mindszenty had been a prominent figure in the Hungarian Catholic Church for decades. He had been arrested by the short-lived Bela Kun communist government in 1919 and then again by the fascist Arrow Cross government in 1944 for protesting the continuation of war on Hungarian soil.[13] Mindszenty came from a small village in western Hungary and had been a rural pastor for almost 30 years but was elevated as the head of the Hungarian church by Pope Pius XII in 1945 in part because of his outspoken anti-communist views.[14] He possessed impeccable anti-fascist credentials, was a sworn enemy of the left, and was primed for a fight. In his inaugural sermon after being named cardinal-prelate, Mindszenty boldly described himself as the 'pontifex, bridge builder, and first dignitary of the state, vested with nine hundred year-old rights'.[15] He was pitted against a formidable figure in his own right—Mátyás Rákosi, the Hungarian communist leader, who was committed to destroying old Hungary and building a new revolutionary order along the Danube.

While the early postwar years church–state relations were relatively peaceful, Mindszenty and the church were increasingly seen as an obstacle to communist political plans and social engineering. This was especially so in terms of the state's desired land reform, nationalization of children's education and the 'de-christianization' of the national press.[16] Mindszenty fought back in various ways. He continually challenged state authority and refused to declare any loyalty to the Hungarian state unless the

---

[13] P.A. Hanebrink (2006) *In Defense of Christian Hungary: Religion, Nationalism and Antisemitism, 1890–1944* (Ithaca: Cornell University Press), p. 228.

[14] His original family name was Pehm—for this reason the communists constantly accused him of being Swabian, and not truly Hungarian. In 1941 he changed his surname to Mindszenty (after the name of the village) to stress his identification with the Hungarian nation and made much of his authentic centuries-long Hungarian heritage in the beginning of his memoirs. Mindszenty, *Memoirs*, 1.

[15] Quoted in H. Stehle (1981, orig. 1975) *Eastern Politics of the Vatican, 1917–1979*, trans. S. Smith (Athens, OH: Ohio University Press), p. 258.

[16] P.C. Kent (2002) *The Lonely Cold War of Pope Pius XII: The Roman Catholic Church and the Division of Europe, 1943–1950* (Montreal: McGill-Queen's University Press), p. 214.

communist regime guaranteed freedom for all Catholic associations; permitted the publication of a daily not just weekly Catholic newspaper; and resumed diplomatic relations with the Vatican.[17] In a 1947 pastoral letter, Mindszenty deployed the new language of human rights to undermine government legitimacy: 'We must publicly declare that no Christian voter can support a party that rules by violence an suppression, and tramples underfoot all natural laws and human rights'.[18] In an effort to re-energize his followers and win back public space and momentum from the communist state's assault, he declared the period between August 1947 and December 1948 the 'Year of the Virgin', with the slogan 'Hungary is the Virgin Mary's Country'.[19] This 1948 Hungarian Marian Festival was designed to inspire the besieged Hungarian Catholic community through mass sermons, parades, marches and festivals of all kinds as a show of solidarity and resistance against state incursions into Catholic cultural territory. By all counts, the festival was a huge success—the clergy led several pilgrimages, with up to some 1.5 million people participating.

Here it is well to remember that the Cult of Mary enjoyed a strange comeback across the continent, punctuated by a rash of religious sightings, roaming faith healers, pilgrimages and miracles that attracted hundreds of thousands of believers in a number of countries.[20] In fact, it was part of an explosion of popular religiosity across Europe that is still largely unremarked in histories of the postwar era. Recalling its real presence offers a very different picture of reconstruction Europe in the decade after the defeat of the Third Reich.[21] The immediate postwar period saw a bounty of miracles and apparitions across Europe, ranging from Spain,

---

[17] 'A Cardinal on Trial: Charges Against the Hungarian Primate', *London Times*, February 3, 1949, 5.

[18] Cited in J. Luxmoore and J. Babuich (1999), *The Vatican and the Red Flag: The Struggle for the Soul of Eastern Europe* (London: Geoffrey Chapman), 43.

[19] The prelate had been inspired by his visit to the World Marian Congress in Ottawa, Canada in 1947, where he took the opportunity to visit Hungarian-American parishes in order to thank them for their generosity toward his countrymen after World War II. G.N. Schuster (1956) *In Silence I Speak: The Story of Cardinal Mindszenty Today and of Hungary's 'New Order'* (London: Victor Gollancz), p. 4.

[20] N. Perry and L. Echeverría (1988) *Under the Heel of Mary* (London: Routledge), pp. 231–257 and J. Baaden (2014) 'Post-War Saints, 1945–1960', *History of European Ideas*, 1–20.

[21] M.E. O'Sullivan (2013) 'West German Miracles: Catholic Mystics, Church Hierarchy and Postwar Popular Culture', *Zeitgeschichtliche Forschungen* iii, 1–14.

France and Ireland to West Germany, Austria and Hungary.[22] It is well known that the modern pattern of Marian witnesses originated in France in the mid-nineteenth century, with the sightings in 1858 at Lourdes as the most famous. Pilgrimages to Lourdes began in 1872 after France's defeat against the Prussians, a defeat that was interpreted as divine judgment for the sins of the Republic and Paris Commune.[23] But a new wave of spectacular apparitions of the Virgin swept across Europe after World War II. Between 1947 and 1954 an average of 13–14 apparitions were reported to church authorities across Europe each year, more than three times the average of either the preceding or following decades. The old pilgrimage sites, such as Lourdes and Marpingen in Germany, maintained their draw after World War II, but now new ones arose in other places too, drawing crowds of tens of thousands for weeks or even months on end, as a new populist Catholic evangelism gripped post-Nazi Europe.[24] Some of these sightings took on Cold War dimensions as well: the famed Fatima apparitions in Portugal in 1917—whose apocalyptic visions had supposed revealed the sorrows and scourges of the Russian Revolution—became a touchstone of Cold War commentary about the heightened fears of the new specter of communism after 1945. An International Pilgrim Virgin Statue—a copy of the Fatima image blessed by the bishop of Leiria of Portugal—embarked on a world tour in 1947, supposedly accompanied by miracles and the presence of doves.[25] The 1947 Oscar-winning film, *Songs of Bernadette*, based on Franz Werfel's 1943 novel of the Lourdes story, was released all over Western Europe in 1948 and was the biggest box office film that year in the Federal Republic.[26] In 1949 there arose in West Germany the Apparition Cult of Heroldsbach-Thurn, a village in

---

[22] D. Blackbourn (1993) *Marpingen: Apparitions of the Virgin Mary in Bismarckian Germany* (London: Clarenden), p. 378.

[23] R. Harris (2000) *Lourdes: Body and Spirit in the Secular Age* (London: Allen Lane), pp. 249–255.

[24] Cited in M. Scheer (2012) 'Catholic Piety in the Early Cold War Years; or, How the Virgin Mary Protected the West from Communism', in *Cold War Culture: Perspectives on Eastern and Western European Societies*, A. Vowinckel, M. M. Payk and T. Lindenberger (ed.) (Oxford: Berghahn), p. 131. See also P.M. Kane (2004) 'Marian Devotion since 1940: Continuity or Casualty?' *Habits of Devotion*, J.M O'Toole (ed.) (Ithaca: Cornell University Press), pp. 89–129 and M. Scheer (2006) *Rosenkranz und Kriegsvisionen* (Tübingen: Tübinger Vereinigung für Volkskunde).

[25] M. Vincent (2013) 'Made Flesh? Gender and Doctrine in Religious Violence in 20th Century Spain', *Gender & History* xxv:3, 678–680.

[26] Scheer, 'Catholic Piety', p. 138.

southeastern Germany, accompanied by widespread reports of daily divine visitations to a group of young female seers. By 1952 some 1.5 million pilgrims had visited the site and its four female visionaries to witness its reported 3000 apparitions, and the town's proximity to the Czech border during the 1948 communist consolidation of power apparently heightened these Cold War fears of apocalyse.[27] Such sightings were not confined to Western Europe. In the Cathedral of Lublin, Poland, it was reported that the statue of Mary began to weep, drawing crowds of up to 100,000 a day.[28] The Pope thus declared the dogma of Assumption in 1950 to harness this new popular religiosity to the church,[29] and the pope's declaration of the International Marian Year in 1954 was broadcast by television to diffuse what Pope Pius XII proudly called the 'further triumphs of Jesus and Mary'.[30] The wider point is that the postwar period—routinely characterized as an era of economic and political miracles—was awash with religious miracles as well.[31]

Communist authorities interpreted Mindszenty's Marian Festival as subversive activism, and local police were instructed to crack down on the faithful.[32] In response Mindszenty brazenly located the Catholic fight against communism in the broader historical Christian struggle against the infidel. The Hungarian Kingdom had long been considered the Eastern flank of Western Christendom, and Mindszenty now updated the struggle to suit Cold War concerns. He made it quite clear that this confrontation with communist authorities was simply the latest round of antagonism with anti-Christian forces that originated in the Hungarian

[27] O'Sullivan, 'West German Miracles', 3.
[28] Scheer, 'Catholic Piety', p. 134.
[29] Blackbourn, *Marpingen* p. 386.
[30] Perry and Echeverría, *Under the Heel* p. 238.
[31] M. Black (2012) 'Miracles in the Shadow of the Economic Miracle: The 'Supernatural '50s' in West Germany', *Journal of Modern History* lxxxiv:4, 836. See too O'Sullivan, 'West German Miracles', 2–5 as well as D. van Melis (2003) '"Strengthened and Purified Through Ordeal by Fire": Ecclesiastical Triumphalism in the Ruins of Europe', in *Life After Death: Approaches to a Cultural and Social History of Europe During the 1940s and 1950s*, R. Bessel and D. Schumann (ed.) (New York), pp. 231–242.
[32] I. Main (2003) 'Weeping Virgin Mary and the Smiling Comrade Stalin', in G.T. Rittersporn, M. Rolf and J. C. Behrends (eds) *Sphären von Öffentlichkeit in Gesellschaften sowjetischen Typs*, (Frankfurt: Peter Lang), 255–278. More generally, W. Christian, Jr. (1984) 'Religious Apparitions and the Cold War in Southern Europe', in *Religion, Power and Protest in Local Communities*, E. Wolf (ed.) (Berlin: De Gruyter), pp. 239–265.

battles against Islam in the sixteenth and seventeenth centuries.³³ Here Mindszenty was following the views of Pope Pius XI and others from the 1930s about the world-historical stakes in this clash between communism and Catholicism.³⁴ The difference was that the cardinal put communist Hungary as the frontline of this new international Christian crusade.

Given his outspoken views and growing defiance, the prelate was arrested on December 26, 1948 as part of the state's banning of all religious orders and then put on trial in early February 1949. 1948 had been a delicate year in Eastern Europe, as Stalin ordered a major wave of arrests of internal enemies in the wake of the Marshall Plan, the Berlin Blockade and the break with Tito. The arrest of Mindszenty took place against the backdrop of fear and paranoia. The communist case against the cardinal was based on several charges. According to the Minister of the Interior, the police confiscated documents written in Mindszenty's hand 'urging Western powers to intervene in Hungary'. Rákosi's government also alleged that the cardinal was the founder of a royalist movement in 1945, and that he was plotting with no other than Otto von Habsburg of the old royal family formerly based in Vienna for 'the overthrow of the democratic State order and the Republic'.³⁵ Most incriminating was supposedly a handwritten confession that he 'expected the restoration of the Monarchy after the conclusion of a third world war by an American victory'.³⁶ And if this were not enough, Mindszenty was accused of embezzling funds and engaging in black marketeering with foreign currency. The Hungarian communist government quickly prepared a so-called *Yellow Book* of documents on the case for the foreign press (both in English and French editions) in order to win over the international community shocked by the cardinal's arrest in December 1948. The *Yellow Book* had the opposite effect, however—the West expressed outrage toward the arraignment, not least because the Hungarian state's published account against the prelate actually appeared a month <u>before</u> the trial began.³⁷ Crude accusatory

---

³³ In the Cold War it was not unusual for commentators to say that Hungary 'in its thousand year history' has 'often spilled its blood for the defense of the West'. Z.K.J. Csáky (1949) *Ich schwöre, daß Kardinal Mindszenty unschuldig ist* (Zurich: Thomas Verlag), p. 106.

³⁴ Hanebrink, *In Defense*, pp. 123–124.

³⁵ J. Közi-Horvath (1979) *Cardinal Mindszenty: Confessor and Martyr of our Time*, trans. Geoffrey Lawman (Williton: Cox, Sons and Co.), 3.

³⁶ *Documents on the Mindszenty Case* (1949) (Budapest: Atheneum), 7–11.

³⁷ The book (and trial) charged him with 'treason, espionage, crimes directed at the overthrow of the Republic, and foreign exchange speculation'. *Documents on the Mindszenty*

photos were included: one featured Mindszenty in unusually dashing clothes about to leave for Rome on an airplane 'put at his disposal by the American mission'; another portrayed his secretary and archivist with an infamous aluminum tube of damning documents supposedly found under the floorboards of the cardinals' residence. The 2-day trial itself was a sham stitch-up, dramatized by the cardinal's puzzling oral confession at the end, after which Mindszenty was sentenced to life imprisonment.

The 1949 trial attracted global attention, and this trial photograph became its most famous illustration (Fig. 12.1). The trial incited a torrent of indignation across the Catholic world, from Western Europe to the USA to South America. The Vatican led the way by protesting that the conviction of a member of the College of Cardinals as an 'act of violence committed against a strenuous defender of the rights of the Church and of the human person'.[38] On February 20, 1949 Pius XII denounced the trial in a speech delivered to some 300,000 people at St. Peter's Square, after which the Holy See decreed that no modus vivendi with communism was possible, so much so that those involved in the trial were formally excommunicated. In the US Cardinal Spellman mounted the pulpit of St. Patrick's Cathedral in New York City for the first time since V-E Day to speak before 3500 listeners. He thundered that the USA needed to unite in prayer and protest for the imprisoned cardinal who was being victimized by the 'world's most fiendish, ghoulish men of slaughter'. For him and other churchmen, the trial was nothing less than 'the crucifixion of mankind'.[39] In London over 6000 Catholics filled Albert Hall to protest the indictment of Mindszenty,[40] and in Paris 'tens of thousands of demonstrators' reportedly demanded the cardinal's release.[41] In Dublin some 150,000 people, including trade union representatives, demonstrated to

Case, 3–4.

[38] Kent, *Lonely Cold War*, p. 228.

[39] Mindszenty followed a similar self-dramatization as Christ-like figure: in recounting his first night in prison, he wrote: 'It was my special cross to be an imprisoned cardinal in the land of the Virgin Mary. Thus there rose before my mind's eye the image of Pilate and his "Ecce homo!"' *Memoirs*, 93.

[40] 'Roman Catholics' Strong Protest: Denial of Human Rights Denounced', *Manchester Guardian*, February 8, 1949, 5.

[41] A pamphlet, *La Verité sur le Cardinal Mindszenty*, drafted by the significantly named 'Comité Internationale de Défense de la Civilisation Chrétienne', challenged the bogus charges against the primate and the anti-religious thrust of the trial. Közi-Horvath, *Cardinal Mindszenty*, p. 57.

Fig. 12.1 Cardinal Mindszenty, Budapest Trial, Feb. 8, 1949

protest the cardinal's imprisonment on May Day.⁴² As far away as India a special mass was held in Catholic Churches throughout country to honor the imprisoned prelate.⁴³ The international Catholic press churned out dozens of pamphlets in outrage toward the trial.⁴⁴ A similar sentiment extended to the Protestant communities across the world; the Anglican archbishop of York, for example, 'denounced the mockery of justice' and felt the trial should be used to bolster a sense of the Atlantic Pact and the Western Union under siege.⁴⁵ Truman himself denounced the trial as 'the infamous proceedings of a kangaroo court'.⁴⁶ That same day the US House of Representatives unanimously adopted a resolution urging the United Nations (UN) to censure the treatment of the cardinal. Protestant leaders and lay parishioners reportedly sent a 'flood of telegrams inundating the State Department and White House'.⁴⁷ *Life* Magazine summed up the sentiment: 'Communists Make Martyr of Cardinal'.⁴⁸ Such shocked outrage found expression among the Eastern European émigré community in the West as well; one book concluded by saying that '[i]n the most stupendous spiritual battle of world history, Cardinal Mindszenty is the embodiment of Western Christianity to-day'.⁴⁹

The case enjoyed great following in mainstream press too. A few statistics provide some perspective of just how big this scandal was at the time: there were some 80 articles on the Mindszenty case in the *New York Times* alone between 1948 and 1949, and another 100 in *Le Monde,* and perhaps even more surprising, some 150 articles about the fate of the cardinal published in *The Manchester Guardian* from 1948 onward. The international press launched defenses and counterattacks, published in French, German and

---

⁴² E. Delaney (2010) 'Anti-communism in Mid-Twentieth Century Ireland', *English Historical Review* CXXVI: 521, 892.

⁴³ Kent, *Lonely Cold War,* p. 229.

⁴⁴ See for example F. Burgess (1949) *The Cardinal on Trial* (London: Richard Madley), M. Derrick, *Cardinal Mindszenty* (1949) (London: Richard Madley) and L. Varga (1955) *Revision of the Sentence Against Cardinal Mindszenty* (New York: Hungarian Catholic League of America).

⁴⁵ 'Atlantic Pact and Western Union' (1948), *The Round Table: The Commonwealth Journal of International Affairs,* 105–110.

⁴⁶ Cited in Luxmoore and Babiuch, *Vatican,* p.95.

⁴⁷ 'The Upper South'; and 'New England: Mindszenty Case Stirs Strong Protest Among All Faiths', both in *NYT,* February 13, 1949, E6.

⁴⁸ 'Communists Make Martyr of Cardinal', *Life,* February 21, 1949, 27.

⁴⁹ The Very Rev. Dr. N. Boer (1949) *Cardinal Mindszenty and the Implacable War of Communism Against Religion and the Spirit* (London: BUE Limited), pp. 9–10.

English, often accompanied by supporting documents.[50] So vociferous was the international attack against the sham trial that the Hungarian government felt compelled to publish a sequel to the *Yellow Book*, dubbed the *Black Book*, to try to refute the Western press's crusade against the trial. The regime's text blasted Western states and the pope for excoriating Hungary in such a manner 'uncommon in international relations and almost forgotten since the fall of Nazi Germany'. Particular ire was provoked by the Western charge of the 'devilish drug'-induced confession, a story whose intellectual level was derided as 'far below an Edgar Wallace murder mystery'.[51] It hoped to defuse Western criticism by portraying the cardinal as a fascist and anti-Semite instead of religious martyr and human rights hero. Different photos were appended to give a different image of the trial, one with the prelate featured in quiet consultation with his counsel and another depicting an orderly trial setting; the Black Book also included relevant legal documents and a final section on 'How a People's Court Works' to underscore that all proceedings were based on law and due process.

Mindszenty also became a common reference point in popular culture in the West, especially in the USA. Numerous television programs in the USA in the late 1940s and early 1950s (e.g., *Studio One*, *Zero-1960*, *Crossroads* and most notably Bishop Fulton Sheen's popular and rabidly anti-communist show, *Life is Worth Leading*) took up the theme of religion and devoted considerable airtime to Mindszenty's trial and his legacy.[52] Cinema also registered the importance of the cardinal's trial. The first film that addressed the cardinal's fate was Felix Feist's 1950 *Guilty of Treason*, which recounted the story of the torturing to death of a Hungarian music teacher who refused to let her primary school class sign petitions for Mindszenty's arrest. Peter Glenville's 1955 film, *The Prisoner*, starring Alec Guinness, was an adaptation of a 1954 West End play by Bridget Boland with the same name. That Guinness (who played the cardinal in the West End productions) converted to Catholicism several weeks after the film's release lent added religious gravity to the film.[53] The

---

[50] Csáky, *Ich schwöre*; and R.P.J. Szalay (1950) *Le Cardinal Mindszenty: Confesseur de la Foi, Défenseur de la Cité* (Paris: Mission Catholique Hongroise de Paris).

[51] *The Trial of Jószef Mindszenty* (1949) (Budapest: Hungarian State Publishing House), pp. 4, 5, 11.

[52] J.F. MacDonald (1978) 'The Cold War as Entertainment in 'Fifties' Television', *Journal of Popular Film* 7:1, 1–27.

[53] T. Shaw (2002) 'Martyrs, Miracles and Martians: Religion and Cold War Cinematic Propaganda in the 1950s', *Journal of Cold War Studies* 4:2, 15–17.

film was a box office success and garnered good reviews in the UK and USA, as it clearly struck a chord with audiences. Not that everyone was happy with the film; the Italian Film Board and the Irish Censorship Board both banned the film because the dramatic breaking of the cardinal's will was deemed too 'anti-Catholic' and 'subtly pro-Communist and tending toward the subversion of public morals'. In any case, such depictions of religion on screen contrasted sharply with Soviet cinema at the time. Soviet cinematic depictions of religion in the 1950s and 1960s were much less draconian or negative than they had been before the World War II. As opposed to the 1920s and 1930s, when priests were characterized as criminal deviants, or during the 1950s, when they were painted as devious Western agents, Soviet cinema of the 1960s (such as Vladimir Skuibin's *The Miracle Worker*, 1960 and Vasilii Ordyskiii's *Clouds over Borsk*, 1960) featured religion in a comparatively positive light. Western films of the period were heading in the opposite direction, fueled as they were by Cold War imperatives to dramatize the battle between Christianity and communism on a global scale.[54]

One of the key effects of this episode was the tightening relationship between Catholicism and politics in the early Cold War, especially between the papacy and European politics. The famed Truman-Pius XII 1947 exchange published in all of the major Western newspapers around the world was widely seen as the spiritual corollary of the Marshall Plan.[55] The Mindszenty trial steeled Pius XII's resolve to abandon the church's usual neutrality and to engage more directly in world politics in order to defend Christian Europe and its institutions.[56] The cardinal's conviction prompted the pope to call upon 'the great nations of the continent' to form a 'large political association' in self-defense,[57] and it was in this context that the Pope even gave his blessing to the Western European Union in a 1948 speech. And on July 1, 1949, for the first time in the twentieth century, the Pope excommunicated members, supporters and followers of communist parties, supposedly even extending the ban to readers of communist newspapers. One commentator was not wrong to say that '[n]ever before had the Pope identified so unequivocably with the western alliance'.[58] Not for nothing was Pius derided by the Soviet Union as the 'Coca Cola Pope' for his ardent support of an American-led Western Europe.

[54] Shaw, 'Martyrs', 9–18.
[55] Kent, *Lonely Cold War*, p. 192.
[56] Quoted in Stehle, *Eastern Politics*, p. 270.
[57] W. Kaiser (2007) *Christian Democracy and the Origins of the European Union* (Cambridge: Cambridge University Press), 181.
[58] Stehle, *Eastern Politics*, p. 296.

The Remaking of Human Rights

Such Cold War politics could be seen in the realm of human rights as well. From the very beginning human rights talk was subject to the ideological rivalry between East and West, wherein issues of poverty and unhappiness were sensationalized on both sides of the Iron Curtain to showcase the superiority of each respective system. On the one hand, Westerners used human rights as a cudgel with which to lambast Soviet despotism behind the Iron Curtain, and in so doing helped consolidate a new Cold War anti-communist consensus. Recall that the Council of Europe was founded in 1949 in response to the 1948 Czech coup and made Council membership conditional upon the respect of human rights and democracy.[59] On the other, the USSR never tired of pointing out the hypocrisy of the West in its own orbit, depicting Western poverty, unemployment and welfare neglect as human rights violations of various kinds. The parlous fate of African-Americans in the US South during the Civil Rights struggle became a favorite communist reference point in drawing attention to American state brutality and racial politics during the 1950s and 1960s.[60]

What is often forgotten is that over the course of the 1940s and early 1950s in Western Europe human rights were being reworked from a decidedly Christian perspective. So much so that human rights were often understood as a distinctly Christian project, as key intellectuals within various Christian churches (especially Catholics in France and Britain) viewed such rights as a blend of spiritualism, individualism and humanism—what was called 'personalism' at the time—that supposedly predated and transcended the world of politics. Catholic intellectuals during the war often used human rights as a language of wartime resistance and anti-fascist solidarity, and much human rights talk just after the war emerged from progressive Catholic circles.[61] Remember, too, that the 1941 Atlantic Charter, a declaration of Anglo-American union of principles based in part on the campaign to protect human rights, ended with a religious service and anthem to Christian soldiers. It has even been argued that Christian lobby groups from the USA and Western Europe were responsible for making

---

[59] M.R. Madsen, '"Legal Diplomacy"—Law, Politics and the Genesis of Postwar European Human Rights', in Hoffmann and Makrakis, p. 67.

[60] M. Dudziak (2000), *Cold War Civil Rights* (Princeton: Princeton University Press), pp. 19–46.

[61] C. Dawson (1942) *The Judgement of the Nations* (London: Sheed & Ward), pp. 185–186 and S. Moyn (2010) *The Last Utopia: Human Rights in History* (Cambridge, Mass: Harvard University Press), 55.

sure that the 1948 UN Declaration of Human Rights provided a robust affirmation of religious liberty at its core.[62] Indeed, the postwar rise of human rights was closely bound with the moral reconstruction of Christian Democratic Western Europe, and such moral politics were very much part of the broader conservative campaign to recast bourgeois Europe in a Christian key.[63] In this rendition, human rights emerged as a key fixture of the anti-collectivist, anti-communist consensus across West Europe.

The language of human rights suffused the West's condemnation of the trial. The Universal Declaration had just been signed a few months before and had received worldwide coverage. The Mindszenty Trial was thus its first real test case and was often framed by this new human rights discourse. At the protest at Albert Hall in London, where some 6000 Catholics had gathered to denounce the trial, the meeting passed a resolution stating that the 'whole proceeding involves a denial of essential human rights and fundamental freedoms as declared in the charter of the United Nations organization and guaranteed in part 2 of the peace treaty with Hungary in 1947'.[64] The Australian House of Representatives adopted a similar resolution and added that Hungary with its signed Peace Treaty must respect stated human rights of freedom of religion and freedom of speech.[65] The British House of Lords discussed the trial's significance as well, joining the international chorus of condemnation.[66]

From the beginning, the Pope drew on the recast understanding of human rights in his February 14, 1949 speech on the trial before the College of Cardinals. In it he intoned that the sad fate of the prelate 'inflicts a deep wound not only on your distinguished college and on the church, but also on every upholder of the dignity and liberty of man'.[67] In a radio address in 1941, the pope had called for an 'international bill' that would recognize that rights flowed from the dignity of the person. The linkage of Catholicism and the 'dignity of the person' had been around since the social encyclicals *Rerum Novarum* (1891) and *Quadragesimo Anno* (1931), and Charles Malik, the Lebanese philosopher of Greek Orthodox faith and a key architect

---

[62] Inboden, *Religion*, p.39.

[63] J. Punt (1987) *Die Idee der Menschenrechte: Ihre geschichtliche Entwicklung und ihre Rezeption durch die modern katholische Sozialverkündigung* (Paderborn: Schoeningh).

[64] 'Roman Catholics' Strong Protest: Denial of Human Rights Denounced', *Manchester Guardian*, February 8, 1949, 5.

[65] 'The Cardinal's Sentence', *London Times*, February 10, 1949, 4.

[66] 'Weapons of the Cold War', *London Times*, March 10, 1949, 6.

[67] Pope Pius XII, 'The Mindszenty Trial', in *Vital Speeches of the Day*, February 15, 1949, 265–266.

of the UN Charter, supposedly drew on these Catholic encyclicals in framing the Universal Declaration, thus underlining the perceived affinity between Christianity and human rights.[68] Catholic publicists were also quick to recognize the power of fusing human rights and Catholic causes.[69] While other non-Europeans present, such as Chinese delegate Peng-Chun Chang, made a strong case for the place of non-Christian Confucian thought in the formation of human rights concepts,[70] the 'Christianization' of human rights talk was undeniable.[71] Others continued with the same rhetoric. Churchmen on both sides of the Atlantic claimed that 'Christianity was the carrier of human freedom and human dignity', that the 'total state [of communism] is the culmination of the enslavement of humanity', and that what was at stake was nothing less than the 'freedom of all humanity and the salvation of the whole Christian world'.[72] The connection between religion and human rights was articulated across the West, from Canada to Western Europe.[73]

Protestants played their part too. Ecumenical movement churches helped lay the groundwork for the 'Commission on a Just and Durable Peace' in 1940, chaired by John Foster Dulles, which saw a necessary linkage of human rights, religious freedom and world peace. Dulles even so far as to say that 'the character of the UN organization [was] very largely determined by the organized Christian forces which worked at San Francisco to complete what had been achieved at Bretton Woods. O. Frederick Nolde, a Lutheran religious educator from Philadelphia, was instrumental in helping establish the Commission on Human Rights as mandated by the UN Charter, and Protestant evangelical groups were also successful drafting Article 18 (on freedom of conscience, religion and belief) in the Universal Declaration.[74]

---

[68] M. A. Glendon (2005) 'Catholicism and Human Rights', in *Believing Scholars: Ten Catholic Intellectuals*, James Heft (ed.) (New York: Fordham University Press), pp. 81–93.

[69] J. Epstein (1947) *Defend These Human Rights: Each Man's Stake in the United Nations: A Catholic View* (New York: The America Press), pp. 47–53.

[70] J. Humphrey (1984) *Human Rights and the United Nations: A Great Adventure* (Epping: Bowker), p. 77.

[71] See C. Malik's introduction to O. F. Nolde (1968) *Free and Equal: Human Rights in Ecumenical Perspective* (Geneva: World Council of Churches).

[72] Csáky, *Ich schwöre*, pp. 106, 114.

[73] G. Egerton (2004) 'Entering the Age of Human Rights: Religion, Politics and Canadian Liberalism', *Canadian Historical Review* lxxxv:3, 451–479.

[74] Canon J. Nurser (2003) 'The 'Ecumenical Movement' Churches, 'Global Order', and Human Rights: 1938-1948', *Human Rights Quarterly* xxv, 841–881. See too R. Traer (1991) *Faith in Human Rights* (Georgetown: Georgetown University Press).

It was not uncommon to link the Nuremberg Trials and Mindszenty trial as egregious violations of human rights. In October 1949 Sir Hartley Shawcross, the British Attorney-General and former judge at the Nuremberg Trials 4 years before, openly condemned the treatment of the cardinal. At a speech before the United Nations in 1949 Shawcross claimed that 'exactly the same techniques were being pursued to-day as the Nazis themselves pursued—in these ex-enemy countries the red flag has been hoisted in place of the swastika'.[75] Even more important in this regard was David Maxwell Fyfe. Maxwell Fyfe had been one of the key British judges at Nuremberg, and then played an instrumental role as one of the framers of the 1950 European Convention on Human Rights. For Maxwell Fyfe, the Mindszenty Affair effectively bridged the two. In April 1949 he was interviewed about the Mindszenty Trial in the British Roman Catholic weekly newspaper, *The Catholic Herald*. In it he made the connection explicit:

> the Hungarian government have no intention of maintaining the guarantees of human rights contained in their own statutes and in their Treaty of Peace. At Nuremberg I learnt in detail and by admission out of their own mouths how human rights and all that civilisation stands for can be destroyed by poisoning the fountain of justice ....When one see the Judiciary failing to hold the scales of justice evenly between the individual and the State, and assuming instead the role of stamping out opposition views and smothering criticism, that is the quintessence of tyranny. That is one of the things that Cardinal Mindszenty fought against in Hungary. May his martyrdom not be in vain.[76]

The human rights issue put the Rakosi's government on the back foot. Special venom was directed at the way that the Western press presented the case as about the 'curtailment of religious freedom', and the cardinal as a 'Catholic martyr' and 'defender of human rights and freedom of speech'. The Hungarian state insisted that no mention was ever made in the trial about any attack on Roman Catholicism or religious rights; according to the Hungarian communist authorities, Mindszenty was simply being

---

[75] 'Loss of Human Rights in East Europe: Sir Hartley Shawcross's Plain Speaking at the UN', *London Times*, October 7, 1949, 4.
[76] 'May Cardinal Mindszenty's Martyrdom not be in Vain Says Sir David Maxwell Fyfe', *Catholic Herald*, April 29, 1949, 1.

treated as an 'ordinary criminal'.[77] That the Hungarian state's *Black Book* version of the trial felt the need to challenge British Foreign Secretary Ernest Bevin's claim that the proceedings of the trial (uttered on the eve of the verdict) was 'utterly repugnant to our conception of human rights and liberties' made clear that the trial had moved onto the human rights plane, however much the Hungarian government had hoped otherwise. That the *Black Book* listed other international notables (Acheson, French Foreign Minister Schuman, Belgian Prime Minister Spaak, Australian Foreign Minister Evatt and of course Pope Pius XII) for their condemnation of the trial on human rights terms suggested the Rakosi government was losing the debate internationally. The issue gathered enough momentum and international media coverage to be brought before the UN for debate. The USA led the way in this regard. The day after the sentencing, US Secretary of State Dean Acheson went to the United Nations to condemn the cardinal's sentence as a 'conscienceless attack upon religious and personal freedom' and a shameless effort to 'remove this source of moral resistance to Communism'.[78] Several months later Acheson delivered another speech at the UN, protesting that the persecution of religious figures in Hungary, Bulgaria and Romania has meant that the Danubian states were in violations of human rights enshrined in the 1947 Peace Treaties with these countries.[79]

The trial also played a mediating role in the creation of the European Convention of Human Rights in 1950, which departed in significant ways from the UN's Universal Declaration of Human Rights of 1948. To recall the UN's 1948 Declaration offered an inclusive raft of rights that went far beyond classic bourgeois rights, which included the right to work, right to education and the right to healthcare, thanks in large measure to the presence of the Soviet Union representatives in these discussions. This more material spirit was in sharp contrast to the new regionalized understanding of human rights enshrined 2 years later in the European Convention, which was expressly conceived as anti-communist in tone, and one in which the social and economic rights featured in the 1948 Universal Declaration were quietly dropped. Enshrined instead were the new hallmarks of a Western understanding of human rights: the sanctity of

---

[77] *Trial*, p. 15.

[78] 'The Cardinal's Sentence', *London Times*, February 10, 1949, 4.

[79] S. D. Kertesz (1949) 'Human Rights in the Peace Treaties', *Law and Contemporary Problems* xiv:4, 627–646.

law, the prohibition of compulsory labor, due process in court, freedom of thought and conscience as well as the new clause 'everyone has the right to respect for his private and family life, his home and correspondence'—all patent liberal values that communists interpreted as bluntly directly at their regimes. The 1950 European Convention was crafted by a group of powerful conservative thinkers, such as Churchill and David Maxwell Fyfe, and was largely an effort to curb both the threat of domestic leftist political parties at home and Soviet hegemony abroad by forging a new West European consensus around an understanding of human rights based on individual liberty.[80] Thus in conservative hands, human rights no longer originated in 1789 but rather was a Christian gift to be defended against the legacy of the French Revolution and the revolutionary threat of communism.[81] The European Convention was thus an articulation of a new West European solidarity that set itself as a bloc against its East European rivals.[82]

There was constant reference in Council of Europe discussions that religious freedom and 'Christian civilization' must serve as a cornerstone of the European Human Rights Convention, and the fate of the captive cardinals in the East—and in particular Mindszenty—dramatized its significance.[83] Mindszenty helped forge the intimate relationship between human rights and anti-communism to such a degree, I would argue, that freedom of religion became the most prominent human rights issue in the West during the early Cold War. One could see this in the way that more secular-minded international organizations shared this view. In April 1949—that is, several months after the trial—the UN debated the convictions of both Cardinal Mindszenty and several Protestant pastors in Bulgaria as violations of the UN's member state obligation to respect human rights for its citizens. The East Bloc representatives at the UN argued passionately that such issues should not come under the UN's

[80] G. D. Cohen, 'The Holocaust and the 'Human Rights Revolution:' A Reassessment', in A. Iriye, P. Goedde, W.I. Hitchcock (2012) eds., *The Human Rights Revolution: An International History* (New York: Oxford University Press), pp. 53–73.

[81] M. Duranti (2013) 'Conservatives and the European Convention on Human Rights', in N. Frei and A. Weinke (ed.), *Toward a New Moral World Order Menschenrechtspolitik und Volkerrecht seit 1945* (Weimar: Wallstein Verlag), pp. 82–93.

[82] This politicized 'regionalization' of human rights spurred similar efforts elsewhere, as noted in the 1970 Latin American Human Rights Convention and the 1981 African Charter of Rights of Peoples and Nations. G. Brunner and E. (1989), *Menschenrechte in der DDR* (Baden-Baden: Nomos Verlag), p. 17.

[83] C. Evans (2001) *Freedom of Religion under the European Convention on Human Rights* (Oxford: Oxford University Press), ch. 9.

purview, but rather should be subject to domestic jurisdiction; the accused, so they continued, were common criminals that had nothing to do with human rights, and thus they rejected the UN's affront to their national sovereignty. But after 10 days of debate, the UN adopted a resolution (30–7) in April 1949 condemning their convictions as egregious human rights violations.[84] The fallout was that Hungary, Bulgaria and Romania were excluded from membership in the UN for breaching 'human rights and fundamental freedoms'.[85] The upshot was that the Mindszenty Affair sealed the tendency for human rights to be increasingly identified 'with the fate of Christianity in a world in which communism claimed to incarnate secularism'.[86]

DRUGS, THE CIA AND THE SPECTER OF 'BRAINWASHING'

There was a scientific aspect of the trial that gained international attention as well. What transfixed the international community the most—whether religious or lay—was the forced confession at the trial, particularly the apparent use of drugs to extract it. This was all the more confusing given that Mindszenty supposedly scribbled down a note on the back of an envelope and handed it to a guard before his interrogation, stating: 'I shall not make a confession. But if despite what I now say you should read that I have confessed or resigned, and even see it authenticated by my signature, bear in mind that it will have been only the result of human frailty. In advance, I declare all such actions null and void'.[87] The Catholic press was the first to break the scandal. The British Roman Catholic weekly, *The Tablet*, reported that 'a tablet of the potent nerve-destroying Actedron' had been used on the cardinal to extract confession.[88] Even before the trial was completed, there were rumors that the cardinal would be drugged to confess.[89]

Particularly unsettling to observers was the specific form of the prelate's confession. After all, it was delivered in a manner unfamiliar to his

---

[84] 'Cardinal Mindszenty and the United Nations', in *The Tablet: The International Catholic News Weekly*, April 30, 1949, 20.

[85] A. Martin (1951) 'Human Rights and World Politics', in *The Year Book of World Affairs 1951* (London: Stevens & Sons), 37–80.

[86] Moyn (2010), pp. 72–73.

[87] Quoted in Schuster, *In Silence*, p. 5.

[88] M. Otterman (2007): *From the Cold War to Abu Ghraib* (Melbourne: Melbourne University Press), p. 14.

[89] 'Plot to Drug Mindszenty Revealed', *Washington Post*, January 23, 1949, M3.

friends—he appeared exhausted and drooling, and his written testimony was riddled with basic grammatical and spelling errors uncharacteristic of him.[90] It was also the appearance of the skittish, wild-eyed archbishop that observers found so disturbing. The famous photograph of the prelate at the trial became one of the most reproduced images of early Cold War culture. The trial was not filmed, though some of it was broadcast live on Hungarian radio; few photos were taken, and thus this photo became emblematic of the whole trial itself. (A dissident Hungarian politician at the time averred that the communists' release of the photograph was 'one of the greatest mistakes of Soviet propaganda'.[91]) In response Cardinal Spellman of New York protested that the prelate had been the victim of 'torture and drugging', and contrasted photographs taken at the trial with those taken of the prelate during his visit in the USA 2 years before.[92] Actedron had supposedly rendered him defenseless and devoid of his faculty of memory and judgment, furnishing sobering 'evidence of his broken, tortured mind, the unhingeing of his mental balance and loss of his self-control'; it was widely reported that the drug induced 'strong persecution mania' and facilitated the breakdown of the victim. One inside contact at the trial remarked that actedron was the 'dread weapon of Communist tyranny against which there is no defense'. In this case, the cardinal's condition proved the 'sad reality of torture and the devilish technique employed by the faithful imitators of Goebbels and Himmler'.[93] One commentator wrote that the trial was an 'unexampled crime of our modern age that the great discoveries of science, psychology and medicine are used for the purpose of war and destruction—instead of progress: not for the service of the health of humanity, but for the spreading of mental and physical disease'.[94]

Public interest in the cardinal's psychological state was closely tied up with two other events that caught the imagination of the Cold War West at the time. The first was the 1949 espionage trial in Budapest of the American ITT business executive, Robert Vogeler, just months after the Mindszenty trial

---

[90] J. Mindszenty (1949) *Four Years Struggle of the Church in Hungary: Facts and Evidence Published by Order of Josef, Cardinal Mindszenty, Prince Primate of Hungary*, trans. Walter C. Breitenfeld (London: Longmans), pp. xi–xiii.

[91] The phrase was attributed to Bela Fábián, former leader of the Hungarian Independent Democratic Party, in 'Mindszenty Trial Held Soviet Farce', *New York Times*, February 10, 1949, 5.

[92] 'Spellman Warns', 1–2.

[93] Boer, *Cardinal Mindszenty*, p. 287.

[94] Boer, *Cardinal Mindszenty*, pp. 288, 290.

had concluded. The Vogeler Affair became an international diplomatic crisis between Hungary and the USA. At the trial Vogeler surprised international followers of the case by pleading guilty of espionage, whereupon he was then sentenced to 15 years in prison, though his sentence was commuted to just 2 years. International journalists roundly condemned the trial as yet another Stalinist 'diabolical puppet show'.[95] What generated huge media attention was Vogeler's tell-all sensationalist memoir of mental torture and drugging published after his return to the USA, called *I Was Stalin's Prisoner* (1952). His story caused an overnight sensation, as his tale of dark psychic manipulation received wide attention in the USA and became fodder for fiction writers, most famously by Paul Gallico in his 1952 novel, *Trial by Terror*.

The second episode was the 1951 publication of Edward Hunter's *Brain-Washing in Red China: The Calculated Destruction of Men's Minds*. Hunter was a CIA propaganda operator and undercover journalist. He coined the term 'brainwashing' as a translation of the Chinese colloquialism *hsi nao* (literally 'wash brain'), which supposedly came from Chinese informants who described its diabolical use. With it Hunter gave a new name to a growing fear of devious communist subterfuge. In his account Hunter aimed to show the sinister ideological makeover of citizens taking place in Mao's China. For Hunter such psychological warfare represented a new and dangerous phase of the Cold War, given that 'what actually is meant by cold war is warfare with unorthodox weapons, with silent weapons such as a leaflet, a hypnotist's lulling instructions, or a self-criticism meeting in Red China'.[96] Anxiety further intensified when the Chinese government launched a propaganda initiative in 1952 that featured recorded statements by American pilots supposedly confessing to having committed war crimes, including germ warfare. By end of the war it was reported that some 70 % of 7190 US prisoners in Korea had either confessed or signed petitions demanding an end to the American war effort in Asia; only 5 % were openly resistant. Worse for many observers is that many of those who had confessed stood by their confessions even once they returned to America after the war. Such uncomfortable stories prompted a new round of US government worry about communist techniques of brainwashing.[97]

[95] Otterman, *American Torture* p.16.

[96] E. Hunter (1951) *Brainwashing in Red China: The Calculated Destruction of Men's Minds* (New York: Vanguard), 12, quoted in D. Seed (1997) 'Brainwashing and Cold War Demonology', *Prospects* xxii, 535–573, 539.

[97] Even the British Ministry of Defence published its own report on the brainwashing of British POWs in the Korean War, though it was mocked in the British press and not taken

The blended media coverage of Mindszenty, Vogeler and the Korean War confessions spurred a range of new scientific studies on psychological manipulation, mostly in the USA. Popular fear about the destruction of individualism was registered in American pulp fiction, ranging from Robert Heinlein's *The Puppet Masters* (1951) to *The Invasion of the Body Snatchers* (1956) to AE van Vogt's *The Mind Cage* (1957), and found cinematic expression as well. Lewis Seiler's 1954 film, *The Bamboo Curtain*, exploited the brainwashing theme related to American involvement in the Korean War. What catapulted the issue into American popular mythology was Richard Condon's bestselling 1959 novel, *The Manchurian Candidate*, which was turned into a box office smash that same year by director John Frankenheimer.[98] In this setting it was no accident that the films on Mindszenty's plight centered on the drugs and brainwashing elements: *Guilty of Treason* drew attention to the cardinal's confession as a form of evil hypnosis,[99] while *The Prisoner* depicted the Interrogator's use of modern psychological techniques as part of communism's dark arts.

Disquiet about the sinister science behind Mindszenty's confession was scarcely confined to film and fiction. The danger of malicious 'truth serums' was extensive enough to warrant attention at the UN, and again the Mindszenty trial was the principal reference. In April 1950 the UN put forward a motion to get such serums banned on the grounds that they represented a breach of human rights. Even if the endeavor failed when the international commission 'could not agree to include 'confessional drugs' under an approved provision forbidding torture or inhuman treatment',[100] its coverage at the UN reflects how prominent the trial was internationally in terms of coupling 'dark psychiatry' with human rights concerns.

Nevertheless, the most ardent student of the trial was arguably the CIA. Like fiction writers and filmmakers, the CIA also believed that brainwashing was a form of modern witchcraft, whose evil spell (and spellbinders) needed to be studied and counteracted. By the late 1940s there was already a concern within the CIA (fueled by defector testimonies) that the USA had fallen behind the USSR in medical intelligence, and the cardinal's confession was widely regarded as indicative of new

---

very seriously. See Ministry of Defence (1954) *The Treatment of British POWs in Korea*, cited in Cyril Cunningham (1973) 'Brainwashing', *RUSI Journal* cxviii: 3, 39–43.

[98] S.L. Carruthers (1998) 'Redeeming the Captives: Hollywood and the Brainwashing of America's Prisoners of War in Korea', *Film History* x, 275–294.

[99] Shaw, 'Martyrs', 16.

[100] '"Truth Serum" Ban is Dropped in UN', *New York Times*, April 1, 1950, 3.

advances in Soviet psychological techniques. CIA operatives were keen to develop new means of both extracting key information quickly from captured Soviet agents, and to help US personnel resist Soviet truth serums or other psychochemical agents if captured by enemies. The cardinal's trial, the detonation of the USSR's first atomic bomb later that same year and the 'confessions' of captured US soldiers in the Korean War all heightened the urgency for the newly created CIA to step up scientific research into psychomedical science.[101] Granted, historians have long made the claim that as early as 1943 the Office of Strategic Services (the forerunner to the CIA) had conducted research into mind-altering drugs and truth serums (based in part on a marijuana derivative) that 'might break down the psychological defenses of Axis agents subjected to questioning by Allied counterintelligence officers'.[102] Some even claimed that the US government analyzed Nazi experimental data collected at Dachau.[103] Whatever the case, Mindszenty's trial had a galvanizing effect. Later interviews with key CIA agents noted that they had looked into the 'vacant eyes' of the Hungarian prelate at his treason trial and 'had been horrified;' they were convinced that his confession had been extracted under the influence of 'some mysterious mind-bending drug' reminiscent of Soviet show trials from 1930s. Senior CIA officials supposedly went to Europe in 1949 to conduct drug experiments on returned prisoners from Eastern Europe.[104]

Internal CIA documents made clear the depth of agency apprehension about Mindszenty's appearance in court. In February the CIA was already gathering reports from abroad about the Mindszenty trial, in particular from the report in the British Roman Catholic weekly, *The Tablet*. Speculation abounded that the cardinal was under the influence of 'some unknown force' and suspected the mysterious power was actedron, the trade name for benzedrine. Worry too was that communist countries were 'further advanced' than previously thought, and that CIA operatives abroad would be in danger. CIA security memoranda proposed that 'a program be initiated which would serve to inform an applicable employee

---

[101] R.E. Doel and A. A. Needell (2008) 'Science, Scientists and the CIA: Balancing International Ideals, National Needs, and Professional Opportunities', *Intelligence and National Security* xii:1, 72.

[102] John M. Crewson, 'Files Show Tests for Truth Drug Began in OSS', *New York Times*, September 5, 1977, 1.

[103] John Marks (1979) *The Search for the Manchurian Candidate: The CIA and Mind-Control* (New York: WW Norton), p. 11.

[104] Marks, *Search*, pp. 24–25.

as to his susceptibility to the various interrogation techniques', and 'if desired, condition him so that he will no longer be susceptible to the various unfriendly interrogation techniques to which he might be exposed'.[105] The Vogeler confession was attributed to the same devious 'truth serum' developed in the Mindszenty case.[106] For the CIA, the cardinal's trial was explicitly linked to the Soviet purge trials of the 1930s, wherein 'overt Russian judicial procedure has been noteworthy for the dramatic trials in which the defendants have exhibited anomalous and incomprehensible behavior and confessions... Most noteworthy and incredible has been the recent "confession" of His Eminence Cardinal Josef (sic) Mindszenty while on trial in the People's Court of Hungary'.[107] By 1952 there was a growing desire to fund a new research program, spurred in large measure by fear of 'Soviet interest and research in this direction, plus rumors, reports and evaluations of possible Soviet use of these techniques in these interrogations', which were seen as 'sufficient to warrant careful investigation of the subject as a basis for evaluating the extent of the threat to US National Security'.[108]

The CIA then set up a new secret program, Project Bluebird, to conduct new scientific research into fending off communist mind-control techniques.[109] The first field operation was set up in 1950 to test-drive behavioral techniques and drug experimentation (sodium amytal and bezendrine) ostensibly in order to induce amnesia on suspected double agents; further tests were carried out on an additional 25 subjects, allegedly North Korean POWs.[110] The project's secret brief was further expanded in its renamed guises as Project Artichoke (1952) and MKULTRA (1953), with virtually no governmental oversight or accountability whatsoever, as first revealed in John Marks's classic 1977 exposé of CIA secret programs, *The Search for the Manchurian Candidate: The CIA and*

[105] Interrogation Techniques of Unfriendly Countries, Security Research Section, CIA, 24 Feb. 1949, 1–3, 184367, MK Ultra Collection, CIA Archives, Washington.

[106] Interrogation Techniques, from Asst. Director, Special Operations to the Chief of Inspection and Security Staff, CIA, Paris, January 16, 1950, Doc 144916, MK Ultra Collection, CIA Archives, Washington.

[107] Communist 'Confession' Techniques, 5, May 17, 1949, Doc 144891, MK Ultra Collection, CIA Archives, Washington.

[108] Proposed Memorandum for the Secretary of Defense, 5, January 20, 1952, Doc 144688, MK Ultra Collection, CIA Archives, Washington.

[109] Marks, *Search*, pp. 24–25.

[110] Marks, *Search*, p. 25.

Mind-Control. Project Artichoke was designed as an 'integrated CIA program' to 'develop special interrogation techniques for the purpose of controlling an individual without his knowledge'.[111] Scientists associated with the shadowy programs were especially keen to study the world of the occult and 'black psychiatry'; though many of their ideas thankfully never left the drawing board, they did dispatch two 'special interrogation teams' to missions in Europe and Asia, and the CIA continued its research into psychochemicals aimed to get people to talk.[112] Over the course of the 1950s CIA director Allen Dulles commissioned various reports on brainwashing, but it was clear that the CIA's obsession with mind control was first sparked by Mindszenty's trial.[113] In one internal 1955 CIA report called 'Out of Soviet Laboratories—Brainwashing', for example, Mindszenty's story opened the history of the new danger of sinister Soviet 'Pavlovian' science, one that posed a 'grave threat to the freedom and independence of all people everywhere'.[114] The trial spurred a new round of worry (and funding) inside the American intelligence community to engineer their own version of what they significantly called at the time the 'Mindszenty effect', leading to the agency's infamous LSD research in the mid-1950s.[115] While a 'mind-control gap' may have been as illusory as 'missile gaps', it was also true that by the early 1950s 'mind control took on a momentum of its own'.[116] This new shotgun marriage of Cold War politics and 'psychological warfare' was not limited to the USA. An International Documentation and Information Center, or Interdoc, was established in February 1963 in The Hague as a kind of French-West German-Dutch joint venture to confront the 'ideological threats' posed by Soviet and Chinese communism and to build Western European intelligence networks outside of

[111] Assistant Director, Scientific Intelligence to Deputy Director for Central Intelligence, Memo: Project Artichoke, February 4, 1952, 1, 5, Doc 144689, MK Ultra Collection, CIA Archives, Washington.

[112] 'Mind-Control Studies Had Origins in Trial of Mindszenty', *New York Times*, August 2, 1977, 16.

[113] Giles Scott-Smith, 'Interdoc and West European Psychological Warfare: The American Connection', *Intelligence and National Security* 26:2–3 (April–June 2011), 361.

[114] 'Out of Soviet Laboratories—Brainwashing', 1955, CIA, 3, 146095, MK Ultra Collection, CIA Archives, Washington.

[115] Crewson, 1; and more generally, M.A. Lee and B. Schlain (1985) *Acid Dreams: The CIA, LSD and the Sixties Rebellion* (New York: Grove Press).

[116] Marks, *Search*, 31.

American control.[117] East European intelligence agencies, such as East Germany's Stasi and Romania's Securitate, also developed new interest in psychology and social science to monitor dissidents and to control informants.[118]

Still, the development of a new Cold War 'covert sphere' to combat the perceived threat of communism's psychological warfare found its most elaborate form in the United States.[119] The Mindszenty trial and Korean War confessions kicked off a cottage industry of new social science studies on Soviet psychology. In part this had to do with what was perceived as the perverse inversion of psychology—where the original impetus behind Freudian psychoanalysis had been to liberate people from their mental afflictions and 'restore their personality' to them, this new 'black psychiatry' was by contrast geared toward manipulation, control and the destruction of individual personality.[120] Western observers were aware of the show trials of the 1930s in which older Bolsheviks were targeted and disgraced by the Party; but these post-1945 trials were very different, given that they featured hardened anti-communists, thereby making their confessions all the more troubling. This was what one observer called the 'Iron Curtain mystery', which warranted serious scientific attention.[121] With it new scholarly interest emerged in how 'Pavlovian' psychology had been turned to new political ends. Examples included George S. Counts and Nucia Lodge, *The Country of the Blind: The Soviet System of Mind Control* (1949), Joost Meerloo, *The Rape of the Mind: The Psychology of Thought Control* (1956), William Sargant, *Battle for the Mind: A Physiology of Conversion and Brainwashing* (1957) and most famously, Robert Jay Lifton's *Thought Reform and the Psychology of Totalism: A Study of 'Brainwashing' in China* (1961).[122] Often the link between the American secret services and social

[117] G. Scott-Smith (2011) 'Interdoc and West European Psychological Warfare: The American Connection', *Intelligence and National Security* xxvi:2/3, 355–376.

[118] Jens Gieseke, *Der Mielke-Konzern: Die Geschichte der Stasi, 1945–1990* (Stuttgart, 2001).

[119] T. Melley (2012) *Covert Sphere: Secrecy, Fiction and the National Security State* (Ithaca: Cornell University Press), esp. pp. 44–72.

[120] S.G. Solomon (1980) 'Reflections on Western Studies of Soviet Science', in Linda Lubrano and SG Solomon (ed.), *The Social Context of Soviet Science* (Boulder, CO: Westview), pp. 1–30.

[121] W.H. Lawrence, 'Why Do they Confess? A Communist Enigma', *New York Times*, 8 May 1949, SM7.

[122] Seed 'Brainwashing'; and on the Hungarian perspective, see I. Rév (2002) 'The Suggestion', *Representations* lxxx:1, 62–98.

science scholarship was an intimate one: for instance, Federal Bureau of Investigation (FBI) director Allen Dulles commissioned a secret intelligence report by two Cornell neurologists, Harold Wolff and Lawrence Hinkle, to study communists techniques of mind control, which eventually was published in an academic journal.[123] By the late 1950s the brainwashing fear had generated hundreds of social science books and articles about the dangers of what later FBI Director J. Edgar Hoover called the communist 'thought control machine'.[124] Not all of this literature was a crass demonization of Soviet psychology and behavioral science: Lifton admitted for example that McCarthyism was an American form of 'totalism' in its blend of 'political religion and extreme opportunism'.[125] Yet the cottage industry of books on the subject reflected growing concern in the 1950s that the Cold War was moving on to new terrain, and the new battleground was the science of suggestion and mind control. In 1953 FBI Director Allen W. Dulles remarked that '[t]he Soviets are now using brain perversion techniques as one of their main weapons in prosecuting the cold war ... Possibly the case that most startled the West was that involving the confession of Cardinal Mindszenty, in Hungary. Here a man of proven courage and outstanding intellect was brought to a point of publicly confessing actions which those who knew this outstanding character could not possibly have attributed to him'. For Dulles, US soldiers in Korea were subject to the same brutal treatment.[126]

By the early 1960s, Mindszenty and his place within Cold War cosmology was becoming more and more marginal. In an era of détente the CIA's secret psychological programs were winding down, as the obsessive 1950s interest in brainwashing and mind-control faded from prominence. Mindszenty's star had also faded among international politicians and even Catholic activists. This may seem strange at first, given that Mindszenty dramatically resurfaced during the Budapest uprising of 1956. After some

---

[123] L.E. Hinkle and H.G. Wolff (1956) 'Communist Interrogation and Indoctrination of 'Enemies of the State;' Analysis of Methods Used by the Communist State Police (A Special Report)', *Archives of Neurology and Psychiatry* lxxvi, 115–174, discussed in Melley, *Covert Sphere*, pp. 58–59.

[124] J. Edgar Hoover, *Masters of Deceit* (New York: Pocket, 1958), p. 75.

[125] R.J. Lifton (1961) *Thought Reform and the Psychology of Totalism: A Study of 'Brainwashing' in China* (New York: WW Norton), p. 457.

[126] Summary of Remarks by Mr. Allen W. Dulles at the National Alumni Conference of the Graduate Council at Princeton University, Hot Springs, VA, April 10, 1953, 1, 8, 11, Doc. 146077, MK Ultra Collection, CIA Archives, Washington.

confusion, Premier Imre Nagy freed the cardinal, allowing him to return to Budapest as a liberated man. Mindszenty wasted little time in praising the insurgents, and delivered speeches and radio addresses calling for the restoration of the Christian public sphere. Yet 4 days later Soviet tanks cracked down on the agitators; the cardinal—seen as the one of the leaders of the counter-revolution—was forced to take refuge in the American Legation in Budapest, where he remained for no less than 15 years in political asylum until 1971, making Julian Assange's asylum in the Ecuadorian Embassy seem rather tame by comparison. His temporary release in 1956—after spending 7 years in jail already—and new confinement in the US Legation attracted a fresh round of worldwide condemnation.[127] So vociferous was the anger that a militant Cardinal Mindszenty Foundation was founded in 1958 in the USA, which elevated the Hungarian cardinal as the masthead martyr of the American anti-communist cause.[128] But within Christian circles in the East Bloc, Mindszenty was seen as a growing liability, in that his inflexible position had made things difficult for the Church and its parishioners both in Hungary and across the East Bloc.[129] Across Eastern Europe Catholics interpreted Mindszenty's showy intransigence as simply a pretext for communist repression, and opted for a more accommodationist line in the name of preserving the church's semi-independence. Heinrich Grüber, a high-ranking authority in the GDR's Evangelical Church, even went so far as to say that Mindszenty 'the schemer, avid for power' was responsible for the bloodshed in 1956.[130]

But even if Mindszenty had not changed, the Cold War had. For one thing, Pope Pius XII died in 1958 and was succeeded by Pope John XXIII. His 1963 encyclical *Pacem in Terris* left open the possibility of cohabitation with communism, setting in train a new Vatican *Ostpolitik*. Over the course of the 1960s the Hungarian church eventually made its peace with the communist state, as had the other churches across the East

---

[127] 'Mindszenty Urges UN to Assist Hungarians', *Milwaukee Journal*, November 12, 1956, 1. See too Alex Last, 'Fifteen Years Holed Up in an Embassy', *BBC News Magazine*, September 6, 2012.
[128] D.L. O'Connor (2006) 'The Cardinal Mindszenty Foundation: American Catholic Anti-Communism and its Limits', *American Communist History* v:1, 40.
[129] Luxmoore and Babiuch, *Vatican*, 43.
[130] V. Conzemius (1998) 'Protestants and Catholics in the GDR, 1945–1990: A Comparison', *Religion, State and Society* xxvi:1, 51–59, here 56.

Bloc.[131] The communist states' relatively successful effort to suppress the transnational dimension of Catholicism both across the Bloc and across the Iron Curtain meant that these churches were fundamentally nationalized. By the late 1960s the new pope, Paul VI, went even further in this spirit of détente: he declared the cardinal 'a victim of history' (significantly, not communism) and annulled the excommunication imposed by Mindszenty on his political opponents. In response, the Hungarian communist government was prepared to let him leave the country. By the early 1970s the USA felt that Mindszenty had become an obstacle and embarrassment to warming Soviet-American relations, since his presence, as the primate himself understood, 'stood in the way of the policy of détente'.[132] In 1971 he was thus pushed out unceremoniously from the American Embassy as 'Nixon's unwanted guest', whereafter the cardinal went into his 'second exile' at the Pázmáneum seminary in Vienna. (Mindszenty was allowed to publish his memoirs abroad in 1974; in it Mindszenty recounted having been drugged and coerced in the form of imposed sleeplessness and being beaten by a rubber truncheon.[133]) Less than a year later five new bishops were approved by the state, the Hungarian Prime Minister, György Lázár, was received privately by Pope Paul VI, and Kadar himself was received at the Vatican in 1977.[134] While his old supporters were hugely disappointed by this humiliating turn of events and expressed anger toward the church for having lost its fighting spirit from the early days of the Cold War,[135] Mindszenty now seemed a relic from a bygone age. And with the suppression of the Catholic public sphere across Eastern Europe until the mid-1970s, the Christian cause (and its martyrs) in Eastern Europe effectively dropped away from Western attention.

Not that he was totally forgotten. His death in 1975 garnered coverage and commemorations across the West and Catholic world. In Cleveland, Ohio, a city square was named after him, and in New Brunswick, New Jersey, home to tens of thousands of Hungarians who fled the country in 1956, a 10-foot statue was unveiled in St. Ladislaus Square bearing his name; a memorial to him stands in Chile as well.[136] The end of communism

---

[131] J. Wildmann (1986) 'Hungary: From the Ruling Church to "Church of the People"', in *Religion in Communist Lands* xiv:2, 160–171.
[132] Mindszenty, *Memoirs*, 235.
[133] Mindszenty, *Memoirs*, 114–116.
[134] Luxmoore and Babiuch, *Vatican*, 169.
[135] Közi-Horvath, *Cardinal Mindszenty*, p. 7.
[136] Közi-Horvath, *Cardinal Mindszenty*, pp. 128–130.

has brought back Mindszenty's case to public awareness. In 1989 his case was re-opened, and the captive cardinal was officially exonerated of all charged in 2012; his bones were reburied in Eszterdom Cathedral and his resting place has become a pilgrimage site, and there is now a movement to canonize him.[137] Two films about Mindszenty were produced in 2010 in Hungary: Gábor Koltay's *The White Martyr* (*A fehér vértanu*) and Zsolt Pozsgai's *I Love You, Faust* (*Szeretlek, Faust*), both of which portray the cardinal as a victim of both German National Socialists and Hungarian Communists, reflect both Mindszenty's enduring popular interest and the new regime's interest in using his story to rewrite Hungarian history.[138]

By the mid-1960s Mindszenty no longer served as a flashpoint of Cold War Catholicism and social science as he had in the late 1940s and early 1950s. However, Mindszenty's legacy found echoes in other ways, especially concerning the history of human rights. Admittedly, Mindszenty's anti-communist view of human rights during the early Cold War was overtaken by global events that shifted the meaning of human rights internationally. This was particularly the case with decolonization, which re-aligned human rights with the causes of national liberation, self-determination and political sovereignty. The cardinal's ignominious end seemed to spell the failure of his crusade to make the affront against religious liberty the supreme violation of human rights worldwide. And yet, Mindszenty's campaign to couch the church's struggle against communism in terms of human rights, 'Christian morality', 'Christian civilization' and Christian Europe was picked up by a new generation of Eastern European Catholic activists in the late 1970s and 1980s.[139] The election of John Paul II in 1978 brought renewed attention to the issue.[140] Nowhere was 'political Catholicism' in the East more pronounced than in Poland, where the budding Solidarity movement wasted little time in coupling political and religious freedom in new and politically explosive ways. In the 1950s Mindszenty became a celebrity martyr of Catholics and anti-communists around the world, whose plight (and strange confession) helped cement a new identity for the Cold War West both in the realm of religion and science. In the first half of the Cold War the captive cardinal's

---

[137] A. Leitner, 'Mindszenty Cleared Posthumously', *The Budapest Times*, March 30, 2012.

[138] I thank Dr. Heléna Tóth for this information.

[139] Hanebrink, *In Defense*, p. 229.

[140] P. Ramet (1987) *Cross and Commissar: The Politics of Religion in Eastern Europe and the USSR* (Bloomington: Indiana University Press), 66.

fate helped shape a new political conscience for Western Europe[141]; in the second half Mindszenty's legacy began to resurface as part of a new conscience of the East. The center of gravity for Catholic activism may have shifted from Yugoslavia and Hungary to Poland and the Polish Pope, but Mindszenty's seemingly failed anti-communist crusade had been raised from the dead.

---

[141] Kertesz, 'Human Rights', 219.

The manufacturer's authorised representative in the EU is Springer Nature Customer Service Centre GmbH, Europaplatz 3, 69115 Heidelberg, Germany. If you have any concerns regarding our products, please contact ProductSafety@springernature.com

Printed and bound by CPI Group (UK) Ltd, Croydon, CR0 4YY
23/03/2026
02076682-0008